ENCYCLOPEDIA OF ORGANIC CHEMISTRY

ENCYCLOPEDIA OF ORGANIC CHEMISTRY

Vol. 4

By

Sananda Chatterjee

DAE, DCA, CMCNet

Scientific Consultant

Institute for Natural Sciences

Kolkata

(West Bengal)

DISCOVERY PUBLISHING HOUSE PVT. LTD.

NEW DELHI-110 002

Published by:
Tilak Wasan
DISCOVERY PUBLISHING HOUSE PVT. LTD.
4383/4B, Ansari Road, Darya Ganj
New Delhi-110 002 (India)
Phone : +91-11-23279245, 43596064-65
Fax : +91-11-23253475
E-mail : parul.wasan@gmail.com
discoverypublishinghouse@gmail.com
web : www.discoverypublishinggroup.com

***First Edition:* 2012**

ISBN: 978-81-8356-875-3 (Set)

Encyclopedia of Organic Chemistry

Printed at:
Shree Balaji Art Press
Delhi

Preface

Organic Chemistry is that branch of chemistry, which deals with the reactions related to the life form, or one of the source of the life form, carbon (we can say the end of it).

This carbon mainly conjugates with different elements such as nitrogen, hydrogen and oxygen sometimes metals and other non-metals also). Organic chemistry studies mainly the extraction, structure, properties, of these compounds.

In this book chapters deals with the different methods of purification, extraction of compounds, some processes were used and devised by Alchemists.

The extraction preservation of natural dyes are examples of alchemical work, these been discussed in the chapter "dyes indicators and pigments".

Dyes is the mal-pronunciation of the damathi word dyum meaning colour, it was, the ancient Egyptians who use different natural pigments to extract different colours.

In this book the organic compounds has been mainly classified into two broad groups, the aromatic and the aliphatic compounds. The extraction of alcohol is definitely an alchemical process'.

The destructive distillation of wood is one of the finest processes devised by the alchemists for the preparation of methanol is one of the important topics of this book.

Looking ahead in to the modem world, the world of future, the subject comes in vision is the biochemistry, the base of which is organic chemistry, the chapter containing carbohydrates, and the chapter, chemical compounds of biological and biochemical interest is the best element of focus for the students stepping foot in the advanced staircase of biochemistry.

The last chapter, chapter 41, discusses the different organic reactions and their suitably mechanisms, this will help those students who are very much afraid of organic formulas and reactions. This chapter also deals with the chirality one of the interest of Dr. Palash Gangopadhyay (author's friend).

Organic Chemistry is not a mere subject, it is the rhythm of life tuned in the essence of carbon, which looks black but carries the dazzleness of diamond in its heart.

SANANDA CHATTERJEE

Contents

CHAPTER

17

Nitro Alkanes Diazoalkanes and Azides

Introduction

Nitrous acid (NHO_2) exists in two tautomeric forms.

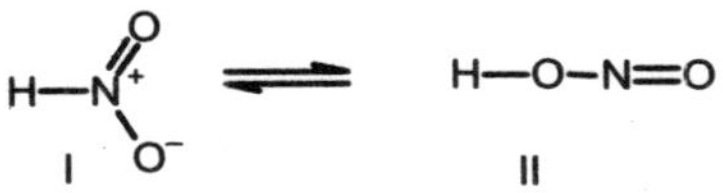

$$\underset{\text{I}}{H-\overset{+}{N}(=O)-O^-} \rightleftharpoons \underset{\text{II}}{H-O-N=O}$$

The alkyl derivatives arriving from I are known as ***Aliphatic nitrate compounds***. Those derived form II are called alkyl nitrites which are in fact esters of nitrous acid and was even discussed in chapter 14.

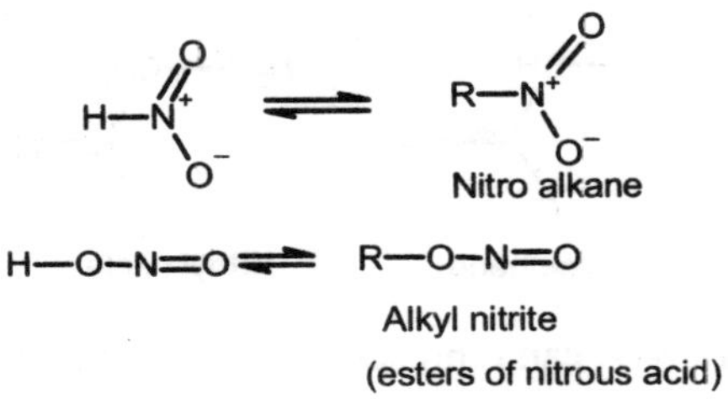

$$H-\overset{+}{N}(=O)-O^- \rightleftharpoons \underset{\text{Nitro alkane}}{R-\overset{+}{N}(=O)-O^-}$$

$$H-O-N=O \rightleftharpoons \underset{\substack{\text{Alkyl nitrite} \\ \text{(esters of nitrous acid)}}}{R-O-N=O}$$

In nitroalkene molecule, nitrogen atom is attached directly to carbon of the alkyl group (C – N) and in nitrites it is attached to carbons through oxygen (C – O – N)

Nitroalkanes are correctly considered as the derivatives of alkanes in which a hydrogen atom is replaced by nitro group – NO_2

$$\underset{\text{Alkane}}{R-H} \rightarrow \underset{\text{nitro alkane}}{R-NO_2}$$

The general formula of nitroalakne is R – NO_2 where the nitro group (– NO_2) is the functional group.

Like alcohols Nitroalkanes are further derived into primary, secondary and tertiary groups according to the nitro group is bonded to a primary (1°) secondary (2°) as a tertiary carbon atom.

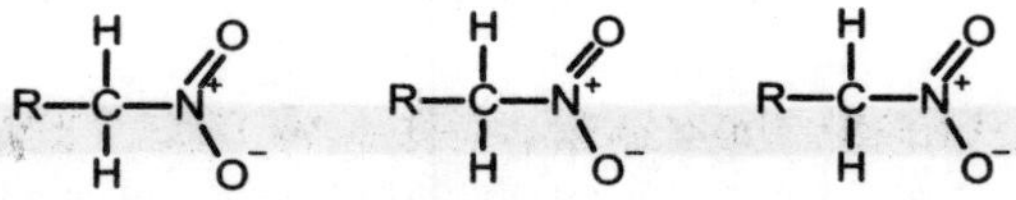

1° nitroalkane   2° nitroalkane   3° nitroalkane

Nomenclature

Aliphatic nitro compounds are slowly named by the IUPAC system. The systematic name of nitro compounds is constructed by prefixing 'nitro' to the alkane in which this – NO_2 group is substituted. The total name emerges as one word, the portion of the – NO_2 group on the carbon chain being indicated by a number. For example.

Formula	IUPAC name
$CH_3 - NO_2$	Nitromethane
$CH_3 - CH_2 - CH_3 - NO_2$	Nitro ethane
$CH_3 - CH_2 - CH_2 - NO_2$	1-Nitropropane

Isomerism

Besides chain and position isomerism nitroalkane also shows functional isomerism with alkyl nitrites. These nitroethane ($C_2H_5NO_2$) is isomeric with ethyl nitrite (C_2H_5ONO)

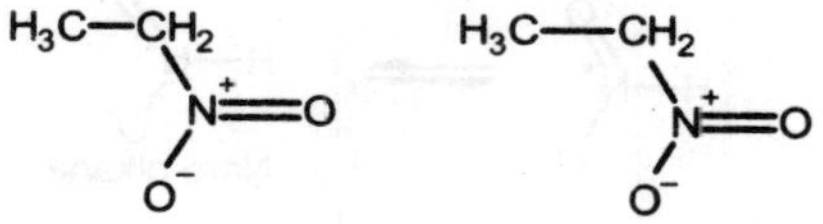

Nitroethane   Ethyl nitrite

The molecular formula $C_4H_4NO_2$ can represent the following nitroalkanes

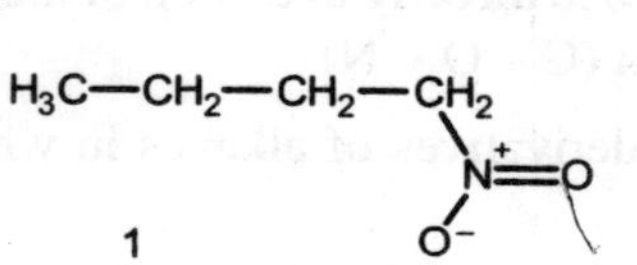

1   3

1-nitrobutane   2-methyl-1-nitropropane

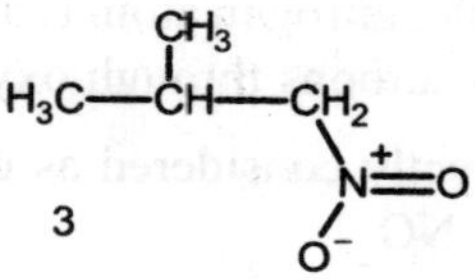

2 $H_3C—CH_2—CH(—N^+(=O)O^-)—CH_3$
2-nitrobutane

4 $H_3C—C(CH_3)_2—N^+(=O)O^-$
2-methyl-2-nitropropane

The structure above represent the chain isomers, while I and II represent position isomers

Structure

In considering the structure of a nitroalkane molecule, nitrogen atom plays the key role. He configuration of the volume shell of nitrogen in different states is as below:

N is ground state	$2s^2$	Sp^1_x	$2py^1$	$2py^1$
N in the enhanced state	$2s^1$	$2p^1_x$	$2py^1$	$2pz^1$

(one slection premoted to $2p_z$ orbital)

N in hybridized state 2 $(sp^2)^x$ $2pz^2$

Pz
sp^2 sp^2 sp^2

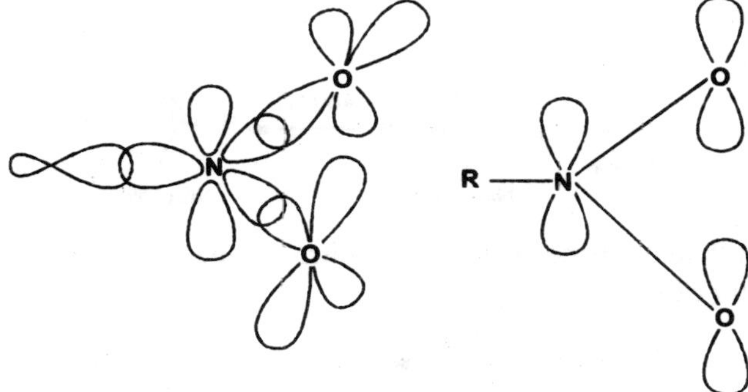

At this stage there is left on nitrogen one hybridized p orbital (2_{p2}) with an unshared electron pair and a half filled p – orbital (2py) on each oxygen atom. The two unused P_2 orbitals of oxygen are equidistant from P_2 orbital of nitrogen and parallel it. These available p orbitals of nitrogen and the two oxygen overlap sidewise producing a colorized cloud which encompasses all these atoms.

In terms of valance bonds, the structure of nitro alkanes may be written as

$R—N(=O)\rightarrow O$ (classical) $R—N^+(=O)—O^-$ (polar)

The polar structure is supported by a large dipole moment of the nitro group. Thus nitro alkane is represented as a resonance hybrid of the canonical forms I and II

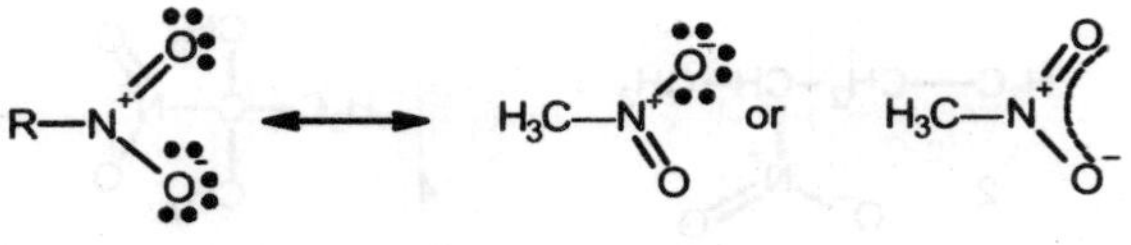

R—N(=O)—O⁻ ⟷ $H_3C—N^+(O^-)=O$ or $H_3C—N^+$(…O)(…O⁻)

I II III

The hybrid structure III suggests the equivalence of the two nitrogen – oxygen bonds. The spectroscopic and diffraction studies of nitro alkanes have shown that the bond distance for each of the N – O bonds is 1.22A, which is less than N – O single bond (1.36Å) AND N = O double bond (1.15Å). This confirms the bond structure and distribution of charge in nitroalkane molecule.

Methods of Preparation

The general methods of preparation of nitro alkanes are listed below:

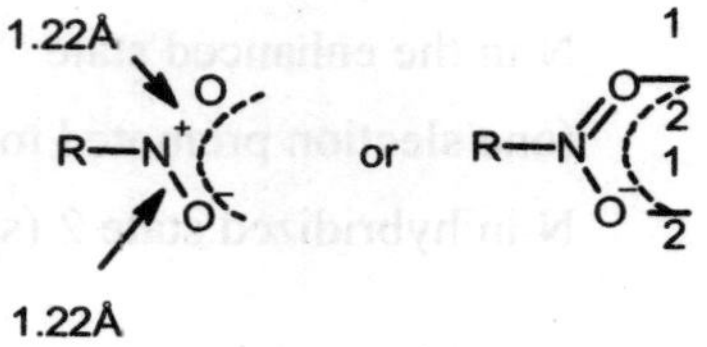

1.22Å, 1.22Å: R—N⁺(O)(O⁻) or R—N⁺(O)(O⁻) (1, 2, 1, 2)

(1) Vapour Phase of Nitration of Alkanes

Nitro alkanes are made industrially by passing a gaseous mixture of nitric acid and alkane, they are allowed to pass through a narrow metal tube at about 400°C

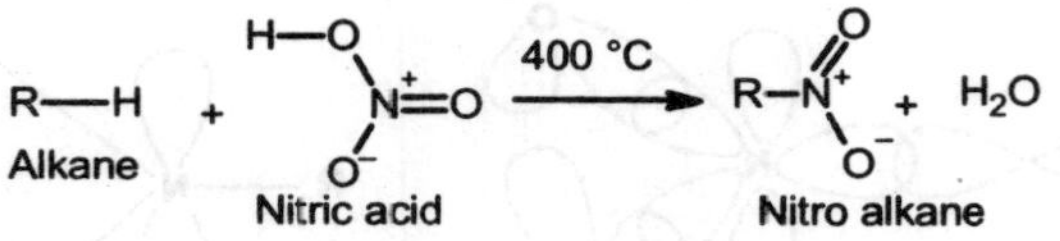

$$R—H + HO—N^+(=O)O^- \xrightarrow{400\,°C} R—N^+(=O)O^- + H_2O$$

Alkane, Nitric acid, Nitro alkane

With alkanes other than methane, a mixture of nitro alkanes is obtained which can be separated by fractional distillation, for example.

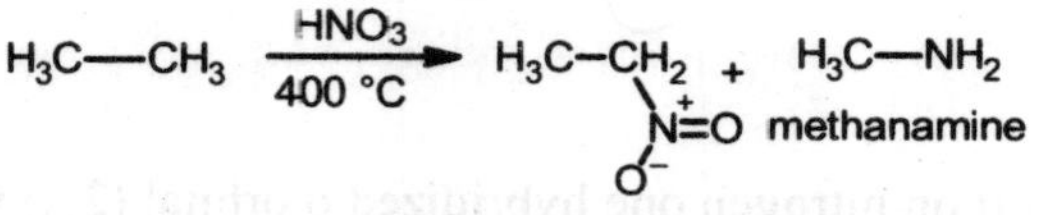

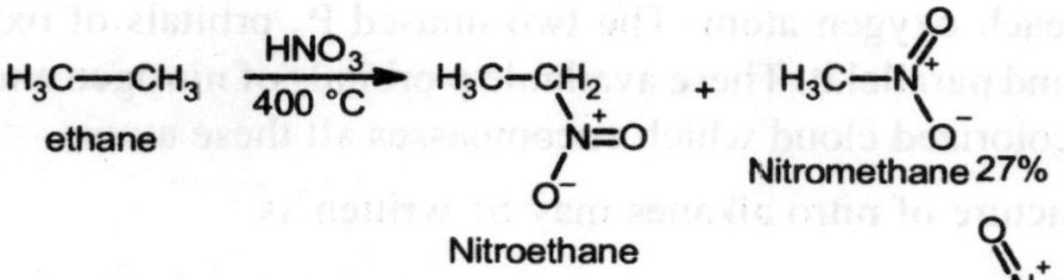

$$H_3C—CH_3 \xrightarrow[400\,°C]{HNO_3} H_3C—CH_2—N^+(=O)O^- + H_3C—NH_2$$

methanamine

$$H_3C—CH_3 \xrightarrow[400\,°C]{HNO_3} H_3C—CH_2—N^+(=O)O^- + H_3C—N^+(=O)O^-$$

ethane, Nitroethane, Nitromethane 27%

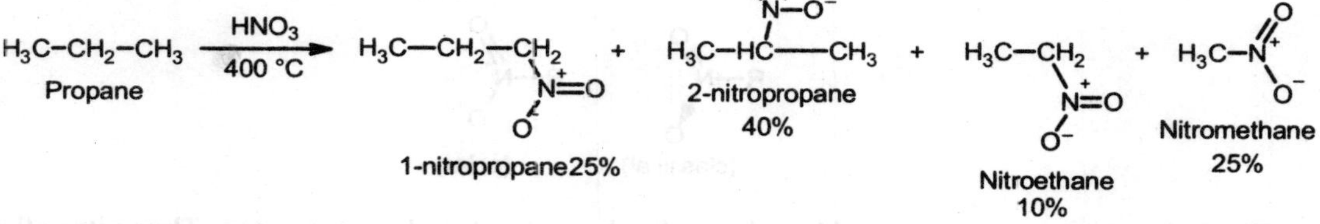

$$H_3C—CH_2—CH_3 \xrightarrow[400\,°C]{HNO_3} H_3C—CH_2—CH_2—N^+(=O)O^- + H_3C—CH(N^+(=O)O^-)—CH_3 + H_3C—CH_2—N^+(=O)O^- + H_3C—N^+(=O)O^-$$

Propane, 1-nitropropane25%, 2-nitropropane 40%, Nitroethane 10%, Nitromethane 25%

The nitroalkane with less carbon atoms are produced by the initial rupture of C – C bonds followed by nitration.

Similarly n – butane yields nitro methane, nitro ethane, 1 – nitro butane and 2 – nitro butane .

(2) Action of Alkyl Halides Metal Nitrites

Nitro alkanes are obtained in the laboratory by the action of primary or secondary alkyl halides (bromides or iodides) on silver nitrite* in ethanol.

$$\underset{\substack{1^\circ \text{ or } 2^\circ \\ \text{alkyl halide}}}{R-Br} + AgNO_3 \rightarrow \underset{\substack{\text{Nitroalkane} \\ 80\%}}{R-NO_2} + AgBr$$

The product contains 80% nitroalkane along with 20% of the isomeric alkyl nitrite (RO – N =O) which is also produced in the reaction. The components of the mixture can be easily separated by fractional distillation as the nitroalkane boils as much higher temperature than the alkyl nitrite.

$$R-Br + NaNO_2 \xrightarrow{\text{N, N dimethylformamide}} R-N^{+}(=O)-O^{-} + NaBr$$

Tertiary halides react with metal nitrites form chiefly alkyl nitrites and alkenes and are therefore not used.

(3) Action of Sodium Nitrite with α—halogeno carboxylic acid.

Sodium nitro carbohydrate produced in the for instance decarboxylates to form nitroalkane; for example

$$\underset{\substack{\text{(chloroacetato-}\kappa\text{ O)sodium} \\ \text{sodium chloroacetate}}}{Cl-CH_2-C(=O)-O-Na} + NaNO_2 \xrightarrow[\text{aq solution}]{\text{boil}} \underset{\substack{\text{(nitroacetato-}\kappa\text{ O)sodium} \\ \text{sodium nitro acetate}}}{O=N^{+}(O^{-})-CH_2-C(=O)-O-Na}$$

$$O=N^{+}(O^{-})-CH_2-C(=O)-O-Na + H_2O \longrightarrow \underset{\text{Nitromethane}}{H_3C-N^{+}(=O)-O^{-}} + NaHCO_3$$

(4) Hydrolysis of α-Nitroalkane

Leavy and Scaife recently developed a method of preparing nitro alkanes by hydrolyzing α-nitroalkane in presence of acids, Thus

$$\underset{\text{2-methyl-1-nitroprop-1-ene}}{H_3C-C(CH_3)=CH-N^{+}(=O)-O^{-}} + H_2O \longrightarrow H_3C-C(CH_3)=O + \underset{\text{Nitromethane}}{H_3C-N^{+}(=O)-O^{-}}$$

---

*silver nitrate is a costly reagent by using sodium nitrite is suitable medium like N, N dimethylformamide nitroalkane can be obtained in 50 to 60% yield.

(5) Oxidation of Oximes

Primary and secondary nitroalkanes are obtained in good yields by oxidizing acetaldehydeoximes and ketoximes, respectively with the help of hydrofluoroperoxyacetic acid.

$$CF_3-CO-O-OH \longrightarrow CF_3-COOH + [O]$$

Trifluoroethaneperoxoic acid → Trifluoroacetic acid

$$H_3C-CH{=}N-OH + [O] \longrightarrow H_3C-CH_2-\overset{+}{N}(=O)-OH \rightleftharpoons H_3C-CH_2-\overset{+}{N}(=O)O^-$$

Acetaldehyde oxime → Ethyl(hydroxy)oxoammonium ⇌ Nitroethane

(6) Oxidation of T- Alkyl amine

Tertiary nitroalkanes are best prepared by oxidizing by oxidizing alkyl amines with aqueous potassium permanganate.

$$R_3C-NH_2 + 3[O] \xrightarrow{KMnO_4} R_3C-\overset{+}{N}(=O)O^- + H_2O$$

t-alkyl amine → 3° nitroalkane

Physical Properties

(1) The lower nitroalkanes are colourless pleasant smelling liquids at ordinary temperature.

(2) Nitromethane is about 10%

(3) Since they are polar molecules, nitroalkanes are useful solvents for polar and ionic compounds

(4) They have abnormally high boiling points. Thus,

Nitromethane	CH_3-NO_2	b.p = 101°C
Nitroethane	$CH_3-CH_2-CH_2$	b.p = 115°C
1 – Nitropropane	$CH_3CH_2CH_2-NO_2$	b.p = 112°C
2 – Nitropropane	$(CH_3)_2CH-NO_2$	b.p = 120°C

Properties of Nitromethane

Boiling point:	114°C
Melting point:	-50°C
Relative density (water = 1):	1.05

Property	Value
Solubility in water, g/100 ml at 20°C:	4.5
Vapour pressure, kPa at 20°C:	2.08
Relative vapour density (air = 1):	2.6
Relative density of the vapour/air-mixture at 20°C (air = 1):	1.03
Flash point:	28°C
Auto-ignition temperature:	414°C
Explosive limits, vol% in air:	4.0
Octanol/water partition coefficient as log Pow:	0.2

This is explained by the fact that nitroalkanes are highly polar compounds as shown by their single dipole moments (3.6D) due to the appreciable electronic attraction between the polar molecules, they need a larger amount of energy in form of heat to separate them.

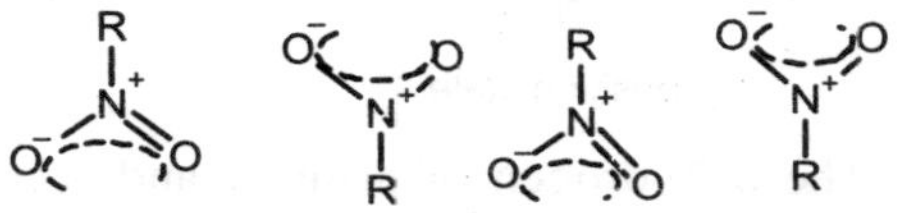

Fig. 17.1 : Intermolecular association in nitroalkane

Nitro alkanes have much higher boiling points than the less polar isomeric

alkyl nitrified $(CH_3 - O - NO)$ bp – 12°C,
$(C_2H_5 - O - NO)$ bp 17°C

They are less toxic than isomeric nitrites and aromatic nitro compounds. Nitro groups of nitro alkanes can be identified by strong infra red bands at 1580 and 1275 cm^{-1}.

Chemical Properties

The structure of nitro group indicates a positive charge on the nitrogen atom. Therefore, it is strongly electron withdrawing group an disable to exert a strong inductive effect (–1) and mesmeric effect (–M). Because of this electron withdraw effect; primary and secondary nitroalkanes possesses active hydrogen atoms on carbon adjacent to NO_2 group. In the presence of a base hydrogen atom is removed and a nucleophile anion results.

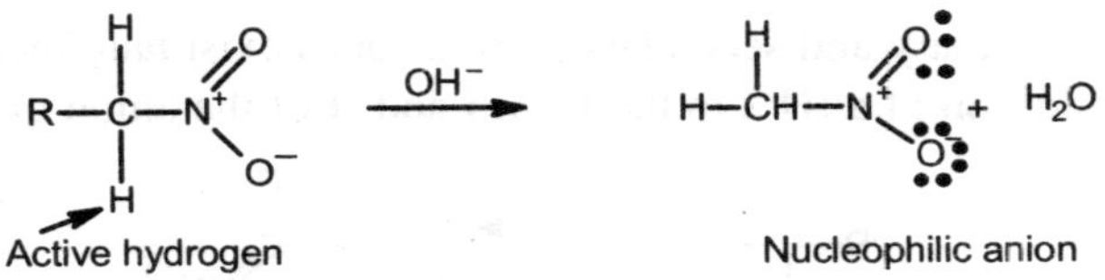

$$R-CH_2-N^+(=O)O^- \xrightarrow{OH^-} H-\bar{C}H-N^+(=O)O^- + H_2O$$

Active hydrogen   Nucleophilic anion

Therefore primary and secondary nitroalkanes are capable of undergoing nucleophilic addition reaction as also exhibiting tautomerism. Their reactions are detailed below:

(1) Action of Heat: Nitro alkanes are decomposed on moderate heating beyond 300°C. Alkanes are formed with the elimination of nitrous acid.

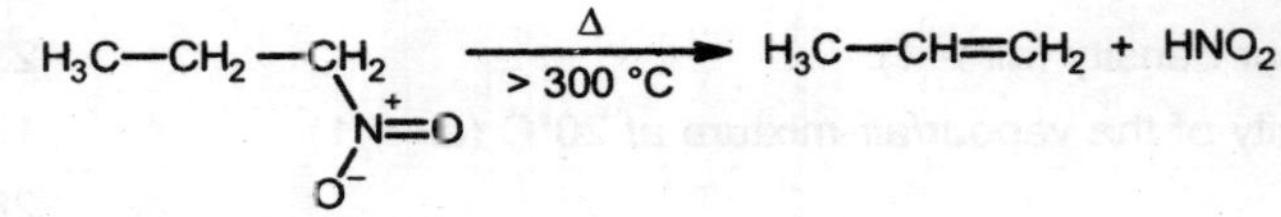

$$H_3C-CH_2-CH_2-\overset{+}{N}(=O)O^- \xrightarrow[> 300\,°C]{\Delta} H_3C-CH=CH_2 + HNO_2$$

Mechanism

The reaction presumably takes place through a cyclic transition state, which results in *cis* elimination.

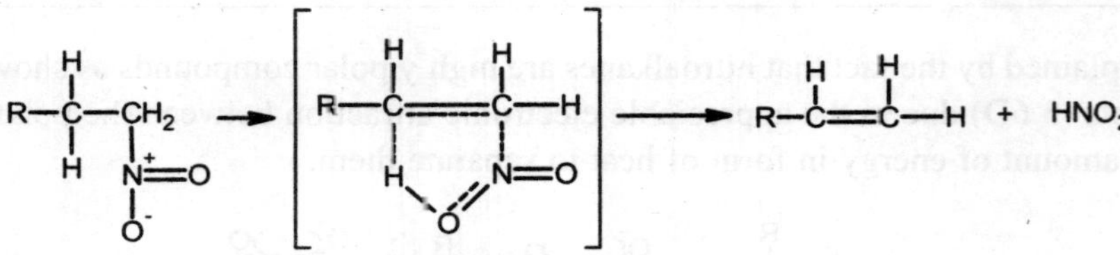

$$R-CH_2-CH_2-\overset{+}{N}(=O)O^- \longrightarrow [\text{Transition state}] \longrightarrow R-CH=CH-H + HNO_2$$

Transition state

(2) Formation of Salts: The α-hydrogen of primary and secondary nitroalkanes are acidic in nature much in the same fashion as the α-hydrogen atom of aldehydes and ketones. This they dissolve in NaOH or KOH forming salts.

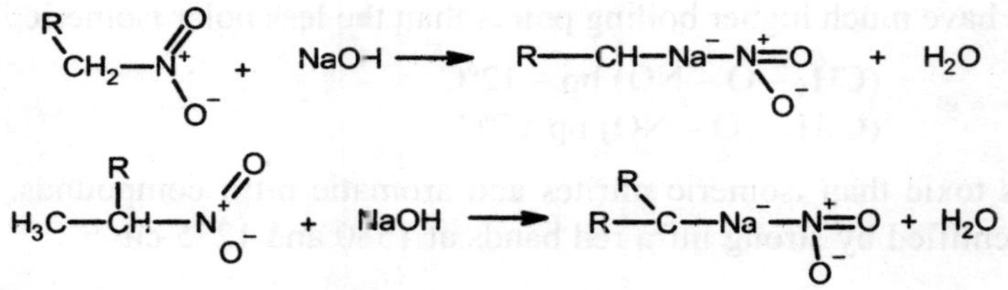

$$R-CH_2-\overset{+}{N}(=O)O^- + NaOH \longrightarrow R-\overset{-}{C}H-Na-\overset{+}{N}(=O)O^- + H_2O$$

$$H_3C-CH(R)-\overset{+}{N}(=O)O^- + NaOH \longrightarrow R-C(R)-\overset{-}{Na}-\overset{+}{N}(=O)O^- + H_2O$$

The acidic behaviour and formation of salt is associated to strongly electron with drawing effect of NO_2 group and resonance stabilization of the following resulting anion.

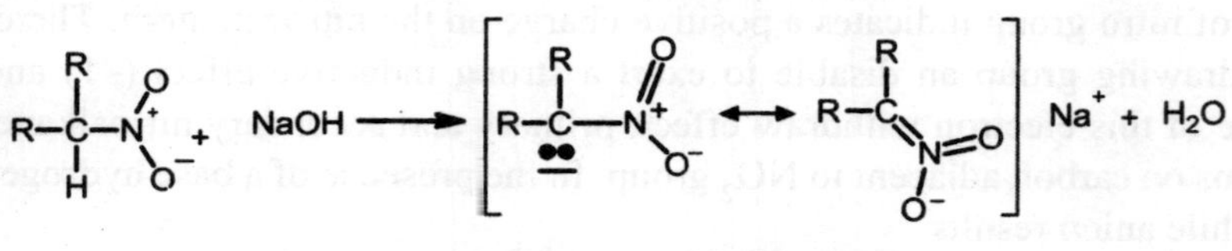

$$R-C(R)(H)-\overset{+}{N}(=O)O^- + NaOH \longrightarrow \left[R-\ddot{C}(R)-\overset{+}{N}(=O)O^- \longleftrightarrow R-C(R)=\overset{+}{N}(=O)O^-\right] Na^+ + H_2O$$

Resonance forms of anion

It is of interest that primary and secondary nitro alkanes exist tautomerism, the tautomeric forms being derived by the addition of a (H^+) to the form I and II of the anion of the salt shown above.

$$R-C(R)(H)-\overset{+}{N}(=O)O^- \rightleftharpoons R-C(R)=\overset{+}{N}(=O)O^-$$

Nitro form or iso form — Nitro ionic or aci form

Tautomeric equilibrium lies almost extremely to the nitro form due to the resonance stabilization of the NO_2 group, that is why nitroalkanes are often called "pseudo acids" the acid behaviour being shown only in presence of strong acids when the (aci form) acids, the aci form is also called nitronic acid.

The existence of the tautomeric mixture has been exhibited by isolation of the ion and aci form the pure state. It is found that iso from dissolves in NaOH solution slowly while the aci form dissolves instantly. Nitro form gives no colour with ferric chloride, while the acidification of sodium salt of nitro alkane produces a brown colour, the characteristic enol structure $(C = C - OH)$. Lastly the acidification of sodium salt of a nitro alkane raises the electrical conductivity of the solution due to the production of the acid form hot at some one the conductivity falls due to the conversion of aci form to unionized nitro form.

(3) Halogenation: Primary and secondary nitroalkanes when treated with halogen (chlorine or bromine) in alkaline solution are readily homogenized. The α-hydrogen are thus replaced by the hydrogen.

$$R-CH_2-N^+(=O)O^- \xrightarrow[NaOH]{+Br_2} R-CHBr-N^+(=O)O^- \xrightarrow[NaOH]{+Br_2} R-CHBr-N^+(=O)O^- \xrightarrow[NaOH]{+Br_2} R-CBr_2-N^+(=O)O^- + NaBr$$

1° nitroalkane 1° monobromo alkane 1° dibromoalkane

$$R-CH_2-N^+(=O)O^- \xrightarrow[NaOH]{+Br_2} R-CHBr-N^+(=O)O^-$$

2° nitroalkane 2° monobromo alkane

$$R_3C-N^+(=O)O^- \xrightarrow[NaOH]{+Br_2} \text{No action}$$

3° nitroalkane

As shown above primary nitroalkane having two replaceable α-hydrogen can form mono and dihydrogen derivatives, while secondary nitroalkanes having one α-hydrogen can form mono-halogen derivative only tertiary nitroalkane with no replaceable hydrogen do not react at all.

In case of nitro alkane, all the three A atom can be successfully replaced by halogen atoms, thus, trichloronitromethane (chloropicrin) an important insecticide is manufactured by the action of Nitromethane with chlorine in presence of sodium hydroxide.

Mechanism

The halogen substitution reactions of nitro alkanes are accounted for as due to the inductive effect of the nitro group.

$$R-CH_2-N^+(=O)O^- + NaOH \longrightarrow R-CH(Na)-N^+(=O)O^- + H_2O$$

$$\left[R-CH_2-\overset{+}{N}(=O)-O^-\right]Na + Br-Br \longrightarrow R-CHBr-N^+(=O)O^- + NaBr$$

α-bromonitro alkane

The above steps are repeated so as to form the dibromo derivation

(4) The reaction with Nitrous Acid: Primary secondary and tertiary nitroalkanes behave

differently when treated with nitrous acid. The reactive hydrogen atoms or to NO_2 group are involved in the reaction :

(a) A primary structure gives a blue nitroso nitroalkane, but in the presence of NaOH solution it produces a red colour due to the formation of sodium salt.

$$R-CH_2-\overset{+}{N}(=O)-O^- + HO-\overset{+}{N}(=O)-O^- \longrightarrow R-CH(N=O)-\overset{+}{N}(=O)-O^- + H_2O$$

1° nitroalkane → Nitroso alkane

$$R-CH(N=O)-\overset{+}{N}(=O)-O^- \rightleftharpoons R-C(=N-O^-)(\overset{+}{N}=O)\ \ (N=O) \xrightarrow{NaOH} R-C(NO)=\overset{+}{N}(O-Na)-O^-$$

Nitrosonitrosoalkane (nitro-form) ⇌ Nitrolic acid (acid form) → Soluble sodium salt (red)

(b) A secondary nitro alkane gives a blue nitroso derivative which no more contains replaceable H – atom and is therefore, insoluble.

$$R_2CH-\overset{+}{N}(=O)-O^- + HNO_2 \longrightarrow R_2C(N=O)-\overset{+}{N}(=O)-O^- + H_2O$$

Isoluble nitroso - nitro alkane (blue)

$$R_3C-\overset{+}{N}(=O)-O^- + HNO_2 \xrightarrow{NaOH} \text{No action}$$

3° nitroalkane

> **The reaction of nitrous acid on the three types (1°, 2°, 3°) of nitroalkanes serves as a test for the identification of nitro alkanes.**

Table 17.1. Identification of 1°, 2°, 3° nitroalkans

Nitroalkane	Reagent	Product	Reagent	Result
$R-CH_2-NO_2$ colourless	HONO →	$R-C(NO)-NO_2$ blue	NaOH →	$R-C(NO)=NO_2^- + Na^+$ Redsolution
R_2CH-NO_2 colourless	HONO →	$R_2C(NO)-NO_2$ blue	NaOH →	Insoluble
R_3C-NO_2 colourless	HONO →	No Change	NaOH →	No change

Since nitroalkanes can be formed from 1°, 2° and 3° alcohols by treatment with HI and then $AgNO_3$, the reaction with nitrous acid also forms the basis of the victor Mayor test for distinguished 1°, 2° and 3° alcohols.

(5) Reaction: Nitroalkanes are reduced to a primary amine with hydrogen on Rancy Nickel and with lithium aluminium hydride.

$$R-N^{+}(=O)O^{-} + 3H_2 \xrightarrow{\text{Rancy Nickel}} \underset{1^\circ \text{ amine}}{R-NH_2} + H_2O$$

$$R-N^{+}(=O)O^{-} \xrightarrow{LiAlH_4} R-NH_2$$

This reaction can also be effected with iron and hydrochloric acid

$$R-N^{+}(=O)O^{-} + 6H^{+} + 6e^{-} \longrightarrow H_3C-NH_2 + 2H_2O$$

Now that nitroalkanes are becoming in large quantities this method can be used for the industrial production of primary amines.

(6) Hydrolysis: (a) Primary nitroalkanes on boiling with concentrated hydrochloric acid or Sulphuric acid are hydrolysed to form a carboxylic acid and hydroxylamine.

$$\underset{1^\circ \text{ nitroamine}}{R-CH_2-N^{+}(=O)O^{-}} + H_2O \xrightarrow[\Delta]{H^+} \underset{\text{Carboxylic acid}}{R-C(=O)-OH} + \underset{\text{Hydrooxylamine}}{H_2N-OH}$$

This reaction proceeds by oxidation of methane group and reduction of nitrogen dioxide group accompanied by C – N bond cleavage. It is used for the commercial production of hydroxylamine.

Mechanism

The reaction is believed to take place through the formation of hydroxamic and acid as.

$$\underset{\text{Nitroalkane}}{R-CH_2-N^{+}(=O)O^{-}} \rightleftharpoons \underset{\text{aci form}}{H_3C-CH=N^{+}(OH)O^{-}} \xrightarrow{+H^+} H_3C-CH=N^{+}(OH)O^{+} \longrightarrow \underset{\text{hydrooxamic acid}}{H_3C-C(OH)=N-OH \;\; \dot{O}H_2}$$

$$\xrightarrow[\text{hydrolysis}]{Al} H_2N-OH + H_3C-C(OH)=O$$

(6) Primary and secondary nitroalkane may be hydrolysed: by first converting then to the salts of their aci form by NaOH, which on boiling with 50% Sulphuric acid, produced aldehydes and ketones reaspectively.

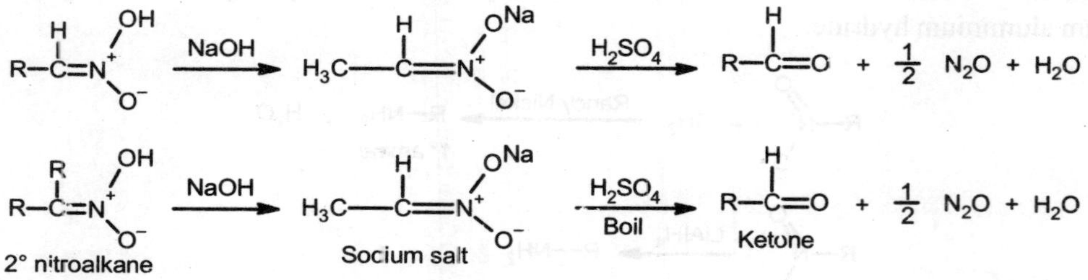

$$R-CH=N^+(OH)(O^-) \xrightarrow{NaOH} H_3C-CH=N^+(ONa)(O^-) \xrightarrow{H_2SO_4} R-CH=O + \frac{1}{2} N_2O + H_2O$$

$$\underset{\text{2° nitroalkane}}{R_2C=N^+(OH)(O^-)} \xrightarrow{NaOH} \underset{\text{Socium salt}}{H_3C-CH=N^+(ONa)(O^-)} \xrightarrow[\text{Boil}]{H_2SO_4} \underset{\text{Ketone}}{R-CH=O} + \frac{1}{2} N_2O + H_2O$$

The above reaction, popularly known as Nef reaction has been successfully employed for the production of aldehydes and ketones

(7) Condensation with Aldehydes and Ketones: Primary and secondary nitroalkanes react with aldehydes and ketones in the presence of dilute alkali. This reaction which yields nitro alcohols is quite similar to aldol condensation, for example

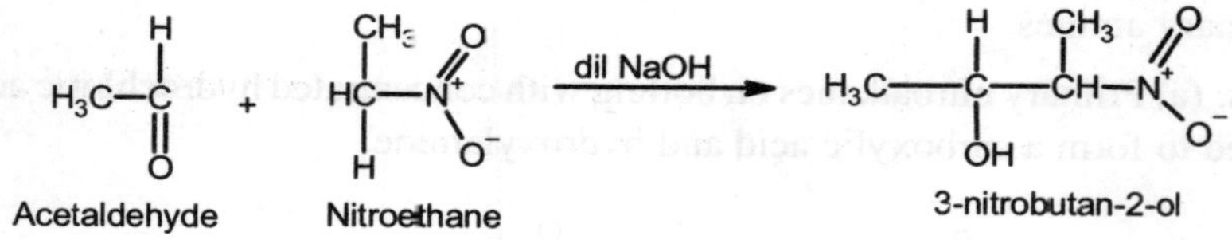

$$\underset{\text{Acetaldehyde}}{H_3C-CHO} + \underset{\text{Nitroethane}}{CH_3-CH_2-N^+(=O)O^-} \xrightarrow{\text{dil NaOH}} \underset{\text{3-nitrobutan-2-ol}}{H_3C-CH(OH)-CH(CH_3)-N^+(=O)O^-}$$

Mechanism

This condensation reaction proceeds by the following steps.

(a) Carbanion formation:

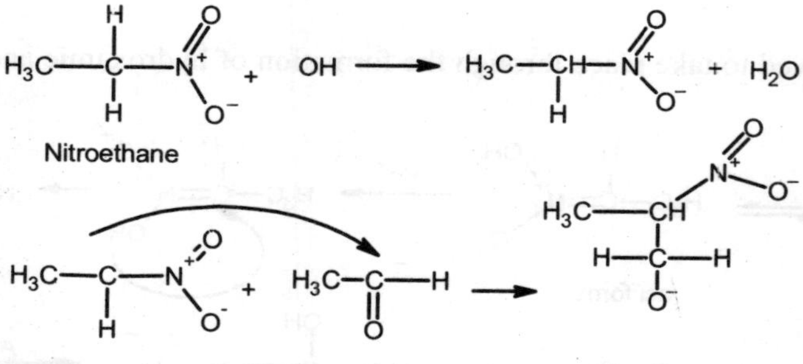

$$\underset{\text{Nitroethane}}{H_3C-CH_2-N^+(=O)O^-} + OH \longrightarrow H_3C-\bar{C}H-N^+(=O)O^- + H_2O$$

$$H_3C-\bar{C}H-N^+(=O)O^- + H_3C-CHO \longrightarrow H_3C-CH(N^+(=O)O^-)-CH(O^-)-H$$

Proton takeup from water:

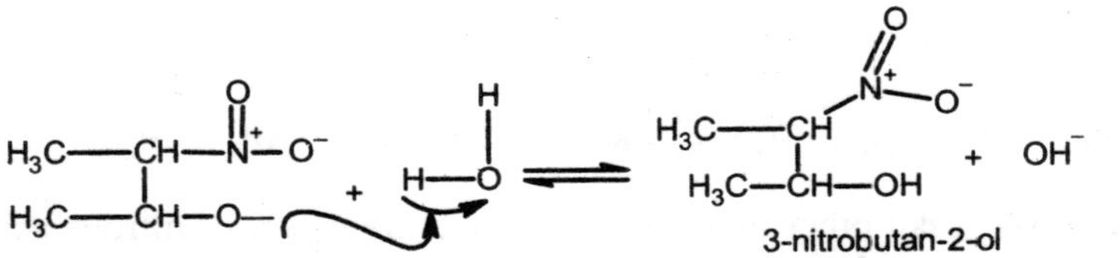

$H_3C—CH—N^+(=O)—O^-$ / $H_3C—CH—O^-$ + $H—O—H$ ⇌ $H_3C—CH(NO_2)—CH(OH)—CH_3$ + OH^-

3-nitrobutan-2-ol
1- nitro -1- methyl 2- propenol
(nitroalcohol)

Uses of Nitro Alkanes

Nitroalkanes can be used in the synthesis of nitroolefines, ß-amino alcohols, aldehydes (Nef Reaction) and nitroesters (Michael Addition). As very polar solvents they are applied in Friedel-Crafts-Reactions.

Nitroalkanes are very important starting materials in the formation of carbon-carbon bond. The nitronate anion can react under basic conditions with saturated carbonyl compounds or with electron poor alkenes leading to the nitroaldol (Henry) reaction or to the Michael reaction, respectively, with the formation of di- or polyfunctionalized nitroderivatives. The most commonly applied protocols to perform the above reactions require the use of basic catalysts under both homogeneous and heterogeneous conditions.

How to distinguish nitroalkane from alkyl nitrites

The two classes of compounds are functional isomers and can be distinguished from each other by specific behaviour

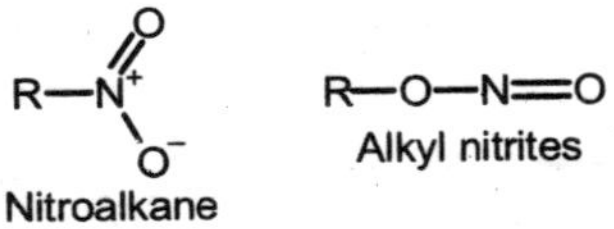

$R—N^+(=O)—O^-$ Nitroalkane $R—O—N{=}O$ Alkyl nitrites

(1) Reduction of nitro alkanes: Nitro alkanes when reduced with hydrogen or Raney's nickel, or $LiAlH_4$ yields amines

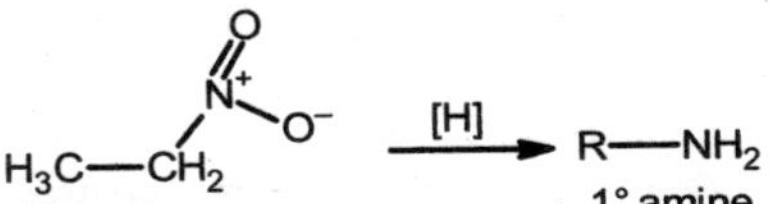

$$H_3C—CH_2—N^+(=O)O^- \xrightarrow{[H]} R—NH_2$$

Nitroethane   1° amine

The fact and amine is so produced demonstrates the presence of C – N bond in the original nitroalkane. The reduction of isomeric alkyl nitrite cannot lead to an amine.

R—O—N=O cannot form $R—NH_2$

Actually alkyl nitrites produce an alcohol on reduction which shows that N is bonded to carbon trough oxygen (C – O – N)

$$H_3C-CH_2-NO_2 \xrightarrow{[H]} H_3C-CH_2-OH + NH_3 + H_2O$$

(2) Basic hydrolysis: **Like other alkyl nitrites which are esters of nitrous acid on boiling which NaOH solution yields the parent alcohol and sodium nitrite.**

$$\underset{\text{Alkyl nitrites}}{R-N^+(=O)O^-} + NaOH \longrightarrow \underset{\text{Alcohol}}{R-OH} + NaNO_2$$

Nitroalkanes have no action with NaOH solution. However 1° and 2°. Nitro alkanes form soluble sodium salts.

$$H_3C-N^+(=O)O^- \rightleftharpoons H_3C-N(OH)O^- \xrightarrow{NaOH} H_2C=N^+(O-Na)O^-$$

(3) Reaction with nitrous acid: **Primary and secondary nitroalkanes when treated with nitrous acid from blue nitroprusso derivatives.**

$$R-CH_2-N^+(=O)O^- \xrightarrow{HONO} R-CH(N=O)-N^+(=O)O^- \xrightarrow{NaOH} \text{Red sodium salt}$$

$$\underset{\text{2° nitroalkanes}}{R_2CH-N^+(=O)O^-} \xrightarrow{HONO} \underset{\text{Nitroso-nitroalkane 2° (blue)}}{R-C(N=O)(CH_3)-N^+(=O)O^-} \xrightarrow{NaOH} \text{Unchanged}$$

One the other hand alkyl nitrites can have no reduction with nitrous acid

$$R-N^+(=O)O^- + HONO \longrightarrow \text{no action}$$

Diazoalkanes

These are derivatives of alkanes in which two hydrogen atoms on the same carbon have been replaced by the divalent diazo group N_2 (di = 2 and azo, azote = nitrogen (see chapter periodic table of elements)

$$H-\underset{H}{\overset{H}{\underset{|}{\overset{|}{C}}}}-H \longrightarrow \begin{matrix}H\\ \end{matrix}\!\!>\!C{=}N{\equiv}N \quad \text{or} \quad CH_2N_2$$

Methane Diazomethane

$$\begin{matrix}H_3C & & H\\ & C & \\ H & & H\end{matrix} \longrightarrow \begin{matrix}H_2C\\ H\end{matrix}\!\!>\!C-\overset{+}{N}{\equiv}N \quad \text{or} \quad CH_3CN_2H$$

Ethane Diazoethane

The diazogroup (>NH_2) is the functional group which is represented as a resonance hybrid of two forms.

$$>C{=}\overset{+}{N}{=}N \longleftrightarrow >C{=}N{\equiv}N$$

Here we will take up a detailed discussion of diazomethane, the simplest member of the classs

DIAZOMETHANE

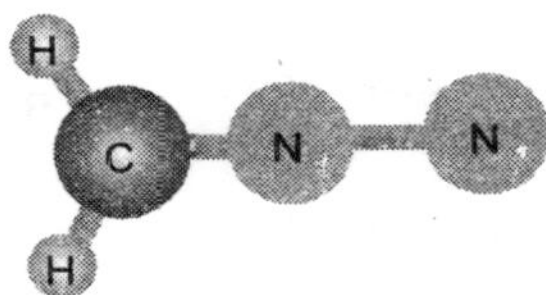

It is most important diazoalkane and has valuable synthetic application.

Diazomethane which is a toxic, explosive yellow gas, is one of the more common diazo compounds. It is usually prepared as a solution in ether and often used in laboratory procedures for converting carboxylic acids into their methyl esters or into their homologues. Diazomethane is prepared in the laboratory at mmol scale from diazomethane precursors such as **Diazald** or N-methyl-N-nitroso-p-toluenesulfonamide and **MNNG** or 1-methyl-3-nitro-1-nitrosoguanidine. Diazald in a solution of diglyme and diethyl ether reacts with a solution of potassium hydroxide in water at elevated temperatures and generated diazomethane is collected by distillation. Diazomethane is liberated from a solution of MNNG in diethyl ether by addition of sodium hydroxide in water at low temperatures. Diazomethane may explode when in contact with ground-glass joints or when heated to about 100°C, and a blast shield should be employed for its use.

Preparation

Diazomethane is an extremely versatile reagent for the preparation of both carbon-carbon and carbon-heteroatom bonds.[1] It is one of the most common methylating reagents for the preparation of methyl esters from the corresponding carboxylic acids.[2] Additionally, diazomethane has found extensive application in the alkylation of phenols, enols, and heteroatoms such as nitrogen and sulfur. Diazomethane has also been used in: cycloalkanone ring expansion,[3] preparation of α-diazo ketones,[4] and pyrazoline formation,[5] and Pd-catalyzed cyclopropanation (**Scheme 1**).[6]

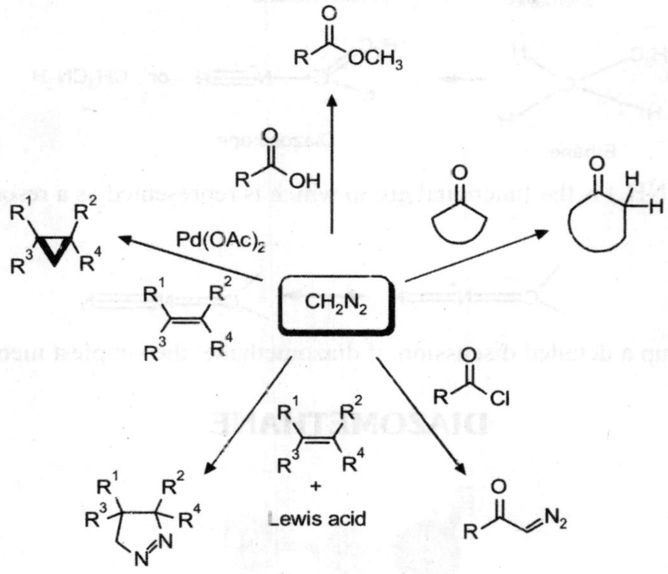
Pd(OAc)$_2$
CH$_2$N$_2$
Lewis acid

Scheme 1

Due to its instability and toxicity, diazomethane is not amenable to storage or transport.[7] Therefore, several diazomethane precursors have been developed over the past few decades. The majority of these precursors contain an *N*-methyl-*N*-nitroso group, which generates diazomethane upon treatment with base.

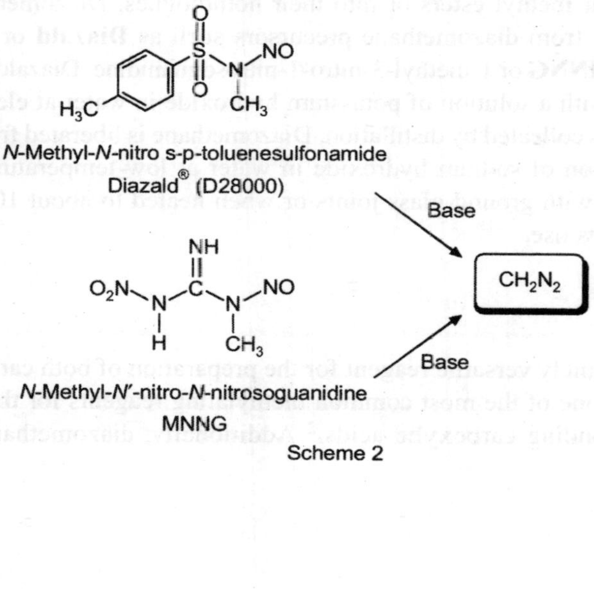
N-Methyl-*N*-nitro s-p-toluenesulfonamide
Diazald® (D28000)
Base
CH$_2$N$_2$
Base
N-Methyl-*N*′-nitro-*N*-nitrosoguanidine
MNNG

Scheme 2

Of the precursors used, *N*-methyl-*N'*-nitro-*N*-nitrosoguanidine (MNNG) and *N*-methyl-*N*-nitroso-*p*-toluenesulfonamide (Diazald) have been the most popular. MNNG is toxic, a severe irritant, a carcinogen and a potent mutagen. Additionally, the use of MNNG has historically been limited to small-scale production of diazomethane (ca. 1 mmol). Because of these hazard concerns, Sigma-Aldrich no longer offers MNNG for sale.

While Diazomethane is also considered to be a severe irritant, it has not demonstrated the acute toxicity of MNNG(methyl nitro nitroso guanidine)rge-scale production (300 mmol) of diazomethane.

Structure

Diazomethane was originally believed to possess a three-membered diazirine ring structure. This was ruled out when electron diffraction and dipole measurement proved that it has a linear structure with H – C – N angle as 120°C

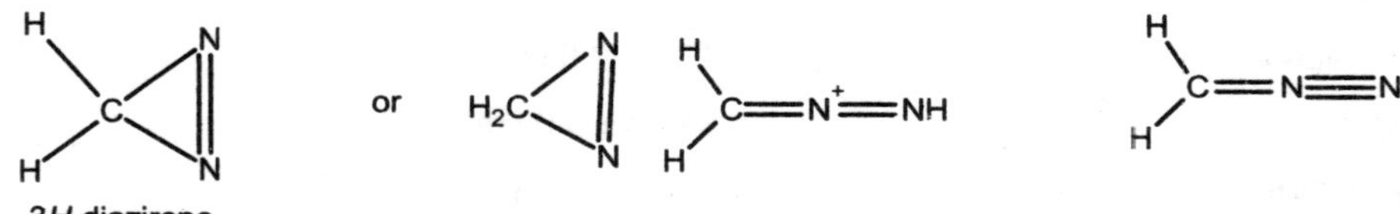

3*H*-diazirene
Diazirine ring structure is incorrent

This is an incorrect double bond

Correct structure of diazomethane

The structure of diazomethane is now represented in two ways: -

(a) As resonance hybrid of the extreme electronic states:

$H_2C{=}\overset{+}{N}{=}\overset{-}{\ddot{N}}\text{:} \longleftrightarrow H_2\overset{-}{C}H{-}\overset{+}{N}{\equiv}N\text{:}$

(b) As Delocalized Orbital Model: Carbon in its sp^2 states of hybridization forms two σ-bonds with H-atoms and one -bond with N – atom in sp state of hybridization. The central

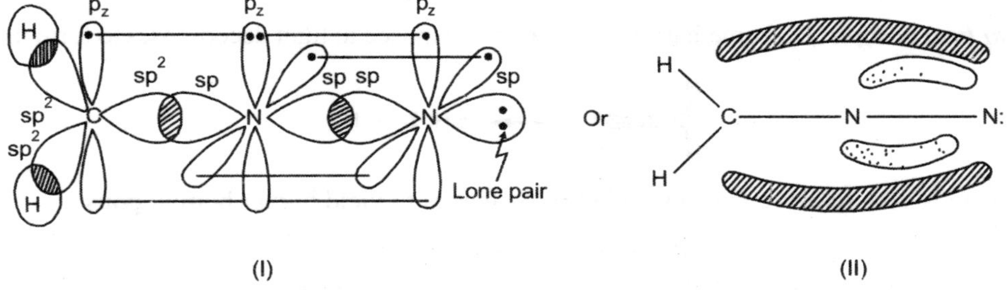

Fig. 17.2 : Molecular Model of Diazomethane

N- atom forms another σ-bond with the second n – atom. This leaves a P_x orbital on carbon a, P_y and P_z orbital and a lone pair in its sp orbital. The various overlaps are shown in fig above (i) and the delocalized π molecular orbital over C – N – N system is shown.

The above orbital structure justifies the linkage of both the nitrogen atoms with the same carbon atom.

Physical Properties

Molecular Formula	= CH_2N_2
Melting Point	= 145 °C
Boiling point	= 23 °C
Formula Weight	= 42.03998
Composition	= C(28.57%) H(4.80%) N(66.64%)
Molar Refractivity	= 11.93 ± 0.5 cm^3
Molar Volume	= 47.9 ± 7.0 cm^3
Parachor	= 112.3 ± 8.0 cm^3
Index of Refraction	= 1.412 ± 0.05
Surface Tension	= 30.2 ± 7.0 dyne/cm
Density	= 0.87 ± 0.1 g/cm^3
Polarizability	= 4.72 ± 0.5 $10^{-24} cm^3$
Monoisotopic Mass	= 42.021798 Da
Nominal Mass	= 42 Da
Average Mass	= 42.040435 Da

Chemical Properties

(I) Action of heat or light: Diazomethane when heated or exposed to light decomposes to form methylene

$$H_2C{=}N{\equiv}N \xrightarrow[\Delta]{h\nu} CH_2\colon + N_2$$

Methylene is very reactive reagent and adds to alkanes to yield higher homologous

$$\underset{\text{alkane}}{R{-}R} + CH_2 \longrightarrow \underset{\text{higher alkane}}{R{-}CH_2{-}R}$$

(II) Reduction: When reduced with sodium amalgam it is converted to methyl hydrazine

$$H_2C{=}N{\equiv}N + 4[H] \xrightarrow{Na/Hg} \underset{\text{methylhydrazine}}{H_3C{-}NH{-}NH_2}$$

(III) Reaction with Mineral Acid: Hydrochloric acid and Sulphuric acid convert methane to diazomium salts which at one decompose to form methyl derivatives and nitrogen thus,

$$H_2C{=}\overset{+}{N}{\equiv}\overset{-}{N} + H{-}Cl \longrightarrow H_3C{-}\overset{+}{N}{\equiv}N\,\overset{-}{Cl} \longrightarrow H_3C{-}Cl + N{\equiv}N$$

1-chloro-2-methyldiazynediium chloromethane nitrogen

(IV) Reaction with carboxylic acids: Diazomethane reacts readily with carboxylic acid to form methyl esters, the H-atom of the – COOH group being converted to methyl group.

$$\underset{\text{carboxylic acid}}{R{-}\overset{O}{\overset{\|}{C}}{-}OH} + CH_2N_2 \xrightarrow{\text{ether}} \underset{\text{methyl ester}}{R{-}\overset{O}{\overset{\|}{C}}{-}O{:}CH_2} + N_2$$

This is an excellent method for converting expensive organic acids to methyl esters. The fact is that the only by product is nitrogen makes it an healthy organic reaction.

Mechanism

The reacticn takes place in two steps (1) the transfer of an organic acid proton to diazomethane at the negative carbon diazomethane generating the nucleophile carboxylate in and a diazomium ion (2) the products of step 1 react to form a new C – O bond giving methyl ester and molecular nitrogen.

(*i*) $$R{-}\overset{O}{\overset{\|}{C}}{-}O{-}H + H_3C{-}\overset{+}{N}{\equiv}N \longrightarrow \underset{\text{Carboxylate ion}}{R{-}\overset{O}{\overset{\|}{C}}{-}O^-} + \underset{\text{Dazomium ion}}{H_3C{-}\overset{+}{N}{\equiv}N}$$

(*ii*) $$R{-}\overset{O}{\overset{\|}{C}}{-}O^- + H_3C{-}\overset{+}{N}{\equiv}N \longrightarrow \underset{\text{methyl ester}}{R{-}\overset{O}{\overset{\|}{C}}{-}O{-}CH_3} + N{\equiv}N$$

(V) Reaction with phenols: Like carboxylic acids phenols are also methylated when treated with dizomethane to form methyl ester. Thus

$$\underset{\text{Phenol}}{C_6H_5OH} + CH_2N_2 \xrightarrow{\text{ether}} \underset{\text{Methyl phenyl ether (anizole)}}{C_6H_5{-}O{-}CH_3} + N_2$$

The reaction mechanism is similar to that of methylation of carboxylic acid

(VI) Reaction with alcohols and Amines: Diazomethane does not react with alcohols and amines ordinarily but in the presence of a catalyst such as BF_3 a Lews acid, the hydrogen of the – OH or >NH group is replaced by a methyl group

$$\underset{\text{Alcohol}}{R-O-H} + CH_2N_2 \xrightarrow{BF_3} \underset{\text{methyl ether}}{R-O-COH_3} + N_2$$

$$R-NH_2 + CH_2N_2 \xrightarrow{BF_3} R-NH-CH_3 + N_2$$

Mechanism

The function of BF_3 is to increase the acidity of the alcohol or amine

The released H^+ promotes methylation as shown below:

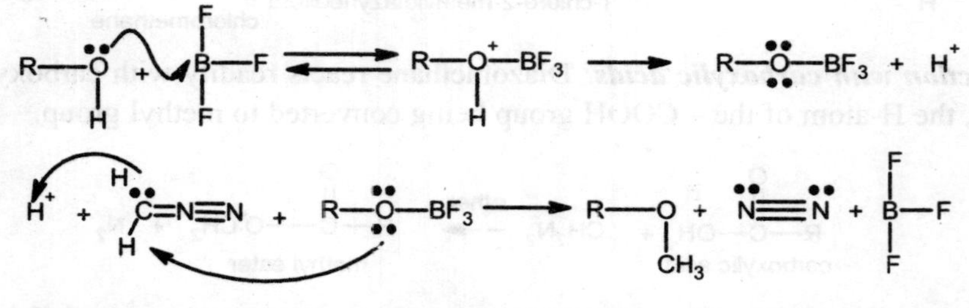

$$R-\ddot{O}(H) + BF_3 \rightleftharpoons R-\overset{+}{O}(H)-BF_3 \longrightarrow R-\ddot{O}-BF_3 + H^+$$

$$H^+ + H_2C=N\equiv N + R-\ddot{O}-BF_3 \longrightarrow R-O-CH_3 + N\equiv N + BF_3$$

(VII) Reaction with carbonyl compounds: Diazomethane converts aldehydes into methyl ketones, while ketones are converted into their higher homologous.

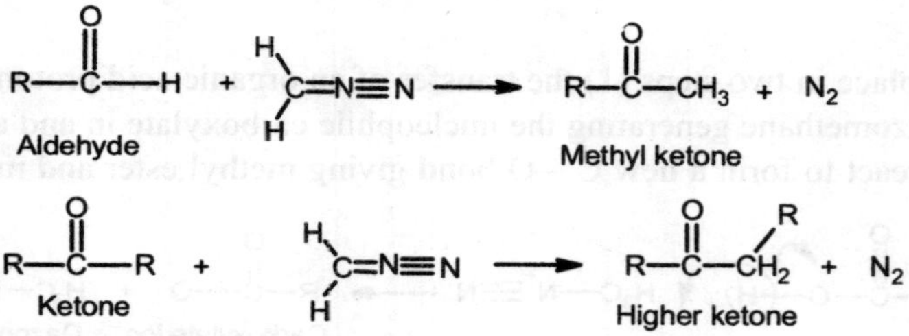

$$\underset{\text{Aldehyde}}{R-CO-H} + H_2C=N\equiv N \longrightarrow \underset{\text{Methyl ketone}}{R-CO-CH_3} + N_2$$

$$\underset{\text{Ketone}}{R-CO-R} + H_2C=N\equiv N \longrightarrow \underset{\text{Higher ketone}}{R-CO-CH_2-R} + N_2$$

Mechanism

The reaction proceeds by nucleophile addition at the carbonyl group. The adjust than looses N_2 by rearrangement.

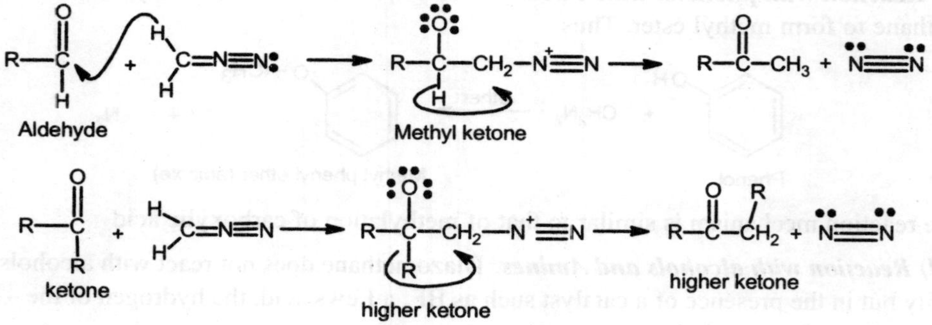

$$\underset{\text{Aldehyde}}{R-CHO} + H_2C=N\equiv N \longrightarrow \underset{\text{Methyl ketone}}{R-CH(O^-)-CH_2-\overset{+}{N}\equiv N} \longrightarrow R-CO-CH_3 + N\equiv N$$

$$\underset{\text{ketone}}{R-CO-R} + H_2C=N\equiv N \longrightarrow \underset{\text{higher ketone}}{R-CR(O^-)-CH_2-\overset{+}{N}\equiv N} \longrightarrow \underset{\text{higher ketone}}{R-CO-CH_2-R} + N\equiv N$$

The most interesting application of this reaction is ring expansion. Thus clyclohexane reacts with diazomethane to form cylcoheptane. The initial product again reacts with diazomethane to yield the corresponding epoxide and higher clycloanones.

Cyclohexanone + Diaomethane ⟶ Cyclooctanone + 1-oxaspiro[2.5]octane +

Mechanism

It involves nucleophiles addition followed by ether rearrangement or ring closure of the epoxide ring.

Cyclohexanone + Diaomethane ⟶ ⟶ Cyclooctanone + N≡N

⟶ + N≡N

(VIII) Reaction with Acid chlorides: Diazomethane reacts with an acid chloride (ROCl) to form a dazomethyl ketone. This when heated with water, in the presence of silver oxide (Ag_2O) yields a carboxylic acid containing one more carbon than the starting acid chloride.

$$\underset{\text{Acid chloride}}{R-\overset{O}{\overset{\|}{C}}-Cl} + \underset{\text{Diaomethane}}{2CH_2N_2} \longrightarrow \underset{\text{Diazomethyl ketone}}{R-\overset{O}{\overset{\|}{C}}-CH_2-\overset{+}{N}\equiv N} + H_3C-Cl + N\equiv N$$

$$R-\overset{O}{\overset{\|}{C}}-CH_2-\overset{+}{N}\equiv N + H_2O \xrightarrow[\Delta]{Ag_2O} H_3C-CH_2-\overset{O}{\overset{\|}{C}}-OH +$$

The complete sequence of reactions of converting a given carboxylic to its next higher homologue (Arudt – Eistert synthesis) is summarized as below:

$$\underset{\text{Acid}}{R-C(=O)OH} \xrightarrow{SOCl_2} \underset{\text{Acid chloride}}{R-\overset{O}{\overset{\|}{C}}-Cl} \xrightarrow{CH_2N_2} \underset{\text{Diazo ketone}}{R-\overset{O}{\overset{\|}{C}}-CH_2\overset{+}{N}\equiv N} \xrightarrow{Ag_2O} \underset{\text{Higher ketone}}{R-CH_2-C(=O)-HO}$$

Mechanism

The two important stages of the reaction of acid chloride with diazo methane leading to the formation of higher acid.

(*i*) Nucleophilic addition of diazomethane at the carboxyl carbon to form diazoketone.

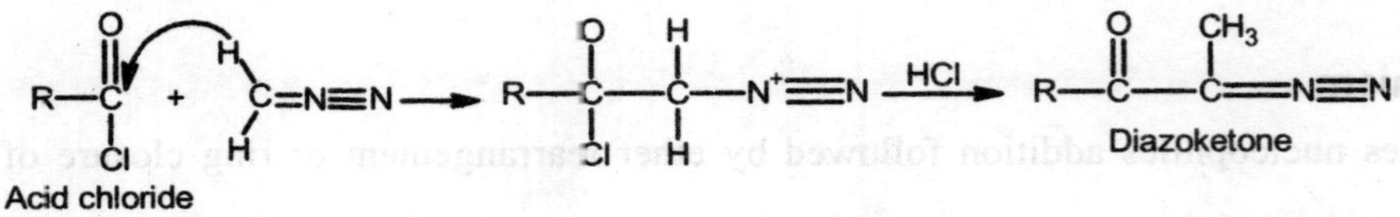

Acid chloride

Diazoketone

(*ii*) Wolf's rearrangement leading to the formation of a ketone that reacts with water to produce the acid.

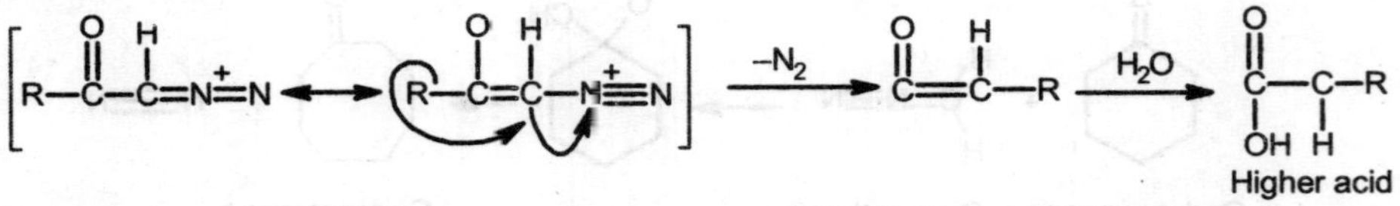

Higher acid

(*iii*) Addition to electrophilic and acetylenic pH bonds

Diazomethane adds to ethylene or acetylenic π-bonds leading to the formation of *heterocyclic compounds

1*H*-pyrazole

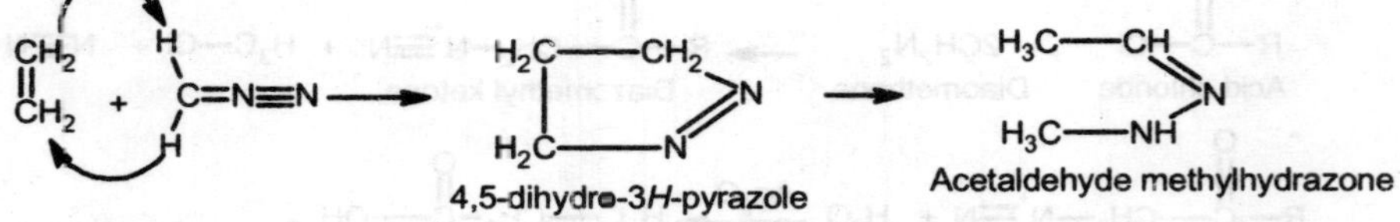

4,5-dihydro-3*H*-pyrazole

Acetaldehyde methylhydrazone

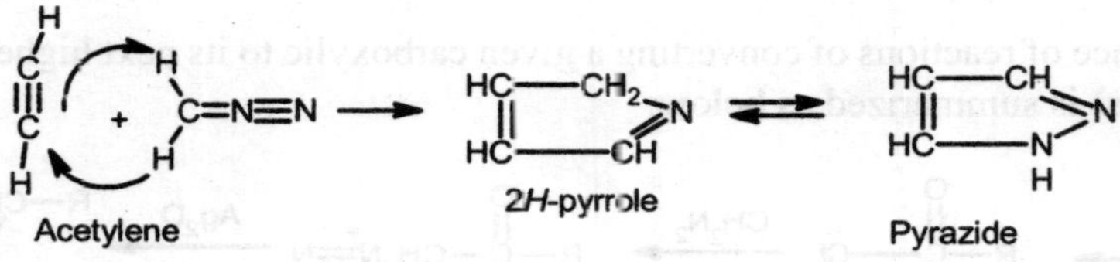

Acetylene

2*H*-pyrrole

Pyrazide

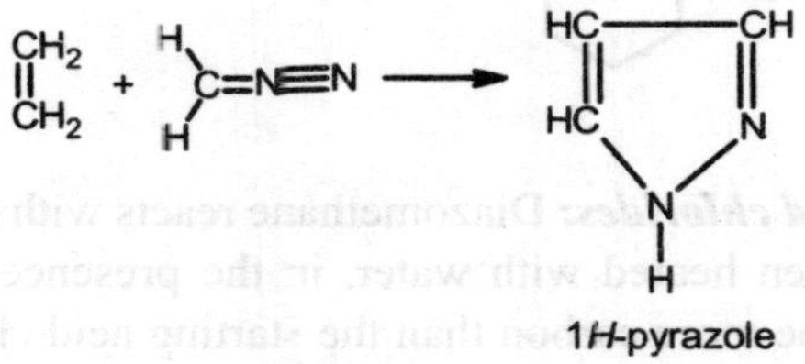

*we will be discussing soon about the heterocyclic compounds in the latter chapters.

AZIDES

These are the derivatives of hydrozoic acid N_3H and are characterized for the presence of the function $N \equiv N = N^-$ (azido group), Azide are named in the some way as halides. The name of the substituent groups is followed by the suffix azide as a separate word. Thus

Acyl azides are also known as azido-alkane

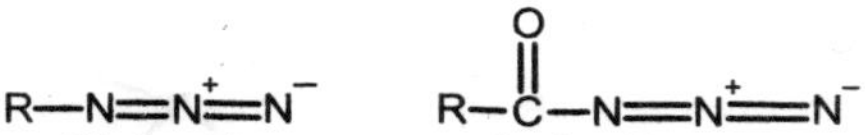

$$R-N=\overset{+}{N}=\overset{-}{N} \qquad R-\overset{O}{\overset{\|}{C}}-N=\overset{+}{N}=\overset{-}{N}$$

Alkyl azide Acyl azide

Structure

Like diazo group, the azido is known as a linear structure. It may be represented as a resonance hybrid of the two extreme bond structure as

$$R=\overset{+}{N}=N=\ddot{N} \longleftrightarrow R=\bar{N}=\overset{+}{N}\equiv\bar{N}$$

Orbital overlaps in alkyl azide leading to delocalized π molecular orbital over the nitrogen atoms is similar to that obtaining in diazo compounds.

Preparation

Alkyl azides may be conveniently obtained by the action of sodium azide (nitrous oxide and sodium azide on alkyl sulphate)

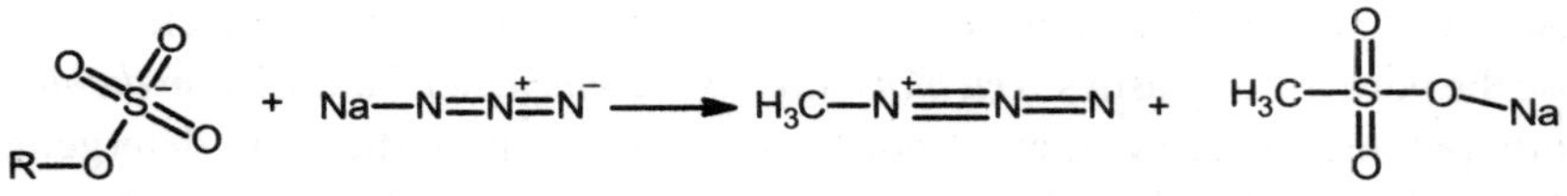

$$R-O-SO_3 + Na-N=\overset{+}{N}=\overset{-}{N} \longrightarrow H_3C-\overset{+}{N}\equiv N=N + H_3C-\overset{O}{\overset{\|}{\underset{\|}{\underset{O}{S}}}}-O-Na$$

(methanesulfonato-κO)sodium

In the above reaction the azide ion acts as the nucleophile bringing about substitution. The azide ion formation form N_2O and $NaNH_3$ may occur as follow:

$$\ddot{N}\equiv N=\ddot{O}: \longrightarrow \underset{H-N-H}{\ddot{N}=N=\ddot{O}:} \rightleftharpoons \underset{N-H}{:\ddot{N}=\overset{+}{N}-OH} \xrightarrow{H_2O} N\equiv N=\ddot{N}:$$

Azide ion

Properties

Azides are explosives as diazo compounds. They possess a very low boiling points including hydrogen bonding in them owing to their great chemical reactivity they are used as synthesi reagent only in the solution state.

(I) Reaction: When reduced with H_2 in presence of platinum or with lithium aluminium hydride, alkyl azides yield primary amines.

$$R-\ddot{N}=\overset{+}{N}=\ddot{N}^{-} + H_2 \xrightarrow{Pt} R-NH_2 + N_2$$

This method of preparation of primary amines is distinctly superior to the ammonolysis of halides, as it gives pure product.

(II) Addition to alkanes, like the diazo compounds, azides add to π-bonds of alkanes (or alkynes) forming heterocyclic compounds.

$$CH_2{=}CH_2 + R-N=N^{+}=N^{-} \longrightarrow \text{1, 2, 3 triazole}$$

1, 2 , 3 triazole

(III) Pyrolysis or Photolysis : By the action of heat or light azides produce very reactive intermediates called nitrines (carbanes form diazo compound)

$$R-N=\overset{+}{N}=\ddot{N}^{-} \longleftrightarrow R-\ddot{N}=N$$

a nitrine

Uses of Azides

Azides have varied uses in the chemical, dye-stuff, plastics, rubber and metal industries. Several compounds are used in wastewater treatment and as chemical intermediates, food additives, and sanitizing agents in dishwashing detergent and swimming pools. *1,1'-Azobis(formamide)* is a blowing agent for synthetic and natural rubber and ethylene-vinyl acetate copolymers. It is also useful as a foaming agent added to increase the porosity of plastics. *Trichlorinated isocyanuric acid* and *sodium dichloroisocyanurate* are used as sanitizing agents for swimming pools and as active ingredients in detergents, commercial and household bleaches, and dishwashing compounds. Sodium dichloroisocyanurate is also used in water and sewage treatment.

Edetic acid (EDTA) has numerous functions in the food, metal, chemical, textile, photography and health care industries. It is an antioxidant in foods. EDTA is used as a chelating agent to remove unwanted metal ions in boiler water and cooling water, in nickel plating and in wood pulping. It also acts as a bleaching agent for film processing in the photography industry, an etching agent in metal finishing and a dyeing agent in the textile industry. EDTA is found in detergents for textiles, industrial germicides, metal cutting fluids, semiconductor production, liquid soaps, shampoos, pharmaceuticals and cosmetics industry products. It is also used in medicine to treat lead poisoning.

Phenylhydrazine, *aminoazotoluene* and *hydrazine* are useful in the dye-stuff industry. Phenylhydrazine is also utilized in the preparation of pharmaceutical products. Hydrazine is a reactant in fuel cells for military uses and a reducing agent in plutonium extraction from reactor waste. It is used in nickel plating, wastewater treatment, and electrolytic plating of metals on glass and plastics. Hydrazine is employed for nuclear fuel reprocessing and as a component of high-energy fuels. It is a corrosion inhibitor in boiler feed water and in reactor cooling water. Hydrazine is also a chemical intermediate and a rocket propellant. *Diazomethane* is a powerful methylating agent for acidic compounds such as carboxylic acids and phenols.

Sodium azide is used in organic synthesis, explosives manufacture and as a propellant in automobile air-bags. Hydrazoic acid is used to make contact explosives such as lead azide.

Other azides, including *methylhydrazine, hydrazobenzene, 1,1-dimethylhydrazine, hydrazine sulphate* and *diazomethane*, are used in numerous industries. Methylhydrazine is a solvent, a chemical intermediate and a missile propellant, while hydrazobenzene is a chemical intermediate and an antisludging additive to motor oil. 1,1-Dimethylhydrazine is used in rocket fuel formulations. It is a stabilizer for organic peroxide fuel additives, an absorbent for acid gases, and a component of jet fuel. Hydrazine sulphate is used in the gravimetric estimation of nickel, cobalt and cadmium. It is an antioxidant in soldering flux for light metals, a germicide and a reducing agent in the analysis of minerals and slags.

Hazards

Diazomethane

Fire and explosion hazards. Either in the gaseous or liquid state, diazomethane explodes with flashes and even at -80°C the liquid diazomethane may detonate.It has been the general experience, however, that explosions do not occur when diazomethane is prepared and contained in solvents such as ethyl ether.

Health hazards. Diazomethane was first described in 1894 by von Pechmann, who indicated that it was extremely poisonous, causing air hunger and chest pains. Following this, other investigators reported symptoms of dizziness and tinnitus. Skin exposure to diazomethane was reported to produce denudation of the skin and mucous membranes, and it was claimed that its action resembles that of dimethyl sulphate. It was also noted that the vapours from the ether solution of the gas were irritating to the skin and rendered the fingers so tender that it was difficult to pick up a pin. In 1930, exposure of two persons resulted in chest pains, fever and severe asthmatic symptoms about 5 hours after exposure to mere traces of the gas.

The first exposure to the gas may not produce any noteworthy initial reactions; however, subsequent exposures in even minute amounts may produce extremely severe attacks of asthma and other symptoms. The pulmonary symptoms may be explained as either the result of true allergic sensitivity after repeated exposure to the gas, particularly in individuals with hereditary allergy, or of a powerful irritant action of the gas on the mucous membranes.

At least 16 cases of acute diazomethane poisoning, including fatalities from pulmonary oedema, have been reported amongst chemists and laboratory workers. In all cases, symptoms of intoxication included irritating cough, fever and malaise, varying in intensity according to the degree and duration of exposure. Subsequent exposures have led to hypersensitivity.

In animals, exposure to diazomethane at 175 ppm for 10 minutes caused haemorrhagic emphysema and pulmonary oedema in cats, resulting in death in 3 days.

Toxicity. One explanation for the toxicity of diazomethane has been the intracellular formation of formaldehyde. Diazomethane reacts slowly with water to form methyl alcohol and liberate nitrogen. Formaldehyde, in turn, is formed by the oxidation of methyl alcohol. The possibilities of liberation in vivo of methyl alcohol or of the reaction of diazomethane with carboxylic compounds to form toxic methyl esters may be considered; on the other hand, the deleterious effects of diazomethane may be primarily due to the strongly irritant action of the gas on the respiratory system.

Diazomethane has been shown to be a lung carcinogen in mice and rats. Skin application and subcutaneous injection, as well as inhalation of the compound, have also been shown to cause tumour development in experimental animals. Bacterial studies show it is mutagenic. The International Agency for Research on Cancer (IARC), however, places it in Group 3, unclassifiable as to human carcinogenicity.Diazomethane is an effective insecticide for the chemical control of *Triatoma* infestations. It is also useful as an algicide. When the ichthyotoxic component of the green alga *Chaetomorpha minima* is methylated with diazomethane, a solid is obtained which retains its toxicity to kill fish. It is noteworthy that in the metabolism of the carcinogens dimethylnitrosamine and cycasin, one of the intermediary products is diazomethane.

Hydrazine and Derivatives

Flammability, explosion and toxicity are major hazards of the hydrazines. For example, when hydrazine is mixed with nitromethane, a high explosive is formed which is more dangerous than TNT. All hydrazines discussed here have sufficiently high vapour pressures to present serious health hazards by inhalation. They have a fishy, ammoniacal odour which is repulsive enough to indicate the presence of dangerous concentrations for brief accidental exposure conditions. At lower concentrations, which may occur during manufacturing or transfer processes, the warning properties of odour may not be enough to preclude low-level chronic occupational exposures in fuel handlers.

Moderate to high concentrations of hydrazine vapours are highly irritating to the eyes, nose and the respiratory system. Skin irritation is pronounced with the propellant hydrazines; direct liquid contact results in burns and even sensitization type of dermatitis, especially in the case of phenylhydrazine. Eye splashes have a strongly irritating effect, and hydrazine can cause permanent corneal lesions.

In addition to their irritating properties, hydrazines also exert pronounced systemic effects by any route of absorption. After inhalation, skin absorption is the second most important route of intoxication. All hydrazines are moderate to strong central nervous system poisons, resulting in tremors, increased central nervous system excitability and, at sufficiently high doses, convulsions. This can progress to depression, respiratory arrest and death. Other systemic effects are in the haematopoietic system, the liver and the kidney. The individual hydrazines vary widely in degree of systemic toxicity as far as target organs are concerned. The haematological effects are self-explanatory on the basis of haemolytic activity. These are dose dependent and, with the exception of monomethylhydrazine, they are most prominent in chronic intoxication. Bone marrow changes are hyperplastic with phenylhydrazine, and blood cell production outside the bone marrow has also been observed. Monomethylhydrazine is a strong methaemoglobin former, and blood pigments are excreted in the urine. The liver changes are primarily of the fatty degeneration type, seldom progressing to necrosis, and are usually reversible with the propellant hydrazines. Monomethylhydrazine and phenylhydrazine in high doses can cause extensive kidney damage. Changes in the heart muscle are primarily of fatty character. The nausea observed with all of these hydrazines is of central origin and refractory to medication. The most potent convulsants in this series are monomethylhydrazine and 1,1-dimethylhydrazine. Hydrazine causes primarily depression, and convulsions occur much less frequently. All hydrazines appear to have some kind of activities in some laboratory animal species by some route of entry (feeding in drinking water, gastric intubation or inhalation). IARC considers them Group 2B, possibly carcinogenic in humans. In laboratory animals, with the exception of one derivative not discussed here, 1,2-dimethylhydrazine (or symmetrical dimethylhydrazine), there is a definite dose response. In view of its Group 2B rating, any exposure of humans should be minimized by proper protective equipment and decontamination of accidental spills.

Phenylhydrazine

The pathology of phenylhydrazine has been studied by means of animal experiments and clinical observations. Information about the effects of phenylhydrazine in humans was obtained from the use of phenylhydrazine hydrochloride for therapy. The conditions observed included haemolytic anaemia, with hyperbilirubinaemia and urobilinuria, and the appearance of Heinz bodies; liver damage with hepatomegalia, icterus, and very dark urine containing phenols; sometimes signs of kidney manifestations occurred. Haematological effects included cyanosis, haemolytic anaemia, sometimes with methaemoglobinaemia, and leucocytosis. Among the more general symptoms were fatigue, giddiness, diarrhoea and lowering of the blood pressure. It was also observed that a student, who had received 300 g of the substance on the abdomen and thighs suffered from cardiac collapse with a coma that lasted for several hours. Individuals with hereditary glucose-6-phosphate dehydrogenase (G6PDH) deficiency would be much more susceptible to the haemolytic effects of phenylhydrazine and should not be exposed to it. With regard to skin damage, there have been reports of acute eczema with vesicular eruption, as well as chronic eczema on the hands and forearms of workers preparing antipyrin. Also described was a case of vesicular dermatosis with the production of phlyctenae on the wrist of an assistant chemist. This appeared 5 or 6 hours after handling and took 2 weeks to heal. A chemical engineer who handled the substance suffered only from a few pimples, which disappeared in 2 or 3 days. Phenylhydrazine is therefore regarded as a potent skin sensitizer. It is very rapidly absorbed by the skin. Because of reports of carcinogenicity of phenylhydrazine to mice, the US National Institute for Occupational Safety and Health (NIOSH) has recommended its regulation as a human carcinogen. A variety of bacterial and tissue-culture studies have shown it is mutagenic. Intraperitoneal injection of pregnant mice resulted in offspring with severe jaundice, anaemia and a deficit in acquired behaviour.

Sodium azide and hydrazoic acid

Sodium azide is manufactured by combining sodamide with nitrous oxide. It reacts with water to produce hydrazoic acid. Hydrazoic acid vapour may be present when handling sodium azide. Commercially, hydrazoic acid is produced by the action of acid on sodium azide.

Sodium azide appears to be only slightly less acutely toxic than sodium cyanide. It may be fatal if inhaled, swallowed or absorbed through skin. Contact may cause burns to skin and eyes. A lab technician accidentally ingested what was estimated to be a “very small amount” of sodium azide. Symptoms of tachycardia, hyperventilation and hypotension were observed. The authors note that the minimal hypotensive dose in humans lies between 0.2 and 0.4 g/kg.

Treatment of normal individuals with 3.9 mg/day of sodium azide for 10 days produced no effects other than a heart-pounding sensation. Some hypertensive patients developed sensitivity to azide at 0.65 mg/day. Workers exposed to 0.5 ppm hydrazoic acid developed headaches and nasal congestion. Additional symptoms of weakness and eye and nasal irritation developed from exposure to 3 ppm for less than 1 hour. Pulse rate was variable and blood pressure was low or normal. Similar symptoms were reported among workers making lead azide. They had definite low blood pressure which became more pronounced during the work day and returned to normal after leaving work. Animal studies showed a rapid but temporary fall in blood pressure from single oral doses of 2 mg/kg or more of sodium azide. Associated haematuria and cardiac irregularities were observed at levels of 1 mg/kg IV in cats. Symptoms observed in animals after relatively large doses of sodium azide are respiratory stimulation and convulsions, then depression and death. The LD_{50} for sodium azide is 45 mg/kg in rats and 23 mg/kg in mice.

Exposure of rodents to hydrazoic acid vapour causes acute inflammation of the deep lung. Hydrazoic acid vapour is about eight times less toxic than hydrogen cyanide, with a concentration of 1,024 ppm being fatal in mice after 60 minutes (compared to 135 ppm for hydrogen cyanide).

Sodium azide was mutagenic in bacteria, although this effect was reduced if metabolizing enzymes were present. It was also mutagenic in mammalian cell studies.

CHAPTER

18

Ureides and Purines

Ureides

Acyl derivatives of urea are called ureides e.g. acetyl urea

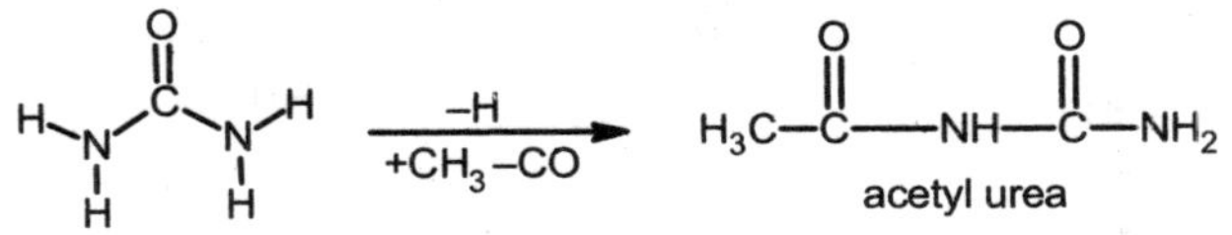

$$H_2N-CO-NH_2 \xrightarrow[+CH_3-CO]{-H} H_3C-CO-NH-CO-NH_2$$

acetyl urea

The ureides are classified (a) simple ureides or open and (b) cyclic ureides

Simple Ureides

They may be prepared by the action of acyl chlorides or acid anhydrides of mono carboxylic acid on urea:

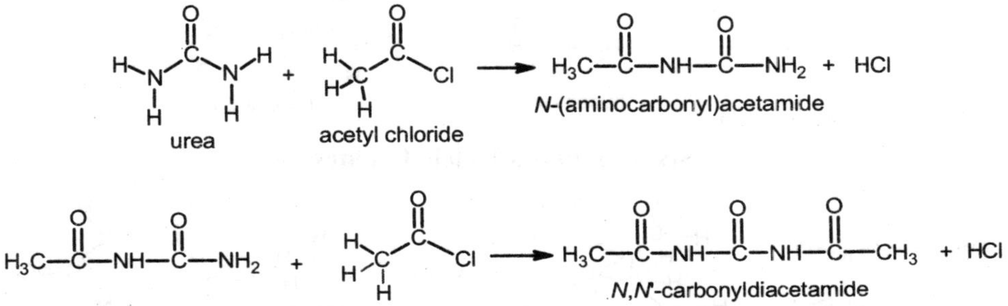

$$H_2N-CO-NH_2 + CH_3-CO-Cl \longrightarrow H_3C-CO-NH-CO-NH_2 + HCl$$

urea, acetyl chloride, *N*-(aminocarbonyl)acetamide

$$H_3C-CO-NH-CO-NH_2 + CH_3-CO-Cl \longrightarrow H_3C-CO-NH-CO-NH-CO-CH_3 + HCl$$

N,N'-carbonyldiacetamide

The simple ureides resemble the amides $(R-CO-NH_3)$ in properties. Many of these simple ureides are useful drugs e.g. bromoural.

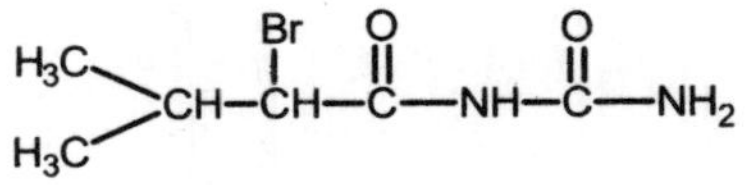

$(CH_3)_2CH-CH(Br)-CO-NH-CO-NH_2$

Cyclic Ureides

They may be prepared by the action of dicarboxylic acid on urea in presence of phosphonyl chloride ($POCl_3$)

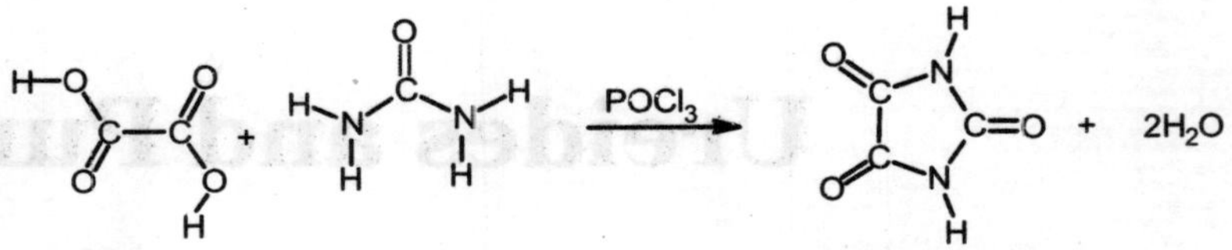

oxalic acid + urea $\xrightarrow{POCl_3}$ oxalylurea (parabinic acid) + $2H_2O$

The cyclic ureides may also be obtained by refluxing a di-ester with urea in ethanoic solution containing sodium e.g. malonic acids forms barbituric acid (malonyl urea)

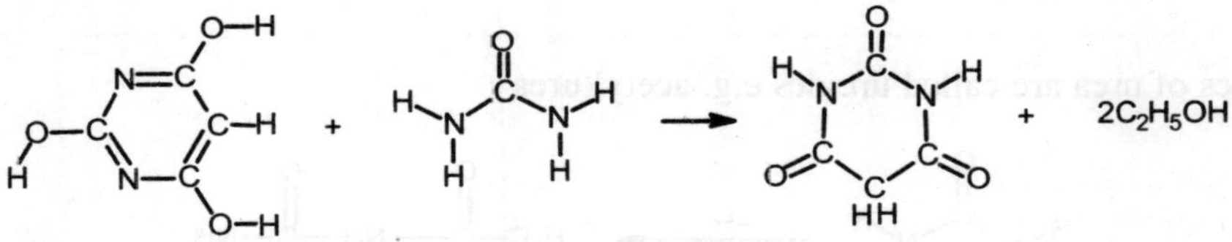

+ urea ⟶ barbituric acid + $2C_2H_5OH$

Types of Cylcic Ureides

There are two types of cyclic ureides, they are: -

Five Membered Cylcic Ureides

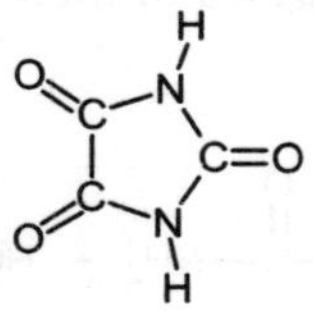

parabinic acid

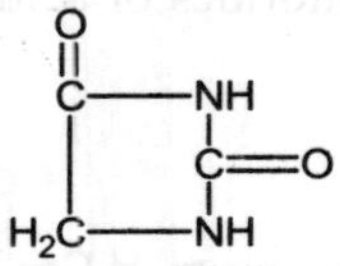

hydantoin

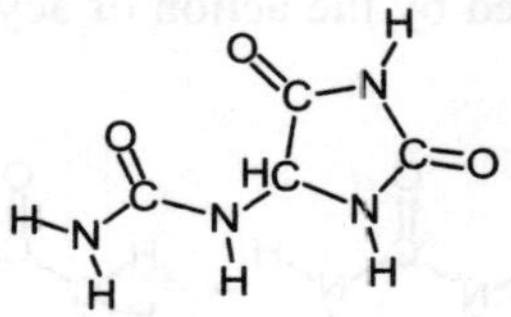

allantoin

Six Membered Cylcic Ureides

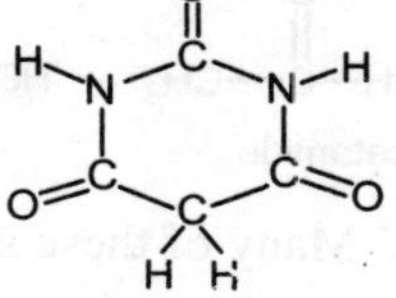

barbituric acid

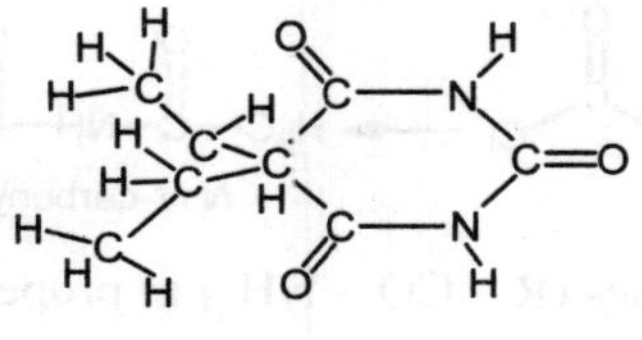

barbitone

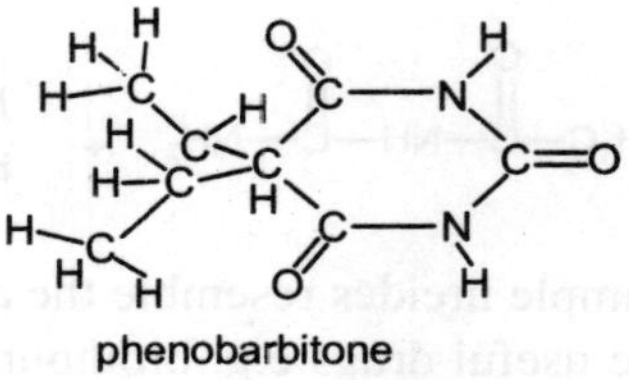

phenobarbitone

alloxan   violuric acid   dilifuric acid 5, nitrobarbituric acid

uramil   dial uric acid   uracil

Ureides are beautifully crystalline substances. They are hydrolysed by alkalies to form the parent acid and urea. The cyclic ureides are acidic owing to isolation and hence they form metallic salts. Many of them are excellent drugs.(the detail discussion of these compounds are not the syllabus of this book, as it comes under biochemistry)

Allantoin

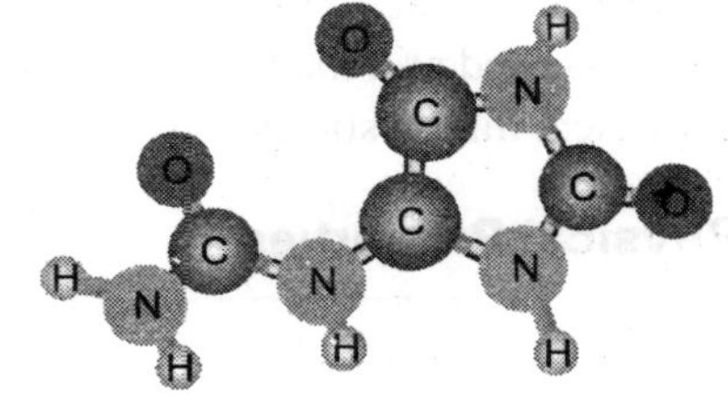

Allantoin is a botanical extract of the comfrey plant and is used for its healing, soothing, and anti-irritating properties. Allantoin helps to heal wounds and skin irritations and stimulate growth of healthy tissue. This extract can be found in anti-acne products, sun care products, and clarifying lotions because of its ability to help heal minor wounds and promote healthy skin.

Its chemical formula is $C_4H_6N_4O_3$. It is also called 5-ureidohydantoin, glyoxyldiureide, and 5-ureidohydantoin. It is a product of oxidation of uric acid. It is a diureide of glyoxilic acid. It is a product of purine metabolism in most mammals except higher apes, and it is present in their urine.

Physical Properties

Molecular Formula	$C_4H_6N_4O_3$
Formula Weight	158.11544
Composition	C(30.38%) H(3.82%) N(35.43%) O(30.36%)
Molar Refractivity	33.42 ± 0.4 cm^3
Molar Volume	95.7 ± 5.0 cm^3
Parachor	288.5 ± 6.0 cm^3
Index of Refraction	1.615 ± 0.03

Surface Tension	82.6 ± 5.0 dyne/cm
Density	1.65 ± 0.1 g/cm^3
Polarizability	13.24 ± 0.5 10^{-24}cm^3
Monoisotopic Mass	158.04399 Da
Nominal Mass	158 Da
Average Mass	158.11681 Da

Uses

The keratolytic effect and abrasive and astringent properties of allantoin are used in skin softening cosmetic preparations. It is also frequently present in toothpaste, mouthwash, and other oral hygiene products, in shampoos, lipsticks, various cosmetic lotions and creams, and other cosmetic and pharmaceutical products.

Uracil

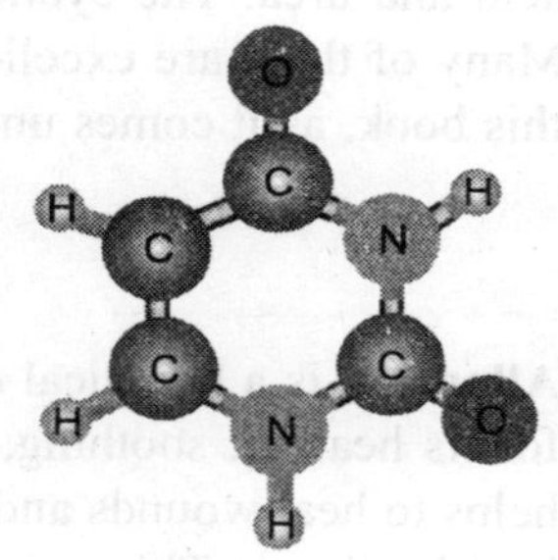

Uracil is one of the four RNA nucleobases, replacing thymine as found in DNA. Just like thymine, uracil can form a base pair with adenine via two hydrogen bonds, but it lacks the methyl group present in thymine. Uracil is only very rarely observed in DNA. Incorporation of uracil coupled with the enzyme, uracil DNA glycosylase (UDG) in polymerase chain reactions (PCR) is used as a method of cross-over contamination prevention, important in clinical diagnostic essays.

Physical Properties

Molecular Formula	$C_4H_4N_2O_2$
Formula Weight	112.08676
Composition	C(42.86%) H(3.60%) N(24.99%) O(28.55%)
Molar Refractivity	25.00 ± 0.3 cm^3
Molar Volume	84.8 ± 3.0 cm^3
Parachor	215.0 ± 6.0 cm^3
Index of Refraction	1.501 ± 0.02
Surface Tension	41.3 ± 3.0 dyne/cm
Density	1.321 ± 0.06 g/cm^3
Polarizability	9.91 ± 0.5 10^{-24}cm^3
Monoisotopic Mass	112.027277 Da
Nominal Mass	112 Da
Average Mass	112.088107 Da

Alloxan

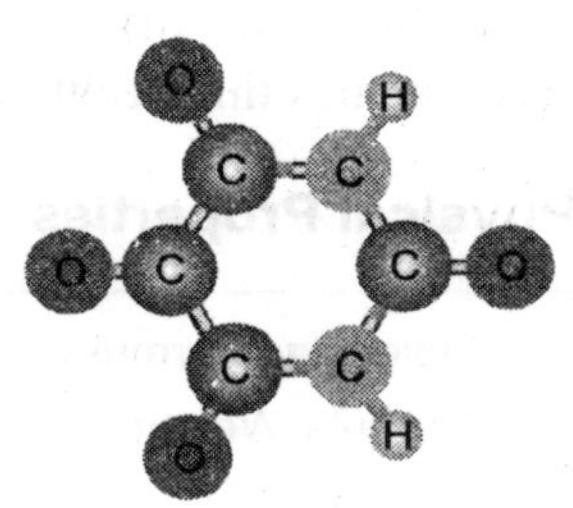

Alloxan or **mesoxalylurea** is an organic compound based on a pyrimidine heterocyclic skeleton. This compound has a high affinity for water and therefore exists as the monohydrate. The compound was discovered by Justus von Liebig and Friedrich Wöhler following the discovery of urea in 1828 and is one of the oldest named organic compounds that exist. The name is derived from allantoin, a product of uric acid excreted by the fetus into the allantois and oxaluric acid derived from oxalic acid and urea, found in urine. The original recipe for Alloxan was by oxidation of uric acid by nitric acid. Alloxan is a strong oxidizing agent and it forms a hemiacetal with its reduced reaction product **dialuric acid** (in which a carbonyl group is reduced to a hydroxyl group) which is called **alloxantin**.

Alloxane is a raw material for the production of the purple dye Murexide. Carl Wilhelm Scheele discovered the dye in 1776. Murexide is the product of the complex *in-situ* multistep reaction of alloxantin and gaseous ammonia. Murexide results from the condensation of the unisolated intermdiate uramil with alloxan, liberated during the course of the reaction. Scheele sourced uric acid from human calculi (such as kidney stones) and called the compound lithic acid. William Prout investigated the compound in 1818 and he used boa constrictor excrement with up to 90% ammonium acid urate. Liebig and Wöhler in the nineteenth century coined the name *murexide* for the dye after the Trunculus Murex which is the source of the Tyrian purple of antiquity. Primo Levi in his world famous novel *The Periodic Table* in chapter *Nitrogen* considers pythons as a source for alloxane on behalf of a lipstick producer but he is turned down by the director of the Turin zoo because the zoo already has lucrative contracts with cosmetics companies (his attempts with chicken dung end in misery).

Alloxan is used as a drug used to induce diabetes in animals used in laboratory experiments. A 2005 website article [1] links the presence of alloxan as an additive in white flour to the occurrence of diabetes and offers vitamine E as a countermeasure. However no scientific evidence to date substantiates this health food claim.

Physical Properties

Molecular Formula	$C_4H_2N_2O_4$
Formula Weight	142.06968
Composition	C(33.82%) H(1.42%) N(19.72%) O(45.05%)
Molar Refractivity	25.66 ± 0.3 cm^3
Molar Volume	84.5 ± 3.0 cm^3
Parachor	237.7 ± 6.0 cm^3
Index of Refraction	1.519 ± 0.02
Surface Tension	62.6 ± 3.0 dyne/cm
Density	1.681 ± 0.06 g/cm^3
Polarizability	10.17 ± 0.5 $10^{-24}cm^3$
Monoisotopic Mass	142.001457 Da
Nominal Mass	142 Da
Average Mass	142.070765 Da

Violuric Acid

Violuric acid, isonitroso- functional group substituted barbituric acid, is used in biological research.

Molecular Formula	$C_4H_3N_3O_4$
Formula Weight	157.08432
Melting Point	240°C
Composition	C(30.58%) H(1.92%) N(26.75%) O(40.74%)
Molar Refractivity	30.99 ± 0.5 cm^3
Molar Volume	71.6 ± 7.0 cm^3
Parachor	235.9 ± 8.0 cm^3
Index of Refraction	1.813 ± 0.05
Surface Tension	117.8 ± 7.0 dyne/cm
Density	2.19 ± 0.1 g/cm^3
Polarizability	12.28 ± 0.5 $10^{-24}cm^3$
Monoisotopic Mass	157.012356 Da
Nominal Mass	157 Da
Average Mass	157.085464 Da

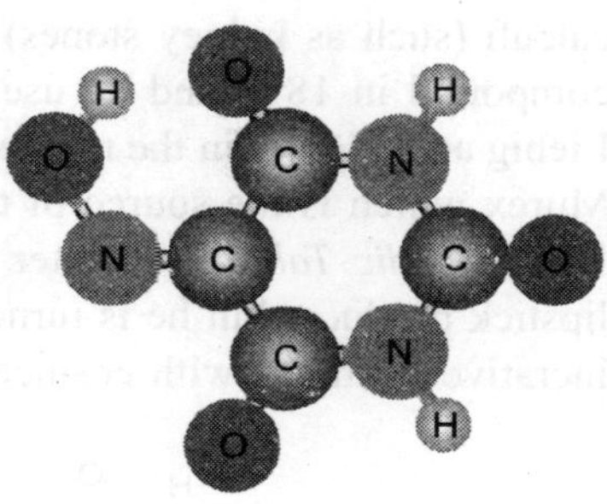

Uramil

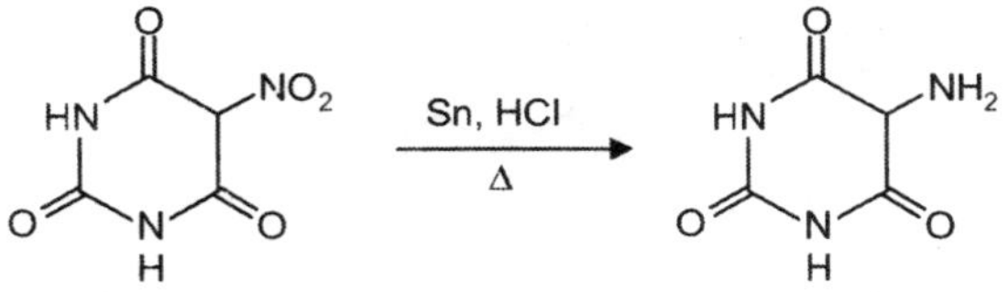

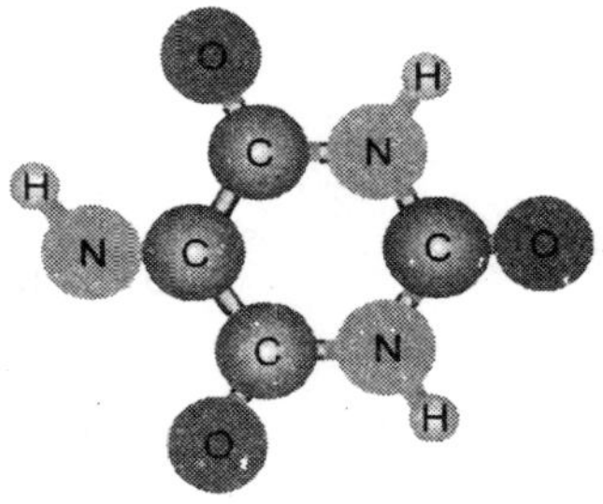

Preparation

In a 5-L. flask are placed 100 g. (0.44 mole) of nitrobarbituric acid and 600 cc. of concentrated hydrochloric acid, and the mixture is heated on a boiling water bath. To the hot mixture are added 250 g. (2.1 gram atoms) of mossy tin and 400 cc. of hydrochloric acid over a period of about thirty minutes; heating is continued until there is no yellow colour in the liquid. The solution is treated with about 3 L. more of concentrated hydrochloric acid and heated until all the white solid is in solution. Norite (decolorizing carbon) is added, and the hot mixture is filtered through a sintered-glass funnel. The filtrate is allowed to stand in an icebox overnight, and then the precipitate of uramil is collected on a filter and washed with liberal quantities of dilute hydrochloric acid and finally with water. The filtrate is concentrated under reduced pressure to about 1 l. and cooled overnight. The additional uramil thus obtained is collected on a Büchner funnel and added to the first product. The uramil is dried in a desiccator over concentrated sulfuric acid, and finally over 40 per cent sodium hydroxide to remove hydrochloric acid Uramil is a fine, white powder which becomes pink to red on standing. The yield of a product which does not melt below 400° is 40–46 g. (63–73 per cent of the theoretical amount).

1. Nitrobarbituric acid forms a yellow aqueous solution, but, as the colour is weak in concentrated hydrochloric acid solution, no trace of it should show at the end of the reaction.
2. If a sintered-glass funnel is not available, the solution may be filtered through a half-inch layer of decolorizing carbon on a double filter paper. After the uramil is once dissolved in the concentrated hydrochloric acid it comes out of solution very slowly, and, if filtered promptly, the solution may be cooled to 60–80° with little loss of product.
3. If, as happens occasionally, the uramil does not crystallize, the solution must be concentrated and cooled again.
4. Unless the product is washed thoroughly it will contain tin salts.
5. To test for uramil an ammoniacal solution is boiled in the air. A positive test is the appearance of a pink color which gradually grows deeper. The reaction proceeds more rapidly in the presence of mercuric oxide.
6. If the material is dried in the air or in an oven, a pink product is almost always obtained. The pink color is produced more rapidly if ammonia or amines are present in the air.

NITROBARBITURIC ACID
[Barbituric acid, 5-nitro-]

fuming HNO_3

1. Procedure

In a 2-L. flask, equipped with a mechanical stirrer and surrounded by an ice bath, is placed 143 cc. of fuming nitric acid (sp. gr. 1.52). Stirring is started, and 100 g. (0.61 mole) of barbituric acid is added over a period of two hours; the temperature is kept below 40° during the addition. The mixture is stirred for one hour after the barbituric acid has been added, and stirring is continued while 430 cc. of water is added and the solution is cooled to 10°. The mixture is filtered, and the residue is washed with cold water and dried on a glass tray at 60–80°. The nitrobarbituric acid is dissolved by adding it to 860 cc. of boiling water in a 2-l. flask and heating the mixture on a boiling water bath while steam is blown in until solution is complete (Note 2). After filtration and cooling overnight, the crystals are removed, washed with cold water, and dried in trays in an oven at 90–95° for two to three hours. The product melts with decomposition at 181–183° when heated rapidly. The yield is 139–141 gm. On drying the product at 110–115° for two to three hours, the yield is 90–94 g. (85–90 per cent of the theoretical amount) of an anhydrous compound which melts with decomposition at 176°.

Physical Properties

Molecular Formula	$C_4H_5N_3O_3$
Formula Weight	143.1008
Composition	C(33.57%) H(3.52%) N(29.36%) O(33.54%)
Molar Refractivity	29.08 ± 0.3 cm^3
Molar Volume	95.5 ± 3.0 cm^3
Parachor	260.2 ± 6.0 cm^3
Index of Refraction	1.520 ± 0.02
Surface Tension	54.9 ± 3.0 dyne/cm
Density	1.497 ± 0.06 g/cm^3
Polarizability	11.53 ± 0.5 $10^{-24}cm^3$
Monoisotopic Mass	143.033091 Da
Nominal Mass	143 Da
Average Mass	143.102111 Da

The Purines

Purine was named by the German chemist **Emil Fischer** in 1884. He synthesised it in 1898. Fischer showed that the *purines* were part of a single chemical family.

Uric acids and other closely related compounds such as caffeine adenine guanine xanthine hypoxanthine form a group of complex cyclic ureides. They are all derived from the same parent substance "purine" and are therefore, named as Purines. Purines may be thought of cyclic diureides urines they could be combined and built from two molecules of urea and one of carboxylic acid.

Purines

The general term **purines** also refer to substituted purines and their tautomers. The below mention figure makes the explanation more clear.

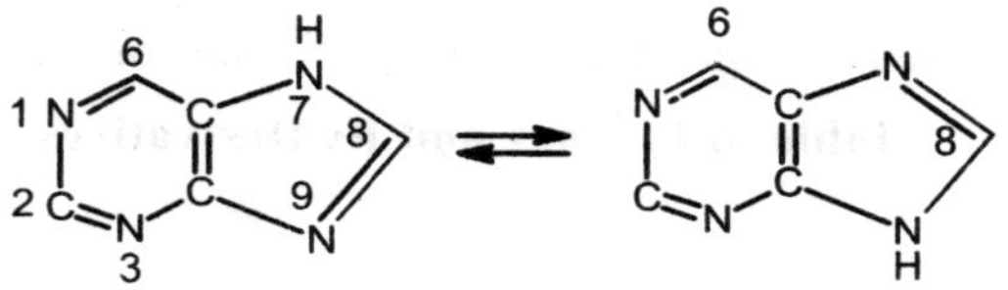

Purine is a colourless solid with a melting point of 214°C. It is slightly soluble in water and has both acidic and basic properties. For the purpose of naming its derivatives, the skeleton of purine is numbered as shown.

Synthesis

Purine does not occur in nature. It can be synthesized by the following methods

(i) Albert and Brown (1954)

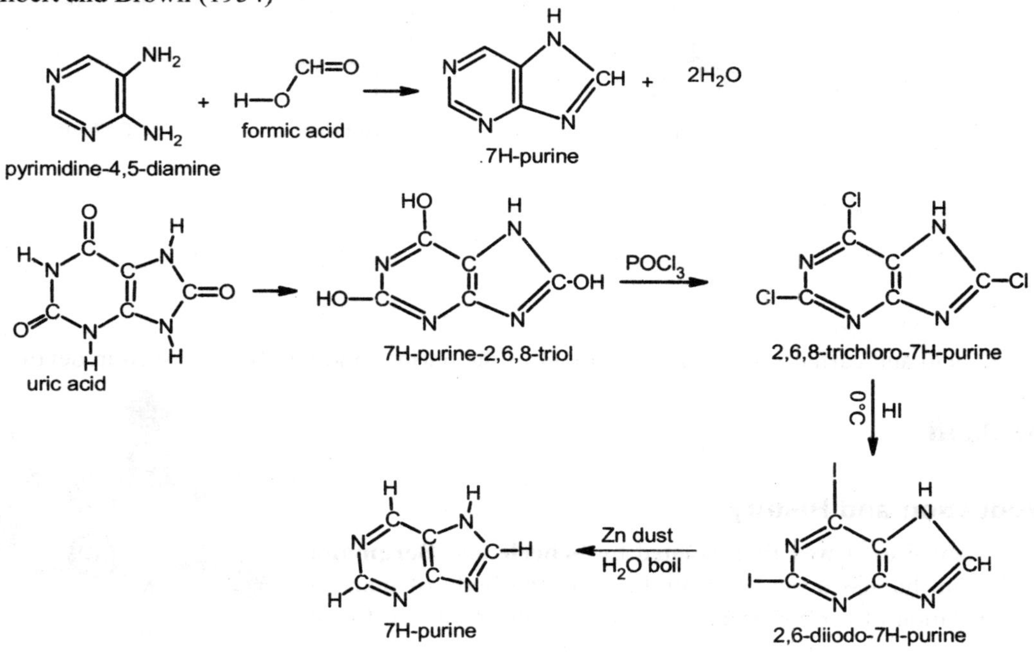

pyrimidine-4,5-diamine + formic acid → 7H-purine + $2H_2O$

uric acid → 7H-purine-2,6,8-triol →($POCl_3$) 2,6,8-trichloro-7H-purine →(HI, 0°C) 2,6-diiodo-7H-purine →(Zn dust, H_2O boil) 7H-purine

Classification of Purine

The natural purines are either the hydroxyl or the amino derivatives of the parent substance purine. Thus they can be divided into two groups: -

(I) Oxypurines: These are the hydroxy derivatives of purines and are named because they can exhibit –enol tautomerism

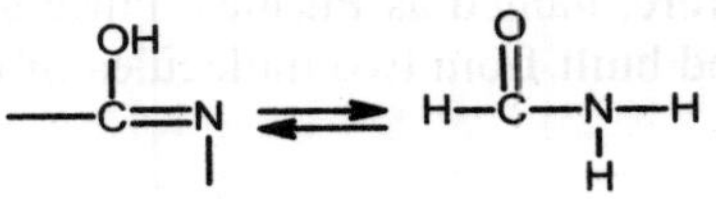

$$-\overset{OH}{\underset{}{C}}{=}N- \rightleftharpoons H-\overset{O}{\overset{\|}{C}}-\underset{H}{N}-H$$

(enol form) (keto form)

Examples of Oxypurines are uric acid, methane and its bases (caffeine, theobrine and hypopanthine)

(II) Aminopurines: These are amino derivatives of purines of e.g. adenine, guanine

Table of Purines and its Derivatives

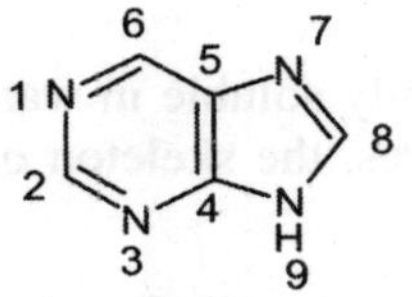

Purine

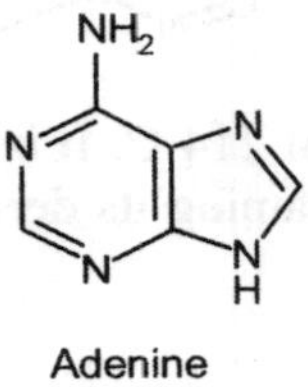

Adenine

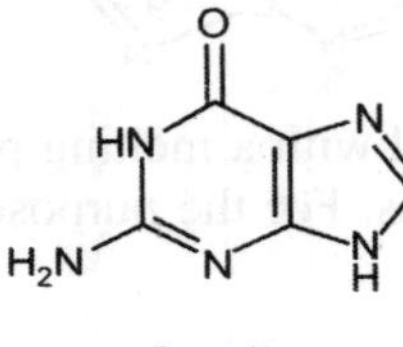

Guanine

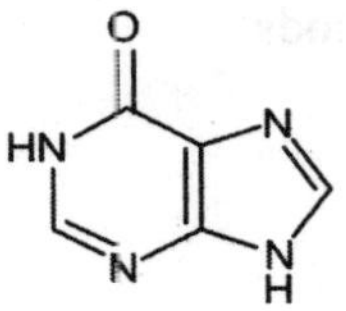

Hypoxanthine

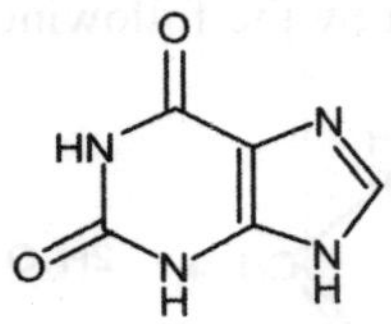

xanthine

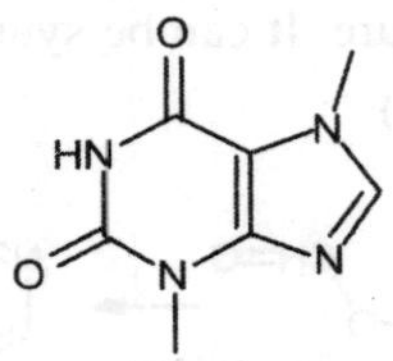

Theobromine

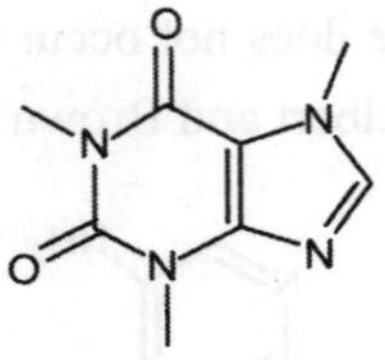

Caffeine

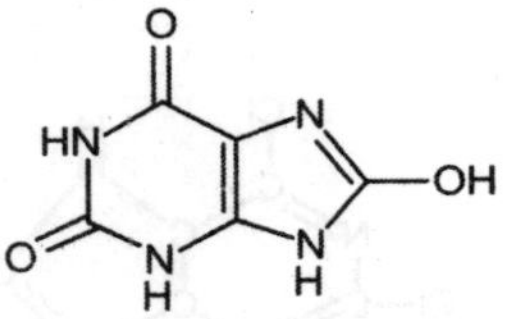

Uric acid

Uric acid and caffeine are by far important of the purines and will be discussed in detail.

Uric Acid

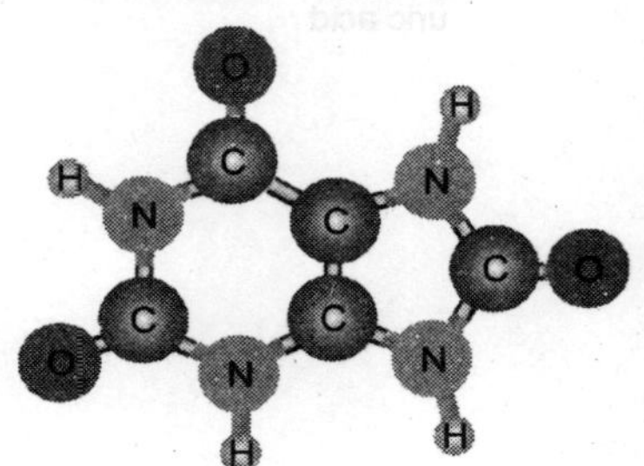

Introduction and History

It is so named as it was first isolated by **Scheele** and **Bergmann** (1776) from human urine. It is produced in the body of man by the degradation of certain protein. Normally only traces are found in the blood and small quantities are excreted in urine owing to its small solubility any temporary excess of uric acid in the blood deposits in the joints (gout) or in the tissues (rheumatism). It may sometimes accumulate in the bladder and in the kidneys forming stones

Uric acid is the chief constituent of the excreta of birds and reptiles. "Guano" which is the excreta of some sea birds is an important source of uric acid.

Preparation

Uric acid may be prepared from human urine or Guano in which it is present as mostly as ammonium salt

***(i)* From Normal Urine**: Uric acid may be isolated from human urine by concentrating it an adding concentrated hydrochloric acid. The crystals of uric acid separates out on standing.

$$\text{Human Urine} \xrightarrow[\text{2.} \qquad \text{cool}]{\text{1. concentration/hydrochloric acid}} \text{Uric Acid crystals}$$

***(ii)* From Guano**: Uric acid is prepared on a large scale from the environment of birds and reptiles. The excrement is powdered and boiled with sodium hydroxide solution, until the evolution of ammonia ceases. The hot solution of sodium urate thus obtained is then filtered and powdered into hydrochloric acid.

$$\text{Amount Urate} \xrightarrow[(^{-}NH_3)]{NaOH} \text{Sodium urate} \xrightarrow[\text{Cool}]{HCl} \text{uric acid} + HCl$$

The uric acid separates as fine crystalline mass on allowing the solution to stand on cold. It is filtered and dried in air.

Physical Properties

Molecular Formula	$C_5H_4N_4O_3$
Formula Weight	168.11026
Composition	C(35.72%) H(2.40%) N(33.33%) O(28.55%)
Melting Point	decomposes
Molar Refractivity	35.51 ± 0.4 cm^3
Molar Volume	89.7 ± 5.0 cm^3
Parachor	279.8 ± 6.0 cm^3
Index of Refraction	1.721 ± 0.03
Surface Tension	94.3 ± 5.0 dyne/cm
Density	1.87 ± 0.1 g/cm^3
Polarizability	14.07 ± 0.5 $10^{-24}cm^3$
Monoisotopic Mass	168.02834 Da
Nominal Mass	168 Da
Average Mass	168.111895 Da

Uric acid is a white crystalline solid having no taste or smell. It decomposes on heating, so that it has no melting point. It is very slightly soluble in water, insoluble in ethanol but soluble in glycol or hot alkali etc.

Chemical Properties

Uric acid behaves as weak tri basic acid due to enolization

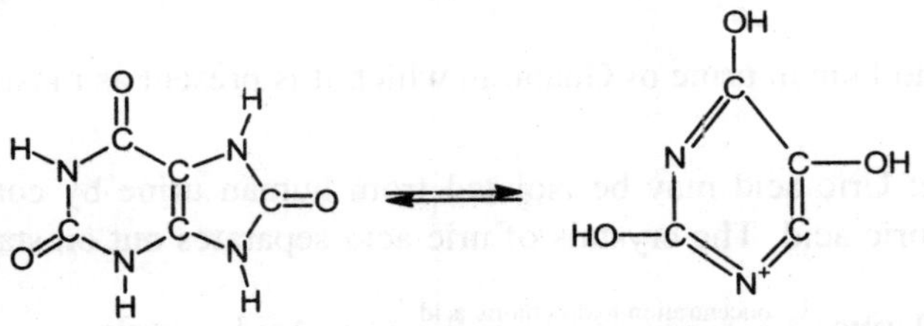

ketonic form ⇌ enolic form (2, 6, 8 trihydroxy purine)

With sodium carbonate, it gives as acidic salt while with sodium hydroxide as normal salt is produced. The acid salts are sparingly soluble in water while normal salts are moderately soluble in water. Lithium salts are freely soluble. That is why lithianated water is often used as remedy in cases of rheumatism and gout to secure elimination of uric acid.

It reacts with phosphorus oxychloride to form 2, 6, 8 trichloropurine indicating there by the existence of a trienolic form of uric acid.

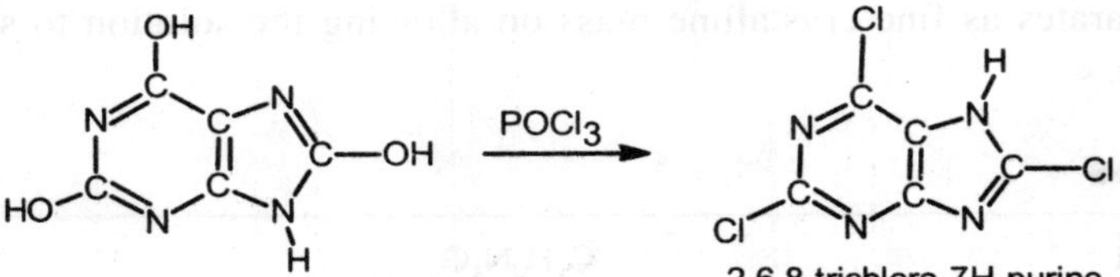

9H-purine-2,6,8-triol
uric acid trienolic form $\xrightarrow{POCl_3}$ 2,6,8-trichloro-7H-purine

Constitution of Uric Acid

(i) Molecular formula of uric acid as deduced from its analytical data and molecular weight determination is $C_5H_4N_4O_3$

(ii) Presence of four imono (>NH) groups: On exhaustive methylation uric acid gives tetramethyluric acid in which all the four hydrogen atoms of uric acid have been replaced by methyl groups. When suggested to hydrolysis with hydrochloric acid, tetramethyluric acid looses all the four nitrogen atoms as methyl amine. It indicates that in tetramethyluric acid all the methyl groups are directly linked to nitrogen atoms. Therefore, it stands to reason that all the four molecules must be attached to nitrogen atoms.

$$C_5O_3(N-H)_4 \xrightarrow{\text{exhaustive}} C_5O_3(N-CH_3)_4 \xrightarrow[\text{distil}]{H_2O} 4CH_3-NH_2$$

This shows that uric acid contains four imino groups (- NH -)

(iii) Presence of Alloxan and Urea Units On oxidation with dilute nitric acid, uric acid forms Alloxan and urea in equimolecular proportions.

$$C_5H_4N_4O_3 + H_2O + [O] \xrightarrow{HNO_3} C_4N_2H_2O_4 + NH_2 - CO - NH_2$$

(iv) Structure of Alloxan When hydrolysed with alkali alloxan produces one molecule of urea and one molecule of mesoxalic acid.

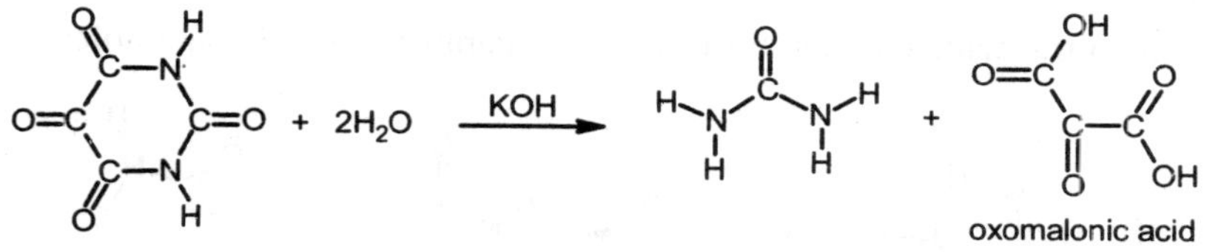

KOH

oxomalonic acid

Since alloxan contains no free amino group or carboxyl group, the products of hydrolysis suggest that alloxan is mesoxyl urea. The structure of alloxan has been confirmed by its synthesis. (Liebig and Wöhler in 1838)

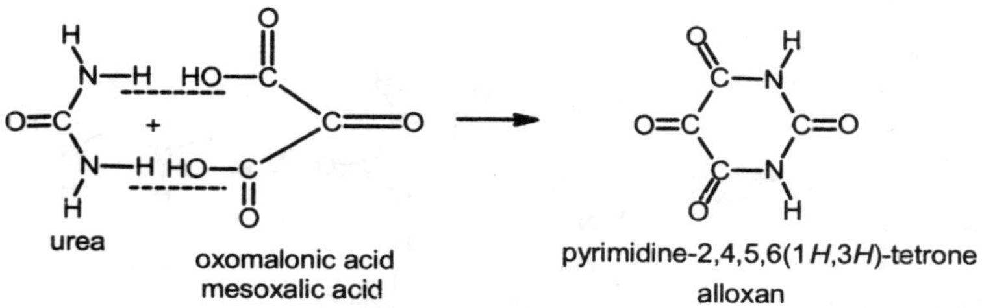

urea

oxomalonic acid
mesoxalic acid

pyrimidine-2,4,5,6(1*H*,3*H*)-tetrone
alloxan

The formation of alloxan from uric acid suggests that the latter contains a six membered ring.

(vi) Presence of Allantoin unit On oxidation with an aqueous suspension of lead dioxide or alkaline potassium permanganate, uric acid forms Allantoin and carbon dioxide.

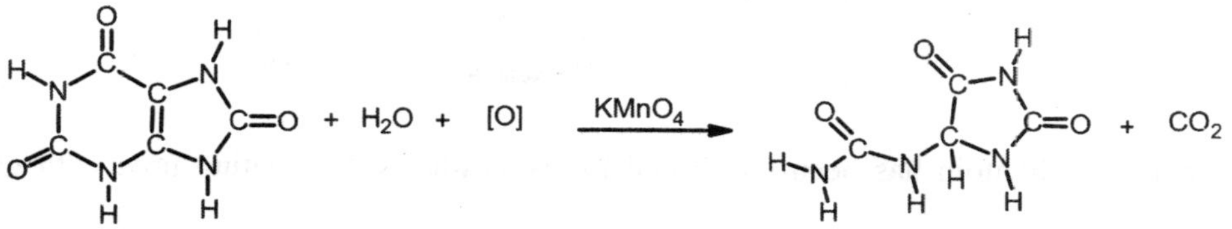

$+ H_2O + [O] \xrightarrow{KMnO_4}$ ... $+ CO_2$

Structure of Allantoin

(*i*) When hydrolysed with alkali, an Allantoin forms two molecules of urea and one molecule of glyoxalic acid.

allantoin + $2H_2O$ $\xrightarrow{OH^-}$ urea + oxoacetic acid (glyoxalic acid)

The formation of these hydrolytic products suggest that Allantoin is a diureides of glyoxalic acid On oxidation with nitric acid, allantoin forms urea and parabanic acid in equimolecular proportions.

allantoin + [O] $\xrightarrow{HNO_3}$ urea + parabinic acid

The reaction of allantoin stated above can, therefore be formulated as follows

urea + parabinic acid $\xleftarrow{HNO_3}$ allantoin $\xrightarrow{HI}$ urea + hydantoin

allantoin $\xrightarrow{2H_2O}$ urea + glyoxalic acid + urea

The structure of allantoin has been confirmed by its synthesis, by heating urea with glyoxalic acid

urea + glyoxalic acid + urea $\xrightarrow{-2H_2O}$ allantoin

(ii) How the structure of uric acid arrived at?

In the formation of allantoin from uric acid (step 5) by oxidation one carbon atom is lost from the latter as carbon dioxide. The problem is then to fit one carbon atom into the allantoin structure, so as to construct the uric acid molecule. Also the structure of uric acid so constructed, must give allantoin on oxidation with nitric acid (step 3). In view of above facts, two structures where proposed for uric acid; one by **Medicus** in 1875 and the second one by **Fittig** in 1878. Both these structure agreed with the facts known upto that time. Medicus formula was found to be correct by Fittig work.

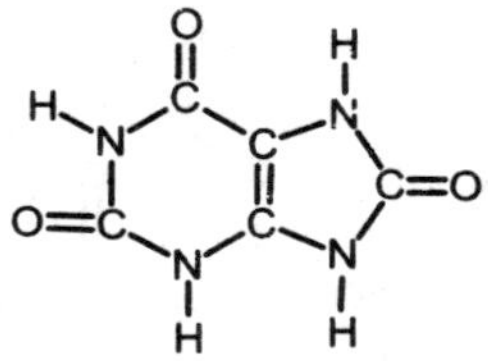

Medicus formula

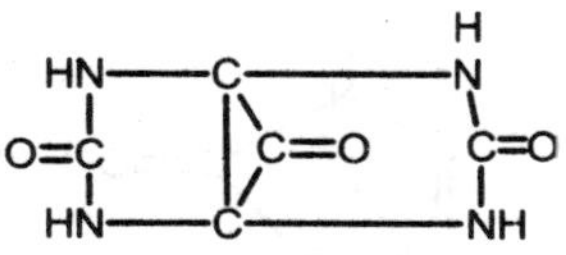

Fittig formula

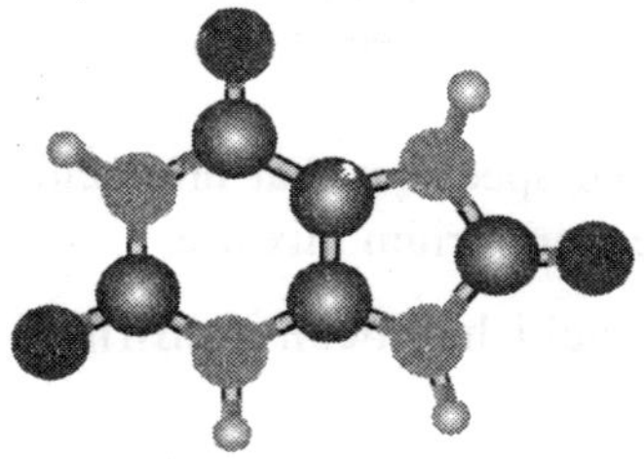

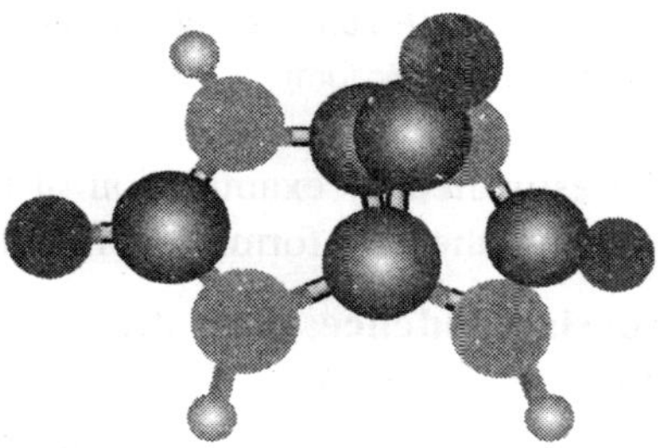

Fischer prepared two isomeric mono methyl uric acid hence this structure is untenable. On the other hand, the Medius formula satisfies the existence of two isomeric monoethyl derivatives.

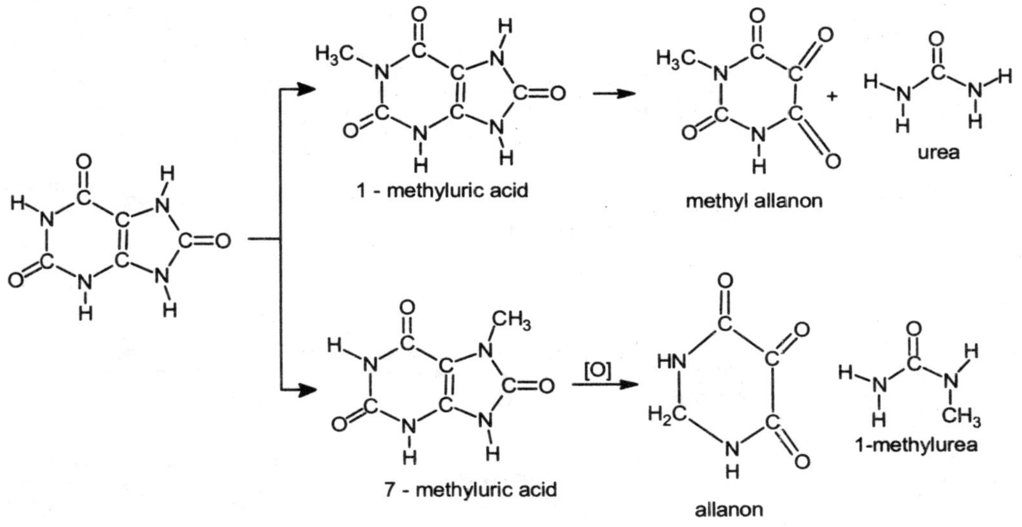

1 - methyluric acid

methyl allanon

urea

7 - methyluric acid

[O]

allanon

1-methylurea

More Medicus formula explains the existence of four monomethyl six dimethyl and four trimethyl derivatives. All of these have been prepared by Fisher, this giving powerful support Medicus formula.

***(iii)*Tautomeric Formula:** As already stated, uric acid reacts with phosphoryl chloride to form tichloro derivative. This shows the presence of three hydroxy groups in the molecule. To explain the presence of three hydroxy group and the acid character uric acid is prepared to be have tautomeric structure.

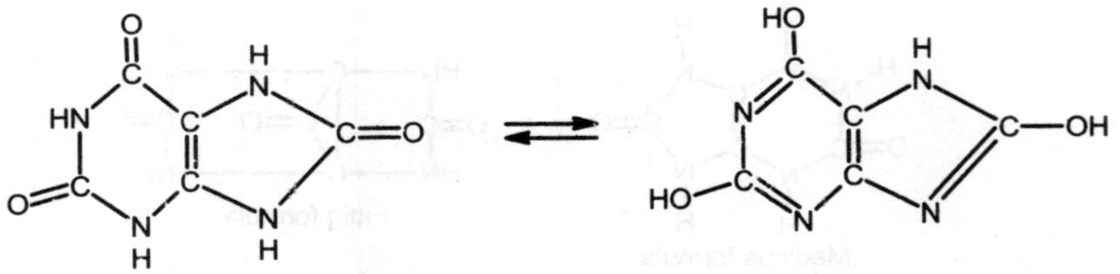

7,9-dihydro-1*H*-purine-2,6,8(3*H*)-trione
ketonic form

7*H*-purine-2,6,8-triol
enolic form

It has been estimated by examination of the infrared spectrum that uric acid exhibits keto-enol tautomerism and that the keto form predominates in the equilibrium mixture.

(*iv*) Synthesis Evidence: The structure of uric acid has been confirmed by its synthesis accomplished by various works:

(1) Behrend and Rosen's Synthesis (1888)

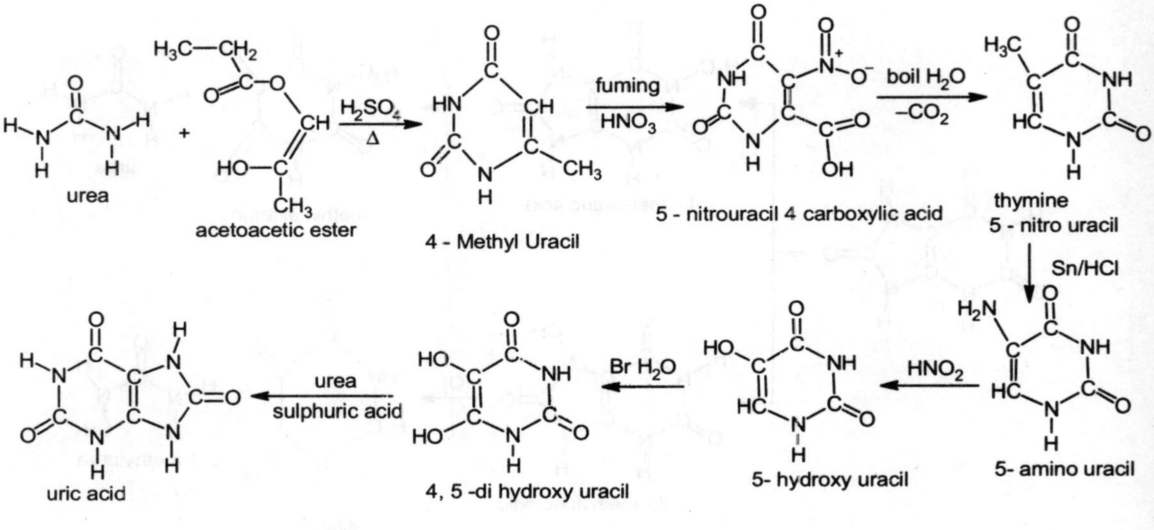

urea + acetoacetic ester $\xrightarrow[\Delta]{H_2SO_4}$ 4 - Methyl Uracil $\xrightarrow[HNO_3]{fuming}$ 5 - nitrouracil 4 carboxylic acid $\xrightarrow[-CO_2]{boil\ H_2O}$ thymine 5 - nitro uracil $\xrightarrow{Sn/HCl}$ 5- amino uracil $\xrightarrow{HNO_2}$ 5- hydroxy uracil $\xrightarrow{Br\ H_2O}$ 4, 5 -di hydroxy uracil $\xrightarrow[sulphuric\ acid]{urea}$ uric acid

(ii) Fischer's Synthesis:

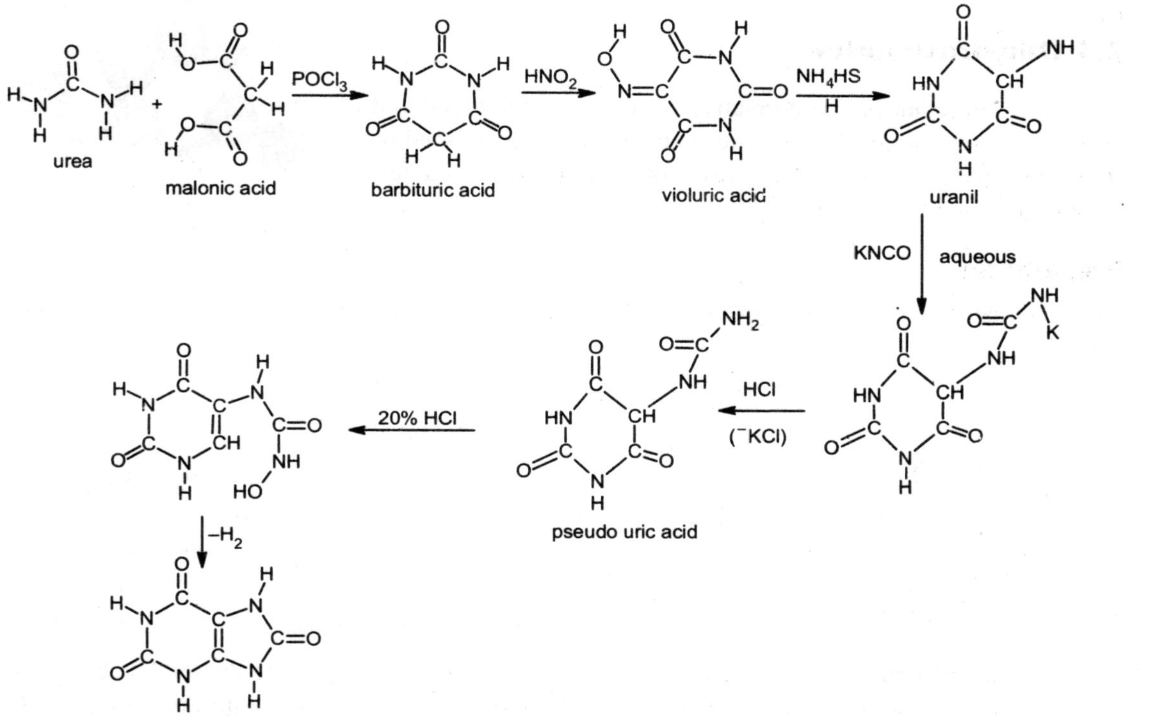

urea

malonic acid

$POCl_3$

barbituric acid

HNO_2

violuric acid

NH_4HS

H

uranil

KNCO aqueous

HCl

(^-KCl)

pseudo uric acid

20% HCl

$-H_2$

***(iii) Traube's Synthesis (1900)*:** It is the most important method as it can be used to prepare any purine derivative. The starting materials are urea and ethyl acetate.

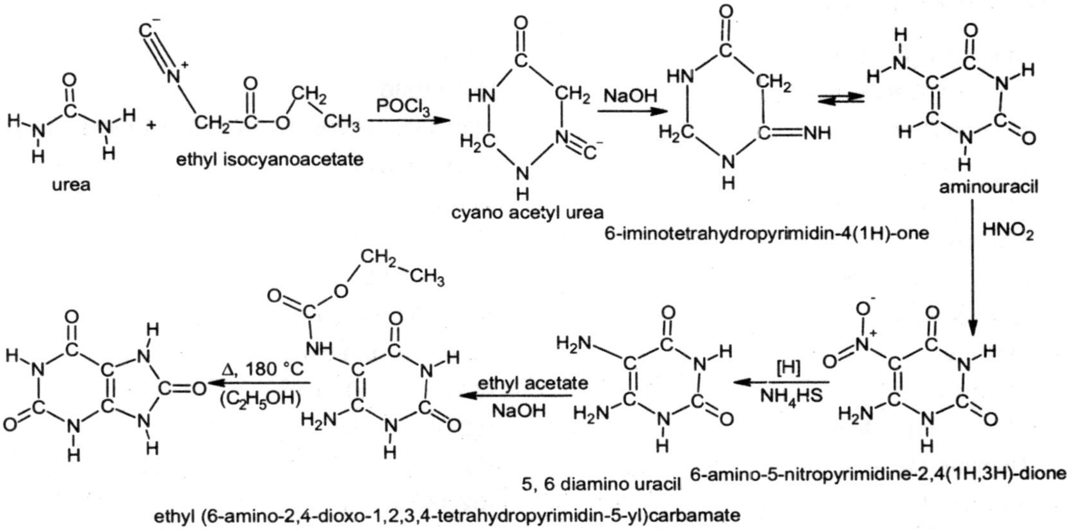

urea

ethyl isocyanoacetate

$POCl_3$

cyano acetyl urea

NaOH

6-iminotetrahydropyrimidin-4(1H)-one

aminouracil

HNO_2

6-amino-5-nitropyrimidine-2,4(1H,3H)-dione

[H]

NH_4HS

5, 6 diamino uracil

ethyl acetate

NaOH

ethyl (6-amino-2,4-dioxo-1,2,3,4-tetrahydropyrimidin-5-yl)carbamate

Δ, 180 °C

(C_2H_5OH)

Xanthine
or
2,4 Dihydroxypurine

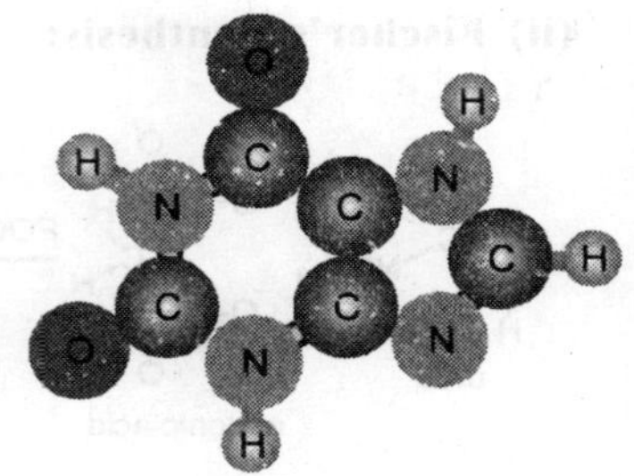

This is an important purine derivative is present in the blood and is excreted in urine. It also occurs in tea extract and emergent seedlings. It is the parent compound of three important bases, caffeine, theobromine and theophylline.

Preparation

(*i*) Xanthine may be prepared from 2, 6, 8 trichloropurine obtained by the action of $POCl_3$ on uric acid.

2,6,8-trichloro-6,9-dihydro-1H-purine $\xrightarrow{C_2H_5Na}$ 8-chloro-2,6-diethoxy-6,9-dihydro-1H-purine $\xrightarrow[\Delta]{HI}$ xanthine

(*ii*) It can also be prepared alternatively by similar method analogous to that of Traube synthesis of uric acid, 4, 5 diaminouracil as obtained in uric acid synthesis is treated with uric formic acid and sodium hydroxide. The sodium salt thus produced is heated at 250°C to give xanthine.

Physical Properties

Molecular Formula	$C_5H_4N_4O_2$
Formula Weight	152.11086
Composition	C(39.48%) H(2.65%) N(36.83%) O(21.04%)
Molar Refractivity	33.29 ± 0.3 cm^3
Molar Volume	92.8 ± 3.0 cm^3
Parachor	276.2 ± 6.0 cm^3
Index of Refraction	1.636 ± 0.02
Surface Tension	78.2 ± 3.0 dyne/cm
Density	1.637 ± 0.06 g/cm^3
Polarizability	13.20 ± 0.5 $10^{-24}cm^3$
Monoisotopic Mass	152.033425 Da
Nominal Mass	152 Da
Average Mass	152.11259 Da

Xanthine crystallizes well. It is very sparingly soluble in water. Chemically it resembles uric acid and forms salts with alkalies and also with hydrochloric acid and nitric acid and when oxidized with potassium chlorate in hydrochloric acid solution xanthine forms alloxan and urea. Like the hydroxy derivatives of purine, xanthine exits tautomerism.

keto form ⇌ enol form

Caffeine

Caffeine, sometimes called **theine** when found in tea, is a xanthine alkaloid found in the leaves and beans of the coffee tree, in tea, yerba mate, guarana berries, and in small quantities in cocoa, the kola nut and the Yaupon holly. In plants, caffeine acts as a natural pesticide that paralyzes and kills many insects feeding upon them.

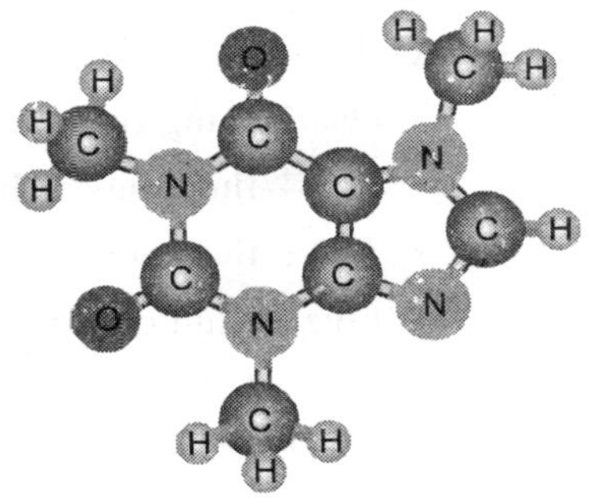

Caffeine is a central nervous system (CNS) stimulant, having the effect of warding off drowsiness and restoring alertness. Caffeine-containing beverages, such as coffee and tea, enjoy great popularity, making caffeine the world's most popular psychoactive substance.

Caffeine is a plant alkaloid, found in numerous plant varieties. The most commonly used of which are coffee, tea, and to some extent cocoa. Other, less commonly used, sources of caffeine include the plants yerba mate and guaraná, which are sometimes used in the preparation of teas and, more recently, energy drinks. Two of caffeine's alternative names, *mateine* and *guaranine*, are derived from the names of these plants.

The world's primary source of caffeine is the bean of the coffee plant, from which coffee is brewed. Caffeine content in coffee varies widely depending on the variety of coffee bean and the method of preparation used, but in general one serving of coffee ranges from about 40 mg for a single shot of espresso to about 100 mg for strong drip coffee. Generally, dark roast coffee has less caffeine than lighter roasts since the roasting process reduces caffeine content of the bean. Arabica coffee normally contains less caffeine content than the Robusta variety.

Tea is another common source of caffeine in many cultures. Tea generally contains somewhat less caffeine per serving than coffee, usually about half as much, depending on the strength of the brew, though certain types of tea, such as black and oolong, contain somewhat more caffeine than most other teas.

Caffeine is also a common ingredient of soft drinks such as cola, originally prepared from kola nuts. Soft drinks typically contain about 10 mg to 50 mg of caffeine per serving. By contrast, energy drinks such as Red Bull contain as much as 80 mg of caffeine per serving. The caffeine in these drinks originates either from the ingredients used or is an additive derived from the product of decaffeination or chemical synthesis.

Chocolate derived from cocoa is a weak stimulant, mostly due to its content of theobromine and theophylline, but it also contains a small amount of caffeine [1]. However, chocolate contains too little of these compounds for a reasonable serving to create effects in humans that are on par with coffee.

Finally, caffeine may also be purchased in most areas in the form of a pill that containing from 50 mg to 200 mg. Caffeine pills are regulated differently among various nations. For example, the European Union requires that a warning be placed on the packaging of any food whose caffeine exceeds 150 mg per litre. In many other countries, however, caffeine is classified as a flavouring and is unregulated.

Caffeine equivalents

In general, each of the following contains approximately 200 mg of caffeine:

- *One* 200 mg caffeine pill (in some countries these are 100 mg, in the UK these are 50 mg)
- *Two* 1-fluid ounce shots of espresso from robusta beans (2 fluid ounces (0,59 dl) total)
- *Two* 8-fluid ounce containers of regular coffee (16 fluid ounces (4.73 dl) total)
- *Five* 8-fluid ounce cups of black tea (40 fluid ounces (1.18 l) total)
- *Five* 12-fluid ounce cans of soda (60 fluid ounces total (1.77 l), although these can vary widely in content)
- *Ten* 8-fluid ounce cups of green tea (80 fluid ounces (2.36 l) total)
- *One and a half* pounds (0,68kg total) of milk chocolate
- *Fifty* 8-fluid ounce cups of decaffeinated coffee (400 fluid ounces (11.82 l) total)

History of caffeine use

Although tea has been consumed in China for thousands of years, the first documented use of caffeine in a beverage for its pharmacological effect was in the 15th century by the Sufis of Yemen, who used coffee to stay awake during prayers. In the 16th century there were coffee houses in Istanbul, Cairo and Mecca, and in the 17th century coffee houses opened for the first time in Europe.

In 1819, relatively pure caffeine was isolated for the first time by the German chemist Friedrich Ferdinand Runge. According to the legend, he did this at the instigation of Johann Wolfgang von Goethe (Weinberg & Bealer 2001).

Extraction of Caffeine

It is very difficult to know the exact amount of caffeine in a particular drink that is not automatically prepared. The amount of caffeine in a single serving of coffee varies considerably due to many variables. Concentration can vary from bean to bean within a given bush; preparation of the raw bean will affect concentration, as well as multiple variables involved in brewing.

Caffeine extraction is an important industrial process and can be performed using a number of different solvents. Benzene, chloroform, trichloroethylene and dichloromethane have all been used over the years but for reasons of safety, environmental impact, cost and flavour, they have been superceded by two main methods:

Water extraction of caffeine

Coffee beans are soaked in water. The water - which contains not only caffeine but also many other compounds which contribute to the flavour of coffee - is then passed through activated charcoal, which removes the caffeine. The water can then be put back with the beans and evaporated dry, leaving decaffeinated coffee with a good flavour. Coffee manufacturers recover the caffeine and resell it for use in soft drinks and medicines.

Supercritical carbon dioxide extraction of caffeine

Supercritical carbon dioxide is an excellent non polar solvent for caffeine (and for many other organic compounds besides) and is harmless. The extraction process is simple: CO_2 is forced through the green coffee beans at temperatures above 31.1°C and pressures above 73 atm. Under these conditions, CO_2 is said to be supercritical: it has gas like properties which allow it to penetrate deep into the beans but also liquid-like properties which dissolve 97-99% of the caffeine. The caffeine-laden CO_2 is then sprayed with high pressure water to remove the caffeine. The caffeine can then be isolated by charcoal adsorption (as above) or by distillation, recrystallization, or reverse osmosis.

Extraction of Caffeine from Tea Leaves

Place 30 g of the tea leaves in a 500 ml beaker. Add 250 mL of distilled water and 5 g of sodium carbonate and stir the contents of the beaker with a glass rod. Boil the contents of the beaker on a hot plate/water bath for 10 minutes. Cool the tea solution to room temperature using an ice-water bath. Filter the cooled solution using glass wool or muslin cloth or cheese cloth. Transfer the filtrate into a separating funnel and extract 3-4 times (4×10^{-25} mL) using dichloromethane. Take care to avoid formation of emulsion for each extraction. Don't shake the separating funnel vigorously, but gently swirl the two immiscible layers for 5 minutes. After each extraction, remove the lower organic layer into 250 mL beaker, leaving any emulsion layer behind. Add anhydrous sodium sulphate to the combined extracts. The sodium sulfate will remove any water and water soluble salts that are retained in the dichloromethane (organic layer) or accidentally transferred during decantation. Filter to remove the solid sodium sulphate and transfer the dry solution to a pre-weighed 250 mL beaker. Evaporate it to dryness by boiling it on a water bath. (It can also be evaporated under vacuum or by

EXTRACTION OF PURE CAFFEINE

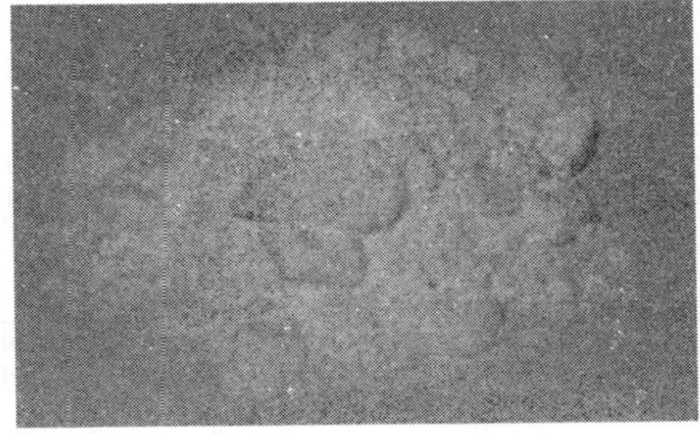

Anhydrous (dry) USP grade Caffeine

blowing dry air or nitrogen gas on the surface of the liquid). The residue will be crude caffeine (usually slight green in colour) (determine its weight-the yield will be 0.5 g approx for 25 g of tea leaves). Purify the crude caffeine either by sublimation or by recrystallization using methanol. Caffeine is reported to sublime at 170°C. Pure caffeine appears as white glistering needles. It is bitter in taste. Determine the melting point using a sealed capillary. The melting point of caffeine is 238°C.

It is very difficult to know the exact amount of caffeine in a particular drink that is not automatically prepared. The amount of caffeine in a single serving of coffee varies considerably due to many variables. Concentration can vary from bean to bean within a given bush; preparation of the raw bean will affect concentration, as well as multiple variables involved in brewing. Caffeine extraction is an important industrial process and can be performed using a number of different solvents. Benzene, chloroform, trichloroethylene and dichloromethane have all been used over the years but for reasons of safety, environmental impact, cost and flavour, they have been superceded by two main methods:

Water extraction of caffeine

Coffee beans are soaked in water. The water - which contains not only caffeine but also many other compounds which contribute to the flavour of coffee - is then passed through activated charcoal, which removes the caffeine. The water can then be put back with the beans and evaporated dry, leaving decaffeinated coffee with a good flavour. Coffee manufacturers recover the caffeine and resell it for use in soft drinks and medicines.

Supercritical carbon dioxide extraction of caffeine

Supercritical carbon dioxide is an excellent nonpolar solvent for caffeine (and for many other organic compounds besides) and is harmless. The extraction process is simple: CO_2 is forced through the green coffee beans at temperatures above 31.1°C and pressures above 73 atm. Under these conditions, CO_2 is said to be supercritical: it has gaslike properties which allow it to penetrate deep into the beans but also liquid-like properties which dissolve 97-99% of the caffeine. The caffeine-laden CO_2 is then sprayed with high pressure water to remove the caffeine. The caffeine can then be isolated by charcoal adsorption (as above) or by distillation, recrystallization, or reverse osmosis.

Synthesis of Caffeine from Uric acid: Uric acid is treated with methyl iodide in alkaline solution to form 1, 3, 7 trimethyl uric acid. This in heating with $POCL_3$ gives chlorocaffeine which on reduction with hydrogen iodide yields caffeine.

uric acid —(CH_3I, NaOH)→ 3-methyl-7,9-dihydro-1H-purine-2,6,8(3H)-trione —($POCl_3$)→ 8-chloro-1,3,7-trimethyl-3,7-dihydro-1H-purine-2,6-dione —(HI)→ caffeine

Traube's Synthesis Method: **W. Traube** formed the nitroso derivative of iminodimethyl barbituric acid action of phosphorus oxychloride on cyanacetic acid and dimethyl urea), and reduced it by ammonium sulphide to 1.3-dimethyl-4.5-diamino-2.6-dioxypyrimidine, the formyl derivative of which, when heated to 250° C., loses the elements of water and yields theophylline (Xanthine). It behaves as a weak base. When oxidized by potassium chlorate and 'hydrochloric acid it yields dimethylalloxan.

1,3-dimethylurea + ethyl isocyanoacetate — $NaNH_2$ / xylene → — HNO_2 → — Zn/H_2SO_4 → — COOH / reflux → — CH_3I / ethanol → caffeine

Physical Properties

Molecular Formula	$C_8H_{10}N_4O_2$
Formula Weight	194.1906
Composition	C(49.48%) H(5.19%) N(28.85%) O(16.48%)
Molar Refractivity	50.38 ± 0.5 cm^3
Molar Volume	133.3 ± 7.0 cm^3
Parachor	364.5 ± 8.0 cm^3
Index of Refraction	1.679 ± 0.05
Surface Tension	55.7 ± 7.0 dyne/cm
Density	1.45 ± 0.1 g/cm^3
Polarizability	19.97 ± 0.5 $10^{-24}cm^3$
Monoisotopic Mass	194.080376 Da
Nominal Mass	194 Da
Average Mass	194.193557 Da

Chemical Properties

Caffeine is a weak base with bitter taste and forma salts with strong acids. On oxidation with potassium chlorate in hydrochloric acid, it gives dimethyl urea and methyl urea

$$\text{caffeine} \xrightarrow{[O]} \text{(dimethyl product)} + H_3C-NH-\overset{O}{\overset{\|}{C}}-NH_2$$

caffeine

1-methylurea

Effects of caffeine

Caffeine has a significant effect on spiders, which is reflected in their web construction.

Caffeine is a central nervous system stimulant, and is used both recreationally and medically to restore mental alertness when unusual weakness or drowsiness occurs. It is important to note, however, that caffeine cannot replace sleep, and should be used only occasionally as an alertness aid.

Caffeine is sometimes administered in combination with medicines to increase their effectiveness, such as with ergotamine in the treatment of migraine and cluster headaches, or with certain pain relievers such as aspirin or acetaminophen. Caffeine may also be used to overcome the drowsiness caused by antihistamines. Breathing problems (apnea) in premature infants are sometimes treated with citrated caffeine, which is available only by prescription in many countries.

While relatively safe for humans, caffeine is considerably more toxic to some other animals such as dogs, horses and parrots due to a much poorer ability to metabolize this compound. Caffeine has a much more significant effect on spiders, for example, than most other drugs do. (1)

Caffeine metabolism

Caffeine is completely absorbed by the stomach and small intestine within 45 minutes of ingestion. It is widely distributed in total body water and is eliminated by apparent first-order kinetics that can be described by a one-compartment open-model system. Continued consumption of caffeine can lead to tolerance. Upon withdrawal, the body becomes oversensitive to adenosine, causing the blood pressure to drop dramatically, which causes headaches and other symptoms.

Caffeine is metabolized in the liver by the cytochrome P450 oxidase enzyme system into three metabolic dimethylxanthines, which each have their own effects on the body:

- Paraxanthine (84%) – Has the effect of increasing lipolysis, leading to elevated glycerol and free fatty acid levels in the blood plasma.
- Theobromine (12%) – Dilates blood vessels and increases urine volume. Theobromine is also the principal alkaloid in cocoa, and therefore chocolate.
- Theophylline (4%) – Relaxes smooth muscles of the bronchi, and is used to treat asthma. The therapeutic dose of theophylline, however, is many times greater than the levels attained from caffeine metabolism.

Each of these metabolites is further metabolised and then excreted in the urine.

Mechanism of Action

The caffeine molecule is structurally similar to adenosine, and binds to adenosine receptors on the surface of cells without activating them. This effect, called competitive inhibition, interrupts a pathway that normally serves to regulate nerve conduction by suppressing post-synaptic potentials. The result is an increase in the levels of epinephrine (adrenaline) and norepinephrine released from the pituitary gland [2]. Epinephrine, the natural endocrine response to a perceived threat, stimulates the sympathetic nervous system, leading to an increased heart rate, blood pressure and blood flow to muscles, a decreased blood flow to the skin and inner organs and a release of glucose by the liver.

Caffeine is also a known competitive inhibitor of the enzyme cAMP-phosphodiesterase (cAMP-PDE), which converts cyclic AMP (cAMP) in cells to its noncyclic form, allowing cAMP to build up in cells. Cyclic AMP participates in the messaging cascade produced by cells in response to stimulation by epinephrine, so by blocking its removal caffeine intensifies and prolongs the effects of epinephrine and epinephrine-like drugs such as amphetamine, methamphetamine, or methylphenidate.

The metabolites of caffeine contribute to caffeine's effects. Theobromine, is a vasodilator that increases the amount of oxygen and nutrient flow to the brain and muscles. Theophylline, the second of the three primary metabolites, acts as a smooth muscle relaxant that chiefly affects bronchioles and acts as a chronotrope and inotrope that increases heart rate and efficiency. The third metabolic derivative, paraxanthine, is responsible for an increase in the lipolysis process, which releases glycerol and fatty acids into the blood to be used as a source of fuel by the muscles (Dews et al. 1984).

With these effects, caffeine is an ergogenic, increasing the capacity for mental or physical labour. A study conducted in 1979 showed a 7% increase in distance cycled over a period of two hours in subjects who consumed caffeine compared to control tests (Ivy et al. 1979). Other studies attained much more dramatic results; one particular study of trained runners showed a 44% increase in "race-pace" endurance, as well as a 51% increase in cycling endurance, after a dosage of 9 milligrams of caffeine per kilogram of body weight (Graham & Spriet 1991). The extensive boost shown in the runners is not an isolated case; additional studies have reported similar effects. Another study found 5.5 milligrams of caffeine per kilogram of body mass resulted in subjects cycling 29% longer during high intensity circuits (Trice & Hayes 1995).

Side effects of caffeine

The minimum lethal dose of caffeines ever reported is 3,200 mg, administered intravenously. The LD_{50} of caffeine is estimated between 13 and 19 grams for oral administration for an average adult. The LD_{50} of caffeine is dependent on weight and individual sensitivity and estimated to be about 150 to 200 mg per kg of body mass, roughly 140 to 180 cups of coffee for an average adult taken within a limited timeframe that is dependent on half-life. The half-life, or time it takes for the amount of caffeine in the blood to decrease by 50%, ranges from 3.5 to 10 hours. In adults the half-life is generally around 5 hours. However, contraceptive pills increase this to around 12 hours, and, for women over 3 months pregnant, it varies from 10 to 18 hours. In infants and young children, the half-life may be longer than in adults. With common coffee and a very rare half-life of 100 hours, it would require 3 cups of coffee every hour for 100 hours just to reach LD_{50}. Though achieving lethal dose with coffee would be exceptionally difficult, there have been many reported deaths from intentional overdosing on caffeine pills.

Too much caffeine, especially over an extended period of time, can lead to a number of physical and mental conditions. *The Diagnostic and Statistical Manual of Mental Disorders, Fourth Edition (DSM-IV) states:* "The 4 caffeine-induced psychiatric disorders include *caffeine intoxication, caffeine-induced anxiety disorder, caffeine-induced sleep disorder, and caffeine-related disorder not otherwise specified (NOS).*"

An overdose of caffeine can result in a state termed *caffeine intoxication* or *caffeine poisoning*. Its symptoms are both physiological and psychological. Symptoms of caffeine intoxication include: restlessness, nervousness, excitement, insomnia, flushed face, diuresis, muscle twitching, rambling flow of thought and speech, paranoia, cardiac arrhythmia or tachycardia, and psychomotor agitation, gastrointestinal complaints, increased blood pressure, rapid pulse, vasoconstriction (tightening or constricting of superficial blood vessels) sometimes resulting in cold hands or fingers, increased amounts of fatty acids in the blood, and an increased production of gastric acid. In extreme cases mania, depression, lapses in judgment, disorientation, loss of social inhibition, delusions, hallucinations and psychosis may occur. [3]

It is commonly assumed that only a small proportion of people exposed to caffeine develop symptoms of caffeine intoxication. However, because it mimics organic mental disorders, such as panic disorder, generalized anxiety disorder, bipolar disorder, and schizophrenia, a growing number of medical professionals believe caffeine-intoxicated people are routinely misdiagnosed and unnecessarily medicated. Shannon *et al* (1998) point out that:

"Caffeine-induced psychosis, whether it be delirium, manic depression, schizophrenia, or merely an anxiety syndrome, in most cases will be hard to differentiate from other organic or non-organic psychoses....The treatment for caffeine-induced psychosis is to withhold further caffeine." A study in the *British Journal of Addiction* declared that "although infrequently diagnosed, caffeinism is thought to afflict as many as one person in ten of the population" (JE James and KP Stirling, 1983).

Because caffeine increases the production of stomach acid, high usage over time can lead to peptic ulcers, erosive esophagitis, and gastroesophageal reflux disease.[*citation needed*] Furthermore, it can also lead to nervousness, irritability, anxiety, tremulousness, muscle twitching, insomnia, heart palpitations and hyperreflexia [4].

Withdrawal

Individuals who consume caffeine regularly develop a reduction in sensitivity to caffeine; when such individuals reduce their caffeine intake, their body becomes oversensitive to adenosine, with the result that blood pressure drops dramatically, leading to an excess of blood in the head (though not necessarily on the brain), causing a headache. Other symptoms may include nausea, fatigue, drowsiness, anxiety and irritability; in extreme cases symptoms may include depression, inability to concentrate and diminished motivation to initiate or to complete daily tasks at home or at work.

Withdrawal symptoms may appear within 12 to 24 hours after discontinuation of caffeine intake, peak at roughly 48 hours, and usually lasts from one to five days. Analgesics, such as aspirin, can relieve the pain symptoms, as can a small dose of caffeine.

Effects on fetuses and newborn children

There is some evidence that caffeine may be dangerous for fetuses and newborn children. In animal studies, caffeine intake during pregnancy has been demonstrated to have teratogenic effects and increase the risk of learning problems and hyperactivity in rats and mice, respectively. The applicability of these results to human infants is disputed since the concentrations involved were high and rodents are more susceptible to most mutagens. In a 1985 study conducted by scientists of Carleton University, Canada, children born by mothers who had consumed more than 300 mg/d caffeine (about 3 cups of coffee or 6 cups of tea) were found to have, on the average, lower birth weight and head circumference than the children of mothers who had consumed little or no caffeine. In addition, use of large amounts of caffeine by the mother during pregnancy may cause problems with the heart rhythm of the fetus. For these reasons, some doctors recommend that women largely discontinue caffeine consumption during pregnancy and possibly also after birth until the newborn child is weaned.The negative effects of caffeine on the developing fetus can be attributed to the ability of caffeine to inhibit two DNA damage response proteins known as Ataxia-Telangeictasia Mutated (ATM) or ATM-Rad50 Related (ATR). These proteins control much of the cells ability to stop cell cycle in the presence of DNA damage, such as DNA single/double strand breaks and nucleotide dimerization. DNA damage can occur relatively frequently in actively dividing cells, such as those in the developing fetus. Caffeine is used in laboratory setting as an inhibitor to these proteins and it has been shown in a study by Lawson et al. in 2004, that women who use caffeine during pregnancy have a higher likelihood of miscarriage than those who do not. Since the dosage rate of self-administration is difficult to control and the effects of caffeine on the fetus are related to random occurrence (DNA damage), a minimal toxic dose to the fetus has yet to be established.

Caffeine pills

Caffeine pills are often used by college students and shift workers as a convenient way to fight sleep, and are often considered harmless. However, like any medication, caffeine can be harmful or deadly in sufficient quantities. Due to the amount of caffeine present in standard pills, it is possible to consume a dangerous amount of caffeine in this form.

Periodically, caffeine pills come under media fire in connection with the death of a college student due to a large overdose of caffeine. One example is the death of a North Carolina student, Jason Allen, who swallowed most of a bottle of 90 such pills [5], equivalent of about 250 cups of coffee. A few other deaths by caffeine overdose have been known, almost always in the case of massive pill consumption.

Identification Test

Murexide test

Add few crystals of caffeine with 3-4 drops of conc. nitric acid in a porcelain dish and evaporate to dryness. Addition of 2 drops of ammonium hydroxide to the residue gives purple colour.

CHAPTER

19

Fats Oils Waxes and Saponification

Introduction

Fat comes from the ancient Greek word Lipos meaning fat, they are generally soluble in organic solvents like chloroform, alcohol and alkaline solution such as sodium and potassium hydroxides. Fats provide a concentrated source of energy in the diet. The building blocks of fats are called fatty acids. These can be either saturated, monounsaturated or poly-unsaturated. Foods rich in saturated fats are usually of animal origin. Vegetable fats are generally unsaturated. Saturated fat raises the level of cholesterol in the blood. Cholesterol is present in animal foods but not plant foods. It is essential for metabolism but is not needed in the diet as our bodies can produce all that is needed. Raised blood cholesterol is associated with an increased risk of heart disease.

Fats and oils are essentially the same. Fats tend to be solid at room temperature whilst oils are liquid. The term lipids include both fats and oils.

Fats are generally constituted in the form of fatty acids, they are found in different form and types.

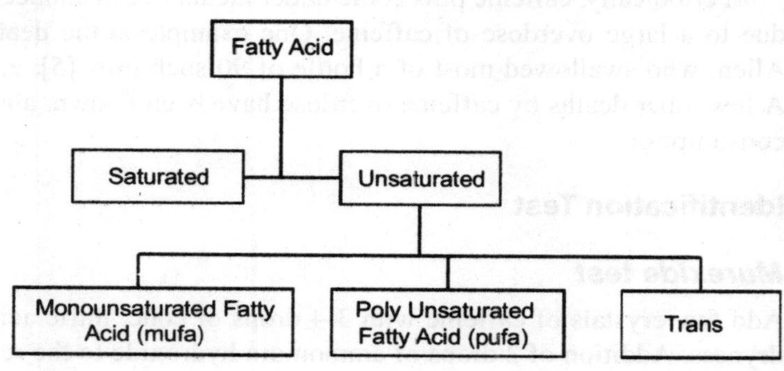

Types of Fatty Acids

Fatty acids when present in food may be either saturated or unsaturated.

Saturated Fatty Acids

If the fatty acid has all the hydrogen atoms it can hold it is said to be saturated (see below)

This type of fat is typically found in larger amounts in foods from animals, e.g. meat, butter, cheese and cream.

Many baked goods such as cakes, biscuits and pastries are also high in saturated fat. Excessive intake of saturated fat can increase blood cholesterol levels, one of the major risk factors for heart disease.

Unsaturated

Some of the carbon atoms are joined to others by a double bond and so are not completely saturated with hydrogen atoms (see below). They could therefore accept more hydrogen atoms:

There are three types of unsaturated fatty acids:-

1. Mono unsaturated fatty acids

If there is one double bond as above, the fatty acid is known as a monounsaturated fatty acid. This is found in significant amounts in most types of nuts, avocado pears, rapeseed oil and olive oil. Monounsaturates do not raise blood cholesterol and evidence shows that they may also help to reduce blood cholesterol levels if they replace saturated fat in the diet.

2. Polyunsaturated fatty acids

If there is more than one double bond, then the fatty acid is known as a polyunsaturated fatty acid (see below). These come mostly from vegetable sources, such as sunflower oil or seeds, but are also found in oily fish such as mackerel and sardines.

Polyunsaturates can actively reduce blood cholesterol levels and are found largely in sunflower and corn oils. Polyunsaturates in oily fish appear to have no effect on blood cholesterol levels, but they do make blood less 'sticky' and therefore less likely to form clots that can block the flow of blood to the heart and cause problems.

3. Trans fatty acids

Trans fatty acids in our diet come mainly from two sources. One is when liquid oils are hardened by partial hydrogenation; the other source of trans fatty acids are meat products and dairy foods. Current scientific evidence suggests that trans fats behave in a similar way to saturated fats, i.e. raising blood cholesterol levels.

For an example of the trans fatty acids configuration see below:

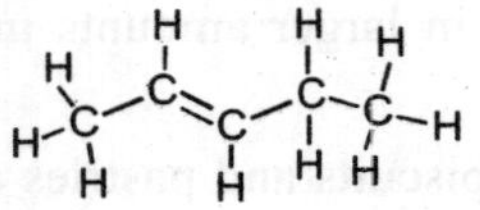

Omega-3s

Omega-3s are considered essential fatty acids because we must get them from our diet and they are found in both plant and animal foods. Plant sources provide only one type of omega-3, called alpha-linolenic acid, and rich sources include flax oil and seeds, walnuts, canola oil, hemp seeds and oil and soybean oil. Animal sources provide two different types of omega-3s, eicosapentaenoic acid (EPA) and docosahexaenoic acid (DHA). The richest sources of EPA and DHA are fish like salmon, lake trout, tuna, herring, mackerel and sardines. Several brands of DHA enriched eggs as well as DHA fortified foods are also now becoming available.

Quantity vs. Quality

When you see how different types of fat can either help or hurt your health, it becomes clear that the quality of fats you choose are equally, if not more important as the quantity you eat. Just keep in mind that fats do provide 9 calories per gram so be careful not to overdo it so you can manage your weight. National guidelines recommend getting less than 30% of your calories from fat.

Changing Your Oil

Oils are made by extracting them from their plant seeds. Sounds simple right? Not exactly. The vegetable oils commonly found on supermarket shelves is most often obtained using chemical extraction, which can reduce the nutrient value, colour and aroma inherent to the oil. However, there are a number of natural ways to get the oil out of a seed but the names and labels can be confusing.

Expeller Pressed Oils

Oils extracted by mechanically expelling the oil from the seeds, just as the name implies.

Cold Pressed Oils

Oils mechanically compressed from the seeds but without using heat, just as the name implies. Extra virgin and virgin olive oils and flax oils are typically cold pressed. Expeller and cold pressed oils can be found at regular grocery stores or natural food stores anywhere. Because these types of oils go through less processing, they tend to be more susceptible to rancidity so we recommend storing them in the refrigerator so they stay fresh.

Common fat-containing foods

Different foods contain different ratios of fatty acids:

- ***Saturated fats*** - sources include fatty cuts of meat, full fat milk and cheese, butter, cream, most commercially baked products such as biscuits and pastries, most deep fried fast foods, coconut and palm oil.
- ***Monounsaturated fats*** - sources include margarine spreads such as canola or olive oil based choices, oils such as olive, canola and peanut oils, avocado, and nuts such as peanuts, hazelnuts, cashews and almonds.
- ***Polyunsaturated fats*** - sources include fish oils, seafood, polyunsaturated margarines, vegetable oils such as safflower, sunflower, corn or soy oils, nuts such as walnuts and brazil nuts, and seeds.

Sources of omega-6 and omega-3 fats

Polyunsaturated fats can be divided into two categories:

- ***Omega-3 fats*** are found in both plant and marine foods and have been found to reduce the risk of heart disease. Food sources include canola and soy oils and canola based margarines. Marine sources include fish especially oily fish such as Atlantic salmon, mackerel, Southern blue fin tuna, trevally and sardines.
- ***Omega-6 fats*** are found primarily in nuts, seeds and plant oils such as corn, soy and safflower.

Benefits of omega fats

Research is ongoing, but the benefits of omega fats in the diet seem to include that they:

- Lower blood cholesterol levels, which reduces an important risk factor in coronary heart disease
- Improve blood vessel elasticity
- Thin the blood, which makes it less sticky and less likely to clot
- Reduce inflammation and boost the immune system
- Contribute to the normal development of the foetal brain.

Plant sterols

Plant sterols are present in all plants. Intakes of 2-3g plant sterols per day have been shown to reduce blood cholesterol levels by an average of 10 per cent. This is because they block the body's ability to absorb cholesterol, which leads to a reduced level of cholesterol in the blood.

It is hard to eat this much from natural sources so there are now plant sterol enriched margarines on the market. Eating 1 to 1½ tablespoons of sterol enriched margarine each day can help to lower blood cholesterol levels.

Energy density

Dietary fat has more than double the amount of kilojoules per gram (37) than carbohydrate or protein (17), making it very energy dense. Some research suggests that saturated fats are more likely to contribute to weight gain (especially around the middle) than polyunsaturated fat and monounsaturated fats, even though they have the same kilojoule content.

Carrying too much body fat is a risk factor in many diseases, including coronary heart disease and type 2 diabetes. Saturated fats should be kept to a minimum.

Saturated fats

Saturated fats tend to raise LDL levels. Most saturated fats raise blood cholesterol. Butter, coconut and palm oil, cottonseed oil, lard, cocoa butter and beef tallow are all high in saturated fats.

These fats are commonly found in many fast foods and in commercial products such as biscuits and pastries.

Trans fatty acids

Trans fatty acids from hydrogenated plant fats appear to act just like saturated fatty acids in the way that they raise blood cholesterol levels. Unlike saturated fats, they also tend to lower HDL cholesterol, so are potentially more damaging. However, they are found in much smaller amounts in the diet than saturated fats and so do not pose as great a risk of heart disease.

Other fats

Other dietary fats that influence blood cholesterol are polyunsaturated fats and monounsaturated fats. They both reduce blood cholesterol levels. You should try to replace saturated fats in your diet with either monounsaturated or polyunsaturated fats.

For example, replace butter in some cooked dishes with olive or canola oil.

Cholesterol in food

People with high blood cholesterol or who are at risk of heart disease should try to limit their intake of cholesterol-rich foods. Dietary cholesterol is only found in animal products such as:

- Full fat dairy products
- Fatty meats
- Egg yolks
- Offal - for example liver, kidney and brains.

The Mediterranean diet

Researchers are investigating the possibility that a diet rich in monounsaturated fats, such as olive oil, may be protective against the development of coronary heart disease. People who have a high consumption of monounsaturated fats from olive oil (for example, in Greece and Italy) tend to have low rates of coronary heart disease, regardless of their body weight. We must remember, though, that the Mediterranean diet contains much more than olive oil. It's possible that the low rate of coronary heart disease in these countries relates to a high intake of vegetables, legumes, fruits and cereals, which are rich in antioxidants. The evidence so far is inconclusive.

Current recommendations

Nutritionists recommend that we limit the amount of fats in the daily diet, particularly saturated and trans fats. Simple suggestions include:

- Use margarine spreads instead of butter or dairy blends.
- Use salad dressings and mayonnaise made from oils such as canola, sunflower, soy and olive oils.
- Use low or reduced fat milk and yoghurt or 'added calcium' soy beverages.
- Try to limit cheese and ice cream to twice a week.
- Have fish (any type of fresh or canned) at least twice a week.
- Select lean meat (meat trimmed of fat and chicken without skin). Try to limit fatty meats including sausages and delicatessen meats such as salami.
- Snack on plain, unsalted nuts and fresh fruit.
- Incorporate dried peas (for example split peas), dried beans (for example haricot beans, kidney beans, three bean mix) or lentils into two meals a week.
- Make vegetables and grain based foods such as breakfast cereals, bread, pasta, noodles and rice the major part of each meal.
- Try to limit take-away to once a week or less.
- Try to limit snack foods such as potato crisps and corn crisps to once a week or less.
- Try to limit cakes, pastries and chocolate or creamy biscuits to once a week or less.
- Try to limit cholesterol-rich foods such as egg yolks and offal like liver, kidney and brains.

Things to remember

- Dietary fat contains more than double the amount of kilojoules per gram than carbohydrate or protein.
- Animal products and some processed foods, especially fried fast food, are generally high in saturated fats, which have been linked to increased blood cholesterol levels.
- Replacing saturated fats with monounsaturated and polyunsaturated fats tends to improve blood cholesterol levels.

Extraction of Fats

(1) ***Rendering.*** Animal fats are recovered from the selected animal material by *Wetrendering* or *Dry Rendering.*

(a) Wet Rendering. The selected chopped material is charged into a cylinder with base. Steam is then blown' through the cylinder for several hours, the pressure being kept constant at about 5 kg/cm2 by releasing steam occasionally. After allowing settlement to the, the floating fat is drawn off leaving the water and exhausted animal matter behind. This process is widely used for the extraction of edible fats such as lard.

The fish oils from 'blubber' or 'liver' are obtained by heating the chopped stuff in a pressure cooker or steam boiler, and running off the floating oil layer.

(b) Dry Rendering. This process is carried out in large steam-heated tanks heating, the chopped animal tissue is churned by metal blades. The fat cells are ruptured and the molten fat so obtained is drained off from the bottom of the tanks. Although dry rendering is cheaper than .wet rendering, the product is somewhat inferior. Hence' this is restricted to the extraction of non-edible fats.

(2) Pressing and Solvent Extraction. In India the most important source of edible oils s such as mustard, groundnut, cottonseed, sesame, niger and sunflower' which has been introduced under experimental conditions. The extraction from seeds is done by a variety of methods right from the village *ghani* to the more sophisticated *Expeller* and modern *Solvent Extraction Plant.* .

(a) Pressing by Expeller. The oil bearing material is screened and crushed by between steel rollers. In order to get the maximum yield the *'meal'* or crushed material cooked at 70-100°C in steam-jacketed vessel to rupture the oil cells. The cooked meal is then pressed by a high pressure *expeller* (Continuous Screw Press) which has now completely replaced the Hydraulic Press. The expeller consists of a perforated cylindrical which a screw shaft moves. As the meal is fed into the expeller

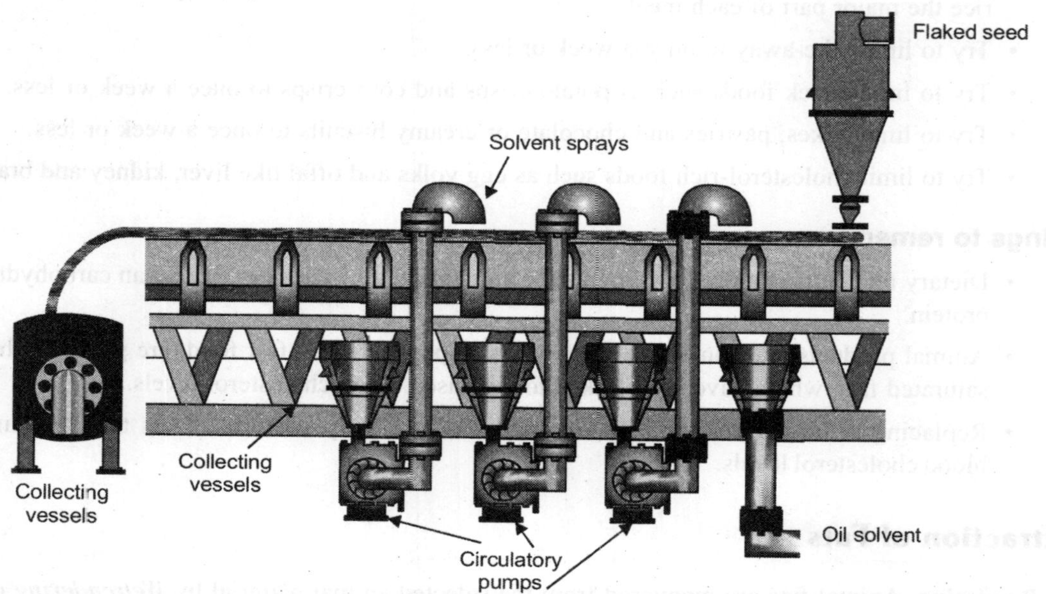

Fig 19.1 : Showing the Extraction of Rolled Oil Seeds by Counter Current Solvent Extraction

tube, it is subjected to increasing pressure by the revolving screw. While the oil expelled from the seeds flows out of the holes in the expeller tube, the pressed oil cake is ejected from the other end of the tube. The exhausted oil cake still contains oil which the modern expeller can reduce to per cent. Most of the exhausted cake is used for cattle feed or subjected to 'solvent In' for the recovery of more oil.

(b) Solvent Extraction. The screened seed is pressed into 'flakes' by passing controllers. The rolled seed is carried along a moving belt of fine wire mesh beneath a succession of solvent sprays (Fig. 19.1)

The oil-bearing solvent, usually petroleum ether, is passed from spray to spray in the opposite direction to the moving belt *(Countercurrent Principle)*. In this way fresh solvent washes the almost extracted cake at the end of the belt. The final solvent spray washing flaked seed at the inlet is heavily loaded with oil. It is then sent to the distillation ('Stripper') when the solvent is recovered and the crudevege table oil is sent to the unit. The exhausted oil cake obtained at the end of the process contain only 0.5 to 1 per cent of the oil and is used as a cattle feed or organic manure.

Refining of Crude Fats and Oils

The crude fats and oils produced by the above techniques contain impurities which not removed would give the product undesirable appearance, taste or odour. These *i* suspended of colloidal matter, free fatty acids, and coloured and odiferous substances.

In the case of high quality fats such as lard, all that is required is filtration to remove suspended matter after coagulation by steam. The vegetable oils need to be given the following treatment for the removal of various impurities.

(a) Filtration. The suspended or colloidal matter is removed by filtration after coagulation with open steam after adding a coagulant such as citric acid.

(b) Neutralisation (Alkali refining). The free fatty acids are removed as soaps by treatment with sodium hydroxide. The oil and alkali are agitated at about 90° until the fatty acids have been saponified. After standing, the soap settles down carrying down suspended and coloured matter. The soapy layer is run off, and the purified oil washed with water and dried by heating in vacuum.

(c) Treatment with Fuller's earth. Remaining traces of colour and by treating the oil with about 1 per cent of Fuller's earth and filtration.

Alternatively the oil is heated under vacuum for several hours using' Palm oil and certain other vegetable oils can be deorised and bleached in this way

Structure and Composition of Fats and Oils

Animal and vegetable fats and oils have similar chemical structure formed from glycerol and long-chain carboxylic acids (often called fatty acids)

glycerol + 3R—C(=O)—OH ⟶ a far or oil (a tryglyceride) + $3H_2O$

glycerol

a far or oil (a tryglyceride)

A triester of glycerol is called a *triglyceride* or *glyceride.* If all the R groups in the general formula are identical, the triester is designated as a Simple glyceride, and if they are not mixed glyceride.

fatty acid groups identical

Simple glyceride

fatty acids groups not identical

Mixed glyceride

Most natural fats and oils are mixed triglycerides having two or three different fatty acid groups.

The carboxylic acids or fatty acids that go to form the fat or oil molecules *(glycerides)* have carbon chains with only even number of carbon atoms. The most common fatty acids have unbranched carbon chains of 14, 16 or 18 carbons. The chains may be saturated or may include one or more double bonds.

The glycerides are referred to as *saturated* or *unsaturated* depending on whether the fatty acid component chains are saturated or contain double bonds.

The most common satarated fatty acids found in fats and oils are myristic acid,$C_{18}H_{27}COOH$, palmitic acid, $C_{15}H_{31}COOH$, and stearic acid, $C_{17}H_{31}COOH$. Amongst the unsaturated fatty acids, oleic acid, $C_{17}H_{33}COOH$. and linoleic acid, $C_{17}H_{31}COOH$, are widely distributed in almost all fats and oils. Oleic acid chain contains one double bond and linoleic acid chain two double bonds. We also know that the presence of a double bond in a fatty acid can cause *cis-trans* isomerism, depending on the configuration of the H atoms attached to the doubly-bonded carbon atoms. Thus oleic acid is the cis-isomer, while its *trans.isomer* is elaidic acid. Linoleic acid has two double bonds and both possess *cis* configuration. Generally speaking, the cis-isomers are found naturally occurring in the unsaturated fatty acid components of fats and oils. . The structure and melting points of some common fatty acids are given in Table below

Composition of Fats and Oils. **As already mentioned, fats and oils are invariably composed of a number of mixed glycerides, *e.g.,***

In the mixed glycerides present in fats and oils a single molecule of glyceride may contain two or three different fatty acids linked by ester bonds to the glycerol. While it is difficult to know exactly

Table of Structure and Melting Points of Different Fatty Acids

Name	Structure	mp°C
A. Saturated:		
Caproic acid	$CH_3-(CH_2)_4-\overset{O}{\overset{\parallel}{C}}-OH$	–2°
Caprylic acid	$CH_3-(CH_2)_6-\overset{O}{\overset{\parallel}{C}}-OH$	17°
Capric acid	$CH_3-(CH_2)_8-\overset{O}{\overset{\parallel}{C}}-OH$	31°
Lauric acid	$CH_3-(CH_2)_{10}-\overset{O}{\overset{\parallel}{C}}-OH$	44°
Myristic acid	$CH_3-(CH_2)_{12}-\overset{O}{\overset{\parallel}{C}}-OH$	54°
Palmitic acid	$CH_3-(CH_2)_{14}-\overset{O}{\overset{\parallel}{C}}-OH$	63°
Stearic acid	$CH_3-(CH_2)_{16}-\overset{O}{\overset{\parallel}{C}}-OH$	70°
B. Unsaturated		
Oleic acid, *cis*-9-octadecanoic acid	$CH_3-(CH_2)_7(H)C=C(H)(CH_2)_7-\overset{O}{\overset{\parallel}{C}}-OH$	16°
Linoleic acid, *is-cis*-9, 12-octadecadienoic acid	$CH_3-(CH_2)_4(H)C=C(H)-CH_2-(H)C=C(H)(CH_2)_7-\overset{O}{\overset{\parallel}{C}}-OH$	–5°

as to which triglycerides are present in a particular fat or oil, the overall percentage composition of fatty acids which make up the fat or oil can be determined by analysis.

$$\begin{array}{l} CH_2-O-\overset{O}{\overset{\|}{C}}-(CH_2)_{12}-CH_3 \\ | \\ CH-O-\overset{O}{\overset{\|}{C}}-(CH_2)_{14}-CH_3 \\ | \\ CH_2-O-\overset{O}{\overset{\|}{C}}-(CH_2)_{16}-CH_3 \end{array}$$

Glyceryl myristopalmitostearate
(*mixed glyceride*)

$$\begin{array}{l} CH_2-O-\overset{O}{\overset{\|}{C}}-(CH_2)_{14}-CH_3 \\ | \\ CH-O-\overset{O}{\overset{\|}{C}}-(CH_2)_{14}-CH_3 \\ | \\ CH_2-O-\overset{O}{\overset{\|}{C}}-(CH_2)_7-CH{=}CH-(CH_2)_7-CH_3 \end{array}$$

Glyceryl diplamitooleate
(mixed glyceride)

Fat or Oil	Fatty Acid			
	Myristic acid	Palmitic acid	Stearic acid	Oleic cid
Oils:				
Olive oil	—	6—10	1—4	83—84
Peanut oil	—	6—9	2—5	50—60
Groundnut oil	—	6—14	2—7	46—72
Cottonseed oil	1—2	17—29	1—4	13—44
Mustard oil	—	1—3	1—3	8—40
Cocoanut oil	1—2	17—29	1—4	13—44
Sunflower oil	—	2—10	1—6	7—42
Soyabean oil	—	7—12	2—6	20—50
Fats:				
Beef etaellow	2—6	24—32	15—25	37—43
Butter fat	7—12	23—30	8—13	30—40
Human fat	3—6	24—26	5—8	40—45
Lard	1—2	25—30	12—18	40—50
Marine Oils				
Whale	5—10	10—20	2—5	33—40
Fish	6—8	10—25	1—3	—

Why are Animal fats solid and Vegetable oils liquid?

It is the degree of unsaturation of the constituent fatty acids which determines whether a triglyceride will be a solid or a liquid. The glycerides in which long-chain saturated acid components predominate tend to be solid or semisolid, and are termed *fats.* On the other hand, oils are glyceryl esters which contain higher proportion of unsaturated fatty acid components. This difference of melting point or consistency is distinctively demonstrated by taking example of glyceryl trioleate and glyceryl tristearate. The former compound contains three unsaturated acid components and is oil (liquid), while the latter compound is made of only saturated acid components and is a fat (solid).

$$
\begin{array}{l}
CH_2-O-\overset{O}{\overset{\|}{C}}-(CH_2)_7-CH{=}CH-(CH_2)_7-CH_3 \\
| \\
CH-O-\overset{O}{\overset{\|}{C}}-(CH_2)_7-CH{=}CH-(CH_2)_7-CH_3 \\
| \\
CH_2-O-\overset{O}{\overset{\|}{C}}-(CH_2)_7-CH{=}CH-(CH_2)_7-CH_3
\end{array}
$$

Glyceryl trioleate (*triolein*);
a oil, mp –5°C

$$
\begin{array}{l}
CH_4-O-\overset{O}{\overset{\|}{C}}-(CH_2)_{10}-CH_3 \\
| \\
CH-O-\overset{O}{\overset{\|}{C}}-(CH_2)_{15}-CH_3 \\
| \\
CH_2-O-\overset{O}{\overset{\|}{C}}-(CH_2)_{16}-CH_3
\end{array}
$$

Glyceryl tristearate (*tristearin*);
a fat, mp 71°

The melting points of mixed glycerides would depend on the extent of unsaturated fatty acid components in the molecule. When two fatty acid components are unsaturated, the glyceride would tend to be an oil, while if two or all the acid components are saturated the triglyceride would tend to be a fat.

Fig. is suggestive as to why the glyceride with mostly saturated acids present in their structure are solids, while unsaturated trigtycerides are oils. A regular saw-tooth

O–C(=O) 1
O–C(=O) 2
O–C(=O) 3

A saturated Triglyceride

O–C(=O)
O–C(=O)
O–C(=O) CIS
6

A partially unsaturated Triglyceride

arrangement of the hydrocarbon chains of a saturated glycerides permit tight packing of. the molecules and this results in a 'pseudocrystalline' solid substance *(fat)*. However, the presence of a *cis* double bond in the chain of the unsaturated fatty acid component causes a big bend at that point, leading to less dense packing of the glyceride molecules. Thus unsaturated glycerides have low melting points and tend to be liquids. If there are two double bonds in the chain of the acid part of a glyceride, there result two bends, giving rise to a much more random conformation. Hence regular packing of molecules of such a glyceride is rendered unlikely. The 'Polyunsaturated' glycerides, therefore, have very low melting point and are liquids (oils).

The terms 'Fat' and 'Oil' are more or less conventional and are now-a-days used in a very general fashion. Chemically common oils and fats are assortment of saturated and un. saturated triglycerides present in varying ratios. The apparent distinguishing difference between the two classes of compounds is their physical state. At ordinary temperature fats are solid or semisolid glycerides, while oils are liquids. But a given sample of glycerides (say *ghee)* may be a 'fat' in winter and an 'oil' in summer. In fact, it would be more advisable to use the term fat for both these classes of substances.

Physical Properties

(1) Oils and fats may be either liquids or noncrystatline solids at room temperature.

(2) When *pure* they are colourless, odourless and tasteless. The characteristic colours, odours, and flavours associated with natural oils and fats are imparted to them by foreign substances. For example, the yellow colour of butter is due to the presence of the pigment carotene; and the taste of butter is due to the following two compounds which are produced by bacteria in the ripening of cream.

$$H_3C-\overset{O}{\overset{\|}{C}}-\overset{O}{\overset{\|}{C}}-CH_3 \qquad H_3C-\overset{O}{\overset{\|}{C}}-\overset{HO}{\overset{|}{CH}}-CH_3$$

biacetyl 3-hydroxybutan-2-one

(3) They are tighter than and insoluble insoluble in water form the upper layer when mixed with it. They are readily soluble in organic solvents like diethyl ether, acetone, alkanes, benzene, chloroform, carbon tetrachloride and carbon disulphide.

(4) They readily form emulsions when agitated with water in the presence of soap, gelatin or other emulsifiers.

(5) They are poor conductors of heat and electricity and, therefore, serve as excellent insulators for the animal body.

Chemical Properties

The reactions of oils and fats are the reactions of triglycerides or triesters of glycerol. Thus they can undergo hydrolysis at all the three ester groups. Also, we know that the chains of the acid components of glycerides may contain one or more double bonds. Therefore the unsaturated glycerides give the addition and oxidation reactions characteristic of alkenes at the seats of these double bonds.

(1) Hydrolysis Triglycerides are easily hydrolysed by enzymes called *lipases.* In the digestive tracts of human beings and mammals to give fatty acids and glycerol. The fatty acids so produced play an important role in the metabolic process in the animal body.

$$\begin{array}{l} CH_2-O-\overset{O}{\overset{\|}{C}}-R \\ | \\ CH-O-\overset{O}{\overset{\|}{C}}-R' \\ | \\ CH_2-O-\overset{O}{\overset{\|}{C}}-R' \end{array} + 3H_2O \longrightarrow \begin{array}{l} CH_2-OH \\ | \\ CH-OH \\ | \\ CH_2-OH \end{array} + \begin{array}{l} R-\overset{O}{\overset{\|}{C}}-OH \\ \\ R'-\overset{O}{\overset{\|}{C}}-OH \\ \\ R''-\overset{O}{\overset{\|}{C}}-OH \end{array}$$

Triglyceride Glycerol Fatty acids

$$\begin{array}{l} CH_2OC(=O)(CH_2)_7CH{=}CH(CH_2)_7CH_3 \\ \quad \text{From oleic acid} \\ CHOC(=O)(CH_2)_{14}CH_3 \\ \quad \text{From palmitic acid} \\ CH_2OC(=O)(CH_2)_7CH{=}CHCH_2CH{=}CH(CH_2)_4CH_2 \\ \quad \text{From linoleic acid} \\ \quad (\textit{A Mixed Glyceride}) \end{array} \xrightarrow{+3H_2O} \begin{array}{l} CH_2OH \\ \\ CHOH \\ \\ CH_2OH \end{array} + \begin{array}{l} CH_3(CH_2)_7CH{=}CH(CH_2)_7{-}C(=O){-}OH \\ \quad \text{Oleic acid} \\ CH_8(CH_2)_{14}{-}C(=O){-}OH \\ \quad \text{Palmitic acid} \\ CH_3(CH_2)_4CH{=}CHCH_2CH{=}CH(CH_2)_7C(=O)OH \\ \quad \text{lionoleic acid} \\ \quad (\textit{Mixture of Fatty Acids}) \end{array}$$

Saponification

The term saponification is the name given to the chemical reaction that occurs when a vegetable oil or animal fat is mixed with a strong alkali. The products of the reaction are two: soap and glycerin. Water is also present, but it does not enter into the chemical reaction. The water is only a vehicle for the alkali, which is otherwise a dry powder. The name saponification literally means "soap making". The root word, "sapo", is Latin for soap. The Italian word for soap is sapone. Soap making as an art has its origins in ancient Babylon around 2500 - 2800 BC. The oils used in modern handmade soap are carefully chosen by the soap maker for the character they impart to the final soap. Coconut oil creates lots of glycerin, makes big bubbly lather, and is very stable. Olive oil has natural antioxidants and its soap makes a creamier lather. Tallow, or rendered beef fat, makes a white, stately bar that is firm and creates abundant lather. Many other oils can be used, each one for a specific reason. Your soap maker will be glad to tell you which oils are used to make her or his soap. The alkali used in modern soap is either potassium hydroxide, which is used to make soft soap or liquid soap because of its greater solubility, or sodium hydroxide, which is used to make bar soap. The common term for the alkali became simply "lye", which curiously is not short for alkali, but originated in the Anglo-Saxon language. Soap made in cottages and on farms in earlier American times became known as "lye soap". That term is now pejorative and derogatory and denotes a harsh soap that would irritate your skin. The old soap got a bad name because it had an excessive amount of caustic. Weighing and measuring techniques were crude, and knowledge of soap chemistry was elementary or non-existent. The true fact is that modern handcrafted soap, though necessarily made with lye to get true soap, has no lye in the final product. It has all been reacted with the oils to form soap and glycerin. A curious fact about modern soap is that most common soap found in the grocery store made in mass-produced factories does have a small amount of excess alkali in it. Also, it has had all of its naturally-occurring glycerin removed so it can be sold as a separate commodity. Why? Greater profit. An important difference between most commercial soap and our Real Handmade soap is that the glycerin is left in Real Handmade Soap and thus it retains its natural moisturizing property.

An example for saponification reaction is

$$\begin{array}{l} CH_2{-}O{-}\overset{O}{\overset{\|}{C}}(CH_2)_{14}CH_3 \\ CH{-}O{-}\overset{O}{\overset{\|}{C}}(CH_2)_{14}CH_3 \\ CH_2{-}O{-}\overset{O}{\overset{\|}{C}}(CH_2)_{14}CH_3 \end{array} + \underset{\text{Sodium hydroxide (or KOH, potassium hydroxide)}}{3\,NaOH} \xrightarrow{\text{Saponification}} \underset{\text{Glycerol}}{\begin{array}{l} CH_2{-}OH \\ CH{-}OH \\ CH_2{-}OH \end{array}} + \underset{\text{A crude soap}}{3CH_3(CH_2)_{14}CO_2Na}$$

A fat

Hydrogenation of Oils (Hardening)

***Definition*: Hydrogenation is the process of forcing hydrogen atoms into the holes of unsaturated fatty acids. This is done with hydrogen gas under pressure with a metal catalyst at a temperature of F (120-210ºC).**

The classical example of a hydrogenation is the attack of hydrogen on unsaturated bonds between carbon atoms, converting alkenes to alkanes. Numerous important applications are found in the pharmaceutical and petrochemical industries.

Hydrogenation of Oleic Acid

$$\underset{\text{Oleic Acid – Unsaturated}}{CH_3(CH_2)_7CH = CH(CH_2)_7\overset{O}{\overset{\|}{C}}{-}OH} + H_2 \longrightarrow CH_3(CH_2)_7CH{-}CH(CH_2)_7\overset{O}{\overset{\|}{C}}{-}OH$$

$$\downarrow$$

$$CH_3(CH_2)_7\overset{H}{\underset{H}{C}}{-}\overset{H}{\underset{H}{C}}{-}(CH_2)_7C{-}OH$$

Hydrogenation reactions also include the reaction between hydrogen and organic sulfur compounds to form gaseous hydrogen sulfide (H_2S). Hydrogenation of that type is very widely used in the petroleum refining and petrochemical industries to desulfurize various final products, intermediate products and process feedstocks by converting sulfur compounds to gaseous hydrogen sulfide which is then easily removed by simple distillation. In those industries, desulfurization process units are often referred to as hydrodesulfurizers (HDS) or hydrotreaters.

Rancidity

Spoilage of fats may occur on storage, particularly if the fats are highly unsaturated and the conditions of storage are conducive to chemical change in the fats. Rancidity is of two types-hydrolytic and oxidative.

Hydrolysis

Hydrolysis is brought about by enzymes that decompose fats into free fatty acids and glycerol. Heating

thoroughly to destroy the lipase enzyme that catalyses the hydrolysis of triglycerides should prevent hydrolytic rancidity. Contaminating microorganisms may also produce lipase and these can similarly be destroyed with sufficient heating.

$$\underset{\text{A butter triglyceride}}{\begin{array}{l} CH_2-O-\overset{O}{\overset{\|}{C}}-CH_2CH_2CH_3 \\ | \\ CH-O-\overset{O}{\overset{\|}{C}}-CH_2CH_2CH_2CH_2CH_3 \\ | \\ CH_2-O-\overset{O}{\overset{\|}{C}}-CH_2CH_2CH_2CH_2CH_2CH_2CH_3 \end{array}} \xrightarrow{+3H_2O} \underset{\text{Glycerol}}{\begin{array}{c} CH_2OH \\ | \\ CHOH \\ | \\ CH_2OH \end{array}} + \begin{array}{c} \underset{\text{Butyric acid}}{CH_3CH_2CH_2-\overset{O}{\overset{\|}{C}}-OH} \\ \underset{\text{Caproic acid}}{CH_3(CH_2)_4-\overset{O}{\overset{\|}{C}}-OH} \\ \underset{\text{Caprylic acid}}{CH_3(CH_2)_6-\overset{O}{\overset{\|}{C}}-OH} \\ (\textit{have offensive odours}) \end{array}$$

Oxidation

Only unsaturated fats and foods which have lipoxygenase are susceptible to oxidative changes. Highly hydrogenated and saturated fatty acids are relatively resistant to oxidation.

$$\underset{\text{Oleic acid}}{CH_3(CH_2)_7-CH\overset{\text{Cleavage}}{=}CH-(CH_2)_7-\overset{O}{\overset{\|}{C}}-OH} \xrightarrow[\text{(air)}]{O_2} \underset{\text{Pelargonic acid}}{CH_3(CH_2)_7-\overset{O}{\overset{\|}{C}}-OH} + \underset{\text{Azelaic acid}}{HO-\overset{O}{\overset{\|}{C}}-(CH_2)_7-\overset{O}{\overset{\|}{C}}-OH}$$

Prevention of rancidity

1. Storage at refrigerator temperature prevents rancidity.
2. Rays of light catalyse the oxidation of fats by the use of coloured glass containers that absorb the active rays, fats can be protected against spoilage. Certain shades of green bottles and wrappers and yellow transparent cellophane wrappers are effective in preventing rancidity.
3. Vacuum packaging also helps to retard the development of rancidity by excluding oxygen.
4. Antioxidants naturally present in the food such as vitamin C, beta carotene and vitamin E protect against rancidity.

Analysis of Fats and Oils

Since fats and oils are obtained from natural sources, their purity and composition is variable. They may contain free fatty acids produced by hydrolysis during storage, and non fatty impurities, Also the suitability of a fat or oil as raw material for the manufacture of soaps, synthetic detergents, paints etc., depends on the carbon chain length and the degree of unsaturation of the acid components in the constituent glycerides. A number of physical and chemical tests have been devised to evaluate a given fat or oil. The usual physical constants that are determined first are melting point, specific gravity and refractive index. The structure of glycerides has also been studied by using the modern physical methods such as X-ray analysis, absorption and mass spectrometery and NMR spectra,

NMR spectra has been particularly used to detect isolated or multiple double bonds in unsaturated Catty acid chains. The fat is subjected to' many analytical tests. Some of the important of these are described here.

(1) Saponification Number. As already discussed, saponification is a term specifically applied to the hydrolysis of an ester when the reaction is carried out in alkaline solution. The saponification number of a fat or an oil is an arbitrary unit that is defined as the Number of milligrams of potassium hydroxide required to saponify one gram of the fat or oil Since there are three ester bonds in a molecule to hydrolyze, three equivalents of potassium hydroxide are needed to saponify one molecular weight of any fat or oil. The following equation depicting the saponification and sample calculation illustrate how saponification value could be determined,

$$\begin{array}{l} CH_2-O-\overset{O}{\overset{\|}{C}}-(CH_2)_{14}-CH_3 \\ | \\ CH-O-\overset{O}{\overset{\|}{C}}-(CH_2)_{14}-CH_3 \\ | \\ CH_2-O-\overset{O}{\overset{\|}{C}}-(CH_2)_{14}-CH_3 \end{array} + \underset{\text{(3×gram formula weight)}}{3KOH} \longrightarrow \underset{\substack{\text{Potassium palmiate} \\ (\textit{a soap})}}{3CH_3(CH_2)_{14}-\overset{O}{\overset{\|}{C}}-\overset{-}{O}\overset{+}{K}} + \underset{\text{Glycerol}}{\begin{array}{l} CH_2OH \\ | \\ CHOH \\ | \\ CH_2OH \end{array}}$$

Glycerol tripalmiate (*mol wt* = 836)

Here 836 grams of the fat require 168, 000 milligrams of potassium hydroxide for saponification. Therefore, one gram of fat will require 168, 000/836 mg of KOH. Hence,

$$\text{Saponification Number of Glyceryl tripalmate} = \frac{168{,}000 \text{ mg of KOH}}{836\text{g fat}} = 208$$

If M be the molecular weight of the fat, the saponification number = 168,000/M. Since the saponification value of a given fat can be determined experimentally, the average molecular weight of the fat 'can be found. The higher the saponification numbers of a fat, the greater the percentage of low-molecular-weight glycerides it contains. As the average molecular weight of the fat depends on the average length of the carbon chain of the fatty acid components, the saponification number also gives an indication of the average length of the carbon chain in the glycerides under examination.

The saponification number of a given sample of fat or oil is determined experimentally as follows. A weighed quantity of the fat is refluxed with excess of standard ethanolic KOH solution. and then titrating the unused alkali against a standard acid solution. Thus, the saponification number (or value) of a given sample of fat while it indicates the average molecular weight of the component glycerides and the chain lengths of the acid portions in them, also gives an estimate of non fatty impurities if present. Further, it tells the amount of alkali which would be actually required by a fat sample for its conversion to soap.

(3) Iodine Number. The extent of unsaturation in a fat or oil is expressed in terms of its **Iodine Number** (or Iodine Value).(see chemical analyser's guide) The iodine number is defined as the Number or grams or Iodine which will add to 100 grams or fat or oil. The following equation and calculation illustrate the definition of Iodine Number.

$$\begin{array}{l} CH_2-O-\overset{\overset{\large O}{\|}}{C}-(CH_2)_7-CH{=}CH-(CH_2)_7CH_3 \\ | \\ CH-O-\overset{\overset{\large O}{\|}}{C}-(CH_2)_7-CH{=}CH-(CH_2)_7CH_3 \\ | \\ CH_2-O-\overset{\overset{\large O}{\|}}{C}-(CH_2)_7-CH{=}CH-(CH_2)_7CH_3 \end{array} + \underset{(6\times126.9)}{3I_2} \xrightarrow{HgCl_2} \begin{array}{l} CH_2-O-\overset{\overset{\large O}{\|}}{C}-(CH_2)_7-\overset{|}{C}H-\overset{|}{C}H-(CH_2)_7CH_3 \\ | \\ CH-O-\overset{\overset{\large O}{\|}}{C}-(CH_2)_7-\overset{|}{C}H-\overset{|}{C}H-(CH_2)_7CH_3 \\ | \\ CH_2-O-\overset{\overset{\large O}{\|}}{C}-(CH_2)_7-\overset{|}{C}H-\overset{|}{C}H-(CH_2)_7CH_3 \end{array}$$

Glyceryl trioleate (triolein)
(*m wt* = 884)

A treble diiodide

The above equation tells that $6 \times 126.9 = 761$.g of iodine will add to 884 g of triolein. The number of grams of iodine that will add to 100 gm of triolein will be $761 \times 100/884$. Therefore

$$\text{Iodine Number of Triolein} = \frac{761.4 \times 100}{884} = 86$$

Obviously, the value of Iodine Number depends on the number of double bonds present in the acid component of the glycerides. A high iodine number indicates that the glycerides contain a large number of double bonds, while a low iodine number implies the presence of a few double bonds. The iodine number of tripalmitin with no double bonds would be zero.

Iodine number of a fat or oil can be experimentally determined by the following procedure *(Hub/'s method):* A weighed amount of the fat or oil dissolved in carbon tetrachloride is allowed to react with a solution of iodine and mercuric chloride (catalyst) in ethanol. The un reacted iodine is titrated against standard thiosulphate solution and by difference the amount of iodine consumed by the weight of fat or oil taken calculated. In another method *(Wijs' method)* the molecular iodine has been replaced by the more reactive iodine monochloride (ICI), the rest of the procedure remaining the same.

The Iodine Numbers as also the Saponification Numbers of a few common oils and fats are given in the Table below:

Fat or Oil	Iodine Number	Saponification Number
Fats:		
Butter	30—40	210—230
Lard	46—70	195—203
Tallow	30—48	190—200
Edible Oils:		
Soyabean oil	127—138	189—195
Cottonseed oil	105—114	190—198
Sunflower oil	140—156	188—194
Nonedible Oils:		
Linseed Oil	170—185	187—195
Tung oil	163—171	190—197

(3) Acid Number. The acid number (or value) of a fat or oil tells the amount of free fatty acids present in it. The acid number is expressed as the Number of milligrams of potassium hydroxide required to neutralize one gram or fat. It is determined by dissolving a weighed, quantity of the fat in ethanol and titrating the solution against standard alkali. The acid number of a fat can give the extent of rancidity in a stored sample.

(4) Reicbert-Meissl Number. The amount of free water soluble, volatile fatty acid (butyric – C_4 to capric - C_{10}) present in a fat. or oil is expressed in terms of *Reichert-Meis Number.* It is defined as the Number of milliliters of 0.1 M potassium hydroxide solution requires to neutralize 5 grams of fat. Reichert-Meissl Number of a fat is determined by treating a known weight of it with ethanolic alkali and distilling the volatile acids. These are titrated again! *M/10* potassium hydroxide and Reichert-Meissl Number calculated.

Manufacture of 'Vanaspatl' or 'Vegetable Ghee'

Margarine, a substitute for butter, is manufactured in USA, UK and other Europeo countries in large quantities by the hydrogenation of vegetable oils and soft fats (Norman, 1902 in India *'Vanaspati'* or *'Vegetable ghee'* was first introduced after the First World War (191! and it has found immense popularity because it resembles natural ghee in appearance. It made industrially by hydrogenation of vegetable oils such as ground nut oil, cottonseed oil sesame oil soybean oil and sunflower oil. The hydrogenation is carried by passing hydrogen gas through the heated oil in the presence of metallic nickel as catalyst. The nickel catalyst required for the process is obtained by mixing a nickel salt (such as nickel formate or nick carbonate) with unsaturated oil and then heating the mixture and passing hydrogen into it Thus the salt is reduced to finely divided nickel dispersed in the oil and is ready for use.

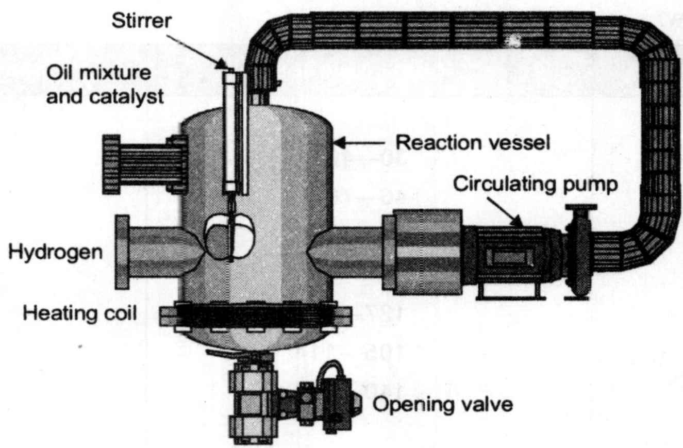

Fig. 19.2 : Manufacture of Vegetable Ghee from oil

The hydrogenation of an oil is actually carried in a hydrogenation tower or converter. It is a tall cylinder made of steel with cone or round base, fitted with a stirring device and also heating and cooling coils. The mixture of oil and nickel Catalyst prepared as described above, is pumped into the converter. Here it is partially mixed by stirring and partially by the flow of hydrogen entering at the base through a perforated pipe. Steam is passed through the heating coils, while the oil mix is continuously pumped to the top of the converter where it is sprayed back down the tower. Since the reaction is exothermic, the steam heating is stopped as the reaction gets going and the temperature is maintained at about, 2000 by passing cooling water through the coils if necessary. The hardening of the oil takes place most readily with a high catalyst concentration and low pressure (30-35 psi). When hardening has taken place to the required degree, the reaction is stopped by lowering the temperature to about 70°, The catalyst is filtered and the product is rebleached and deodorised under vacuum. The *Vonaspati* so prepared has melting point 31-37°C and has the texture of ghee which would melt like ghee when placed on the tongue.

Composition of Vanaspati

Hydrogenation of oil involves the saturation of the double bonds present saturation of the double bonds present in the acid components of glycerides. Thus the fatty acids with both single and two or more double bonds are saturated, However, due to the selective adsorption of the more unsaturated fatty acids on the catalyst surface, oleic acid with a single soluble bond is hydrogenated at a later stage than linoleic acid with two double bonds. For the same reason, hydrogenation of a chain containing two or more double bonds tends to stop on reaching the single double-bond stage. The reactions which take place on hydrogenation include acid to yield 9 – or 12 oleic acid. Also *cis* double bonds are transferred to *trans* configuration.

linolic acid component of a poly-unsaturated glyceride

$$H_3C—O—\overset{O}{\overset{\|}{C}}—(CH_2)_7—CH{=}CH—CH_2–CH{=}CH–(CH_2)_4—CH_2–CH_3$$

$$\downarrow Ni \mid H_2$$

$$H_3C–O—CH_2–(CH_2)_7—CH_2–CH_2–CH_2–CH{=}CH—(CH_2)_4—CH_3$$

12- oleic acid component (haft saturated linoleic component)

The composition of the fatty acids and iodine value of vanaspati and the original groundnut oil is illustrative.

It may be noted that linoleic acid content has fallen markedly as also the entire Cisnonene has been converted to *trans* form. Thus *Vanaspati* as made by hydrogenation contains a large proportion of linoleic acid converted to *trans-nonene* which is biologically antagonistic to human system.

There has been evidence in recent years that the presence of large amounts of saturated fats in the diet may lead to an increase in the level of *cholesterol* in the blood, while the high oil content of the diet tends to diminish cholesterol level in blood. The excess of cholesterol in blood causes *arteriosclerosis* (hardening of the arteries) and consequent heart diseases, the most serious of which are coronary thrombosis and paralytic strokes. As a result of these findings, sunflower oil and other polyunsaturated oils are now increasingly used as cooking medium in preference to solid fats. For the same reason, the Government of India has made it mandatory since April 1972 to blend *vanaspati* with 2.5% sunflower and 7.5% sesame oil in order to raise the content of Polyunsaturated acids, especially linoleic acid in the *cis* form. Vitamins A and D are also added to vanaspati as these are not present in the original oil and are necessary for good health of human body.

WAX

Wax has traditionally referred to a substance that is secreted by bees (beeswax) and used by them in constructing their honeycombs.

In modern terms, **wax** is an imprecisely defined term generally understood to be a substance with properties similar to beeswax, namely

- plastic (malleable) at normal ambient temperatures
- a melting point above approximately 45°C (which differentiates waxes from fats and oils)
- a relatively low viscosity when melted (unlike many plastics)
- insoluble in water
- hydrophobic

Waxes may be natural or artificial. In addition to beeswax, carnauba (a vegetable wax) and paraffin (a mineral wax) are commonly encountered waxes which occur naturally. *Ear wax* is an oily substance found in the human ear. Some artificial materials that exhibit similar properties are also described as wax or waxy.

Chemically, a wax may be an ester of ethylene glycol (ethan-1,2-diol) and two fatty acids, as opposed to a fat which is an ester of glycerin (propan-1,2,3-triol) and three fatty acids. It may also be an ester of a fatty acid with a fatty alcohol. It is a type of lipid.

Types of Waxes

1 Animal and insect waxes

2 Vegetable waxes

3 Mineral waxes

4 Petroleum waxes

5 Synthetic waxes

Animal and insect waxes

- **Beeswax** - produced by honeybees
- **Chinese wax** - produced by scale insects *Coccus ceriferus*

Vegetable waxes

- **Bayberry wax** - from the surface of the berries of the bayberry shrub
- **Candelilla wax** - from the Mexican shrubs *Euphorbia cerifera* and *E. antisyphilitica*

- **Carnauba wax** - from the leaves of the Carnauba Palm
- **Japan wax** - a vegetable tallow (not a true wax), from the berries of *Rhus* and *Toxicodendron* species
- **Castor wax** - catalytically hydrogenated castor oil
- **Jojoba oil** - pressed from the seeds of the jojoba bush, a replacement for spermaceti
- **Ouricury wax** - from the Brazilian Feather Palm
- **Rice bran wax** - obtained from rice bran

Mineral waxes

- **Montan wax** - extracted from lignite and brown coal
- **Ozocerite** - found in lignite beds

Petroleum waxes

- Paraffin wax - made of long-chain alkane hydrocarbons
- Microcrystalline wax - with very fine crystalline structure

Bee Wax

Historical use

Beeswax is used since ancient history; traces of it were found in the paintings in the Lascaux cave and in Egyptian mummies. Ancient Egyptians used it in shipbuilding as well. In the Roman period, beeswax was used as waterproofing agent for painted walls and as medium for the Fayum mummy portraits. In the Middle Ages beeswax was considered valuable enough to become a form of currency. More recently it found use as a modelling material, a component of sealing wax, and in cosmetics.Beeswax is a product from a bee hive. Beeswax is secreted by honeybees of a certain age in the form of thin scales. The scales are produced by glands of 12 to 17 days old worker bees on the ventral (stomach) surface of the abdomen. Worker bees have eight wax-producing glands on the inner sides of the sternites (the ventral shield or plate of each segment of the body). Wax is produced from abdominal segments 4 to 7. The size of these wax glands depends on the age of the worker.Honeybees use the beeswax to build honey comb cells in which the young are raised and honey and pollen are stored. For the wax-making bees to secrete wax the ambient temperature in the hive has to be 33 to 36°C (91 to 97°F). Approximately eight pounds of honey is consumed by bees to produce one pound of beeswax (8 kg/kg). Estimates are that bees fly 150,000 miles to yield this one pound of beeswax (530,000 km/kg). When beekeepers go to extract the honey, they cut off the wax caps from each honeycomb cell. Its color varies from yellowish-white to brownish depending on purity and the type of flowers gathered by the bees. Wax from the brood comb of the honeybee hive tends to be darker than wax from the honey comb. Impurities accumulate more quickly in the brood comb. Due to the impurities, the wax has to be rendered before further use. The leftovers are called **slumgum.** The wax may further be clarified by heating in water and may then be used for candles or as a lubricant for drawers and windows or as a wood polish. As with petroleum waxes it may be softened by dilution with vegetable oil to make it more workable at room temperature, whence it may be used to create sculpture and jewelry models for use in the *lost wax* casting process.

Physical characteristics

It is a tough wax formed from a mixture of several compounds including: hydrocarbons 14%, monoesters 35%, diesters 14%, triesters 3%, hydroxy monoesters 4%, hydroxy polyesters 8%, acid esters 1%, acid polyesters 2%, free acids 12%, free alcohols 1%, unidentified 6% .The main components of beeswax are palmitate, palmitoleate, hydroxypalmitate and oleate esters of long-chain (30-32 carbons) aliphatic alcohols, with the ratio of triacontanylpalmitate

$CH_3(CH_2)_{29}O\text{-}CO\text{-}(CH_2)_{14}CH_3$ to cerotic acid $CH_3(CH_2)_{24}COOH$, the two principal components, being 6:1.

Beeswax has a high melting point range, of 62 to 64°C (144 to 147°F). It does not boil in air, but continues to heat until it bursts into flame at around 120°C (250°F). If beeswax is heated above 85°C (185°F) discoloration occurs. Density at 15°C is 0.958 to 0.970 g/cm^3.

Bee wax can be classified gênerally into European and Oriental types. The ratio of saponification value is lower (3-5) for European beeswax, and higher (8-9) for Oriental types. Hydroxyoctacosanyl hydroxystearate can be used as a beeswax substitute as a consistency regulator and emulsion stabilizer. Japan wax is another substitute.

Uses as a product

Beeswax is used commercially to make fine candles, cosmetics and pharmaceuticals including bone wax (cosmetics and pharmaceuticals account for 60% of total consumption), in polishing materials (particularly shoe polish), as a component of modelling waxes, and in a variety of other products. It is commonly used during the assembly of pool tables to fill the screw holes and the seams between the slates. Beeswax candles are preferred in most Eastern Orthodox churches because they burn cleanly, with little or no wax dripping down the sides and little visible smoke. Beeswax is also prescribed as the material (or at least a significant part of the material) for the Paschal candle ("Easter Candle") and is recommended for other candles used in the liturgy of the Catholic Church.It is also used as a coating for cheese, to protect the food as it ages. While some cheesemakers have replaced it with plastic, many still use beeswax in order to avoid any unpleasant flavours that may result from plastic. The burning characteristics beeswax candles differ from those of paraffin. Beeswax has negative ionization, which binds particulate matter to clear the air. A beeswax candle flame has a "warmer," more yellow color than that of paraffin, and the colour of the flame may vary depending on the season in which the wax was harvested.

Beeswax is also an ingredient in moustache wax, and was used in the manufacturing of the cylinders used by the earliest phonographs.

Chinese wax is a white to yellowish-white, gelatinous, crystalline water-insoluble substance obtained from the wax secreted by certain insects.It resembles spermaceti but is harder, more friable, and with a higher melting point. It is deposited on the branches of certain trees by the scale insect Ceroplastes ceriferus, common in China and India, or a related scale insect, Ericerus pela, of China and Japan. The insects and their secretions are harvested and boiled with water to obtain the wax.

Uses

Used chiefly in the manufacture of polishes, sizes, and candles.

Japan wax is a pale-yellow, waxy, water-insoluble solid with a gummy feel, obtained from the berries of certain sumacs native to Japan and China, such as Rhus verniciflua (Japanese sumac tree) and Rhus succedanea (Japanese wax tree)..Japan wax is a byproduct of lacquer manufacture. It is not a true wax but a fat that contains 10-15% palmitin, stearin, and olein with about 1% japanic acid. Japan wax is sold in flat squares or disks and has a rancid odor. It is extracted by expression and heat, or by the action of solvents.

Uses

Japan wax is used chiefly in the manufacture of candles, furniture polishes, floor waxes, wax matches, soaps, food packaging, pharmaceuticals, cosmetics, pastels, crayons, buffing compounds, metal lubricants, adhesives, thermoplastic resins, and as a substitute for beeswax.

Candelilla wax is a wax derived from the leaves of a small shrub native to northern Mexico and the southwestern United States, *Euphorbia cerifera* and *Euphorbia antisyphilitica*, from the family *Euphorbiaceae*. It is yellowish-brown, hard, brittle and opaque to translucent.

Composition

Candelilla wax consists of mainly hydrocarbons (about 50%, chains with 29-33 carbons), esters of higher molecular weight (20-29%), free acids (7-9%), and resins (12-14%, mainly triterpenoid esters). It is insoluble in water, but soluble in many organic solvents (acetone, chloroform, benzene).

Manufacture

The wax is obtained by boiling the leaves and stems with diluted sulfuric acid and skimmed from the surface and further processed. Its melting point is 67-79°C. It is mostly used mixed with other waxes to harden them without raising their melting point.

Uses

As a food additive, candelilla wax has the E number E902 and is used as a glazing agent. It also finds use in cosmetic industry, as a component of lip balms and lotion bars. One of its major uses was a binder for chewing gums.

Candelilla wax can be used as a substitute for carnauba wax and beeswax. It is also used for making varnish.

Castor wax, also called hydrogenated castor oil, is a hard, brittle, vegetable wax. It is produced by the hydrogenation (chemical combination with hydrogen) of pure castor oil, in the presence of a nickel catalyst. It is odorless and insoluble in water.

Uses

Castor wax is used in polishes, cosmetics, electrical capacitors, carbon paper, lubrication, and coatings and greases where resistance to moisture, oils and other petrochemical products is required.

Carnauba is a wax derived from the leaves of a plant native to northeastern Brazil, the Carnauba Palm (*Copernicia cerifera* Mart). It is known as "queen of waxes" and usually comes in the form of hard yellow-brown flakes. It is obtained from the leaves of the Carnauba Palm by collecting them, beating them to loosen the wax, then refining and bleaching the wax.

Composition

Carnauba wax contains mainly esters of fatty acids (80-85%), fatty alcohols (10-15%), acids (3-6%) and hydrocarbons (1-3%). Specific for carnauba wax is the content of esterified fatty diols (about 20%), hydroxylated fatty acids (about 6%) and cinnamic acid (about 10%). Cinnamic acid, an antioxidant, may be hydroxylated or methoxylated.

Uses

Carnauba wax is a prominent ingredient is cosmetic formulas: lipsticks, eyeliners, mascara, eye shadows, foundations, blushers, skin care preparations, sun care preparations, etc.

Carnauba wax can produce a glossy finish and as such is used in automobile waxes, shoe polishes and floor and furniture polishes, especially mixed with beeswax. Use for coatings of paper is the most common application within the United States.

It is also the main ingredient in surfboard wax, when combined with coconut oil.

As a glazing agent in foods, it finds use particularly in shiny-shelled candies such as M&M's and Tic Tacs, and in some chocolates. Swedish Fish are made out of Carnauba Wax. It is also used in cosmetic industry as a component of creams and lipsticks, and in the pharmaceutical industry as a tablet coating agent.

Technical information

Its INCI name is *Copernicia Cerifera (Carnauba) Wax*

Its E Number is E903.

Its melting point is 78-85°C, among the highest of natural waxes.

Its relative density is ca. 0.97

It is among the hardest of natural waxes.

It is practically insoluble in water, soluble on heating in ethyl acetate and in xylene, practically insoluble in alcohol.

Ouricury wax is a brown-colored wax obtained from the leaves of a Brazilian Feather Palm *Syagrus coronata* or *Cocos coronata* by scraping the leaf surface.

Harvesting

Harvesting ouricury wax is more difficult than harvesting carnauba wax, as ouricury wax does not flake off the surface of the leaves.

Properties and Uses

The physical properties of ouricury wax resemble carnauba wax, so it can be used as a substitute where light colour is not required, e.g. in carbon paper inks, molding lubricants and polishes. Its melting point is 81-84°C.

Castor wax, also called hydrogenated castor oil, is a hard, brittle, vegetable wax. It is produced by the hydrogenation (chemical combination with hydrogen) of pure castor oil, in the presence of a nickel catalyst. It is odorless and insoluble in water.

Uses

Castor wax is used in polishes, cosmetics, electrical capacitors, carbon paper, lubrication, and coatings and greases where resistance to moisture, oils and other petrochemical products is required.

Properties

- Melting point = 60 C
- Acid number = 2
- Saponification value = 179
- Iodine number = 4

Rice bran wax is the vegetable wax extracted from the bran oil of rice (*Oryza sativa*).

Chemical Composition

The main components of rice bran wax are aliphatic acids (wax acids) and higher alcohol esters. The aliphatic acids consist of palmitic acid (C16), behenic acid (C22), lignoceric acid (C24), other wax acids, (C26) etc. The higher alcohol esters consist of ceryl alcohol (C26), melissyl alcohol (C30), etc. Rice bran wax also contains unsaponifiable constituents such as free fatty acids (palmitic acid), squalene and phospholipids.

Uses

Rice bran wax is used in paper coating, textiles, explosives, fruit & vegetable coatings, pharmaceuticals, candles, moulded novelties, electric insulation, textile and leather sizing, waterproofing, carbon paper, typewriter ribbons, printing inks, lubricants, crayons, adhesives, chewing gum and cosmetics.

In cosmetics, rice bran wax is used as an emollient, and is the basis material for some exfoliation particles. It may also serve as a substitute for Carnauba wax.

Physical Properties

Melting point = 77 - 86 °C

Saponification value = 75 -120

Iodine number = 10

Color: Off-white to moderate orange/brown

Odor: typical fatty, crayola-ish

Rice bran wax bleaches and deodorizes readily

Jojoba Oil is the liquid wax produced in the seed of the Jojoba (*Simmondsia chinensis*) plant.

Physical Properties

Unrefined jojoba oil appears as a clear golden liquid at room temperature with a slightly fatty odor. Refined jojoba oil is colourless and odorless. The melting point of jojoba oil is approximately 10°C[1] and the iodine value is approximately 80[2]. Jojoba oil is relatively shelf-stable when compared with other vegetable oils. Jojoba oil has an Oxidative Stability Index of approximately 60[3], which means that it is more shelf-stable than oils of safflower oil, canola oil, almond oil or squalene but less than castor oil, macadamia oil and coconut oil.

Chemical Structure

Jojoba oil is a straight chain wax ester, 36 to 46 carbon atoms in length. Each molecule consists of a fatty acid and a fatty alcohol joined by an ester bond. Each molecule has two points of cis-unsaturation, both located at the 9th carbon atom from either end of the molecule. There is no triglyceride component to jojoba oil.

Uniqueness

Unlike common triglyceride vegetable oils, jojoba oil is chemically very similar to human sebum. Jojoba oil is a straight chain wax ester which is liquid at room temperature, making jojoba and its derivatives good botanical alternatives to spermaceti (whale oil) and its derivatives, such as cetyl alcohol.

Uses

Most jojoba oil is consumed as an ingredient in cosmetics and personal care products, especially skin care and hair care. Jojoba oil is often listed in the ingredient declaration of cosmetic products as either Jojoba Oil or Jojoba esters.

Montan wax

History

The earliest production of montan wax is recorded in Germany, which still supplies the majority of world's production.Montan wax, also known as lignite wax, is a hard wax obtained by solvent extraction of certain types of lignite or brown coal.

Properties

Its color ranges from dark brown to light yellow when crude, or white when refined. Its composition is non-glyceride long-chain (C24-C30) carboxylic acid esters (62-68 weight %), free long-chain organic acids (22-26%), long-chain alcohols, ketones and hydrocarbons (7-15%) and resins; it is in effect a fossilized plant wax. Its melting point is 79-90°C.

Uses

It is used for making car and shoe polishes, paints, and phonograph records, and as lubricant for molding, paper and plastics. About a third of total world production is used in car polishes. Formerly the largest consumer used to be carbon papers.

Unrefined montan wax contains asphalt and resins, which can be removed by refining. Montan wax in polishes improves scuff resistance and increases water repellence, and imparts high gloss.

Technical information

As a food additive, it has the E number E912. It is used for coating of citrus fruits.

Ozokerite or **ozocerite** is a naturally-occurring odouriferous mineral wax or paraffin found in many localities.

Sources

Specimens have been obtained from Scotland, Northumberland and Wales, as well as from about thirty different countries. Of these occurrences the ozokerite of the island (now peninsula) of Cheleken, near Turkmenbashi, and the deposits of Utah in the US, deserve mention, though the last-named have been largely worked out. The sole sources of commercial supply are in Galicia, at Boryslaw, Dzwiniacz and Starunia, though the mineral is found at other points on both flanks of the Carpathians.

Ozokerite deposits are believed to have originated in much the same way as mineral veins, the slow evaporation and oxidation of petroleum having resulted in the deposition of its dissolved paraffin in the fissures and crevices previously occupied by the liquid. As found native, ozokerite varies from a very soft wax to a black mass as hard as gypsum.

Properties

Its specific gravity ranges from -85 to -95, and its melting point from 58 to 100°C. It is soluble in ether, petroleum, benzene, turpentine, chloroform, carbon disulfide, &c. Galician ozokerite varies in color from light yellow to dark brown, and frequently appears green owing to dichroism. It usually melts at 62°C. Chemically, ozokerite consists of a mixture of various hydrocarbons, containing 85-7% by weight of carbon and 14-3% of hydrogen.

Mining

The mining of ozokerite was formerly carried on in Galicia by means of hand-labor, but in the modern ozokerite mines owned by the Boryslaw Actien Gesellschaft and the Galizische Kreditbank, the workings of which extend to a depth of 200 metres, and 225 metres respectively, electrical power is employed for hauling, pumping and ventilating. In these mines there are the usual main shafts and galleries, the ozokerite being reached by levels driven along the strike of the deposit. The wax, as it reaches the surface, varies in purity, and, in new workings especially, only hand-picking is needed to separate the pure material. In other cases much earthy matter is mixed with the material, and then the rock or shale having been eliminated by hand-picking, the "wax-stone" is boiled with water in large coppers, when the pure wax rises to the surface. This is again melted without water, and the impurities are skimmed off, the material being then run into slightly conical cylindrical moulds and thus made into blocks for the market. The crude ozokerite is refined by treatment first with Nordhausen oil of vitriol, and subsequently with charcoal, when the ceresine or cerasin of commerce is obtained. The refined ozokerite or ceresine, which usually has a melting-point of 61 to 78°C, is largely used as an adulterant of beeswax, and is frequently colored artificially to resemble that product in appearance. On distillation in a current of superheated steam, ozokerite yields a candle-making material resembling the paraffin obtained from petroleum and shale-oil but of higher melting-point, and therefore of greater value if the candles made from it are to be used in hot climates. There are also obtained in the distillation light oils and a product resembling vaseline. The residue in the stills consists of a hard, black, waxy substance, which in admixture with india-rubber is employed under the name of okonite as an electrical insulator. From the residue a form of the material known as heel-bail, used to impart a polished surface to the heels and soles of boots, is also manufactured.

Mining of ozokerite fell off after 1940 due to competition from paraffins manufactured from petroleum, but as it has a higher melting point than most petroleum waxes, it is still favoured for some applications, such as electrical insulators and candles, or in extra-soft paper tissues.

Microcrystalline waxes are a type of wax produced by de-oiling petrolatum, as part of the petroleum refining process. In contrast to the more familiar paraffin wax which contains mostly unbranched alkanes, microcrystalline wax contains a higher percentage of isoparaffinic (branched) hydrocarbons and naphthenic hydrocarbons. Microcrystaline wax is generally darker, more viscous, denser, tackier and more elastic than paraffin waxes, and has a higher molecular weight and melting point. The elastic and adhesive characteristics of microcrystalline waxes are related to the non-straight chain components which they contain. Typical microcrystalline wax crystal structure is small and thin, making them more flexible than paraffin wax.

Microcrystalline Wax is a wax derived from petroleum and characterized by the fineness of its crystals in contrast to the larger crystal of paraffin wax. It consists of high molecular weight saturated aliphatic hydrocarbons. Microcrystalline Wax is commonly used in cosmetic formulations.

- **Paraffin** is a common name for a group of high molecular weight alkane hydrocarbons with the general formula C_nH_{2n+2}, where n is greater than about 20, discovered by Carl Reichenbach.
- In the UK, as well as in most Commonwealth countries, the fuel known in the U.S. as kerosene is called paraffin oil (or just *paraffin*), and the solid forms of paraffin are called paraffin wax.
- **Paraffin** is also a technical name for an alkane in general, but in most cases it refers specifically to a linear, or *normal* alkane, while branched, or *iso*alkanes are also called *iso*paraffins. Compare olefin. (Latin *parum* (= barely) + *affinis* with the meaning here of "lacking affinity", or "lacking reactivity")

Physical and chemical properties

It is mostly found as a white, odourless, tasteless, waxy solid, with a typical melting point between about 47°C and 65°C. It is insoluble in water, but soluble in ether, benzene, and certain esters. Paraffin is unaffected by most common chemical reagents, but burns readily.

Pure paraffin is an extremely good electrical insulator, with a electrical resistance of 10^{17} ohm meter. This is better then nearly all other materials except some plastics (notably teflon).

Liquid paraffin

Liquid paraffin has a number of names, including nujol, mineral spirits, adepsine oil, alboline, glymol, liquid paraffin, paraffin oil, saxol, or USP mineral oil. It is often used in infrared spectroscopy, as it has a relatively uncomplicated IR spectrum. When the sample to be tested is made into a mull (a very thick solution), liquid paraffin is added so it can be spread on the disks to be tested.

Uses

- Candlemaking
- Coatings for waxed paper or cloth.
- Coating for many kinds of hard cheese, like Edam cheese.
- As anticaking-, moisture repellent- and dustbinding coatings for fertilizers.
- Preparing specimens for histology.
- Solid propellant for hybrid rockets
- Sealing jars, cans, and bottles
- In dermatology, as an emollient (moisturiser)
- Surfing, for grip on surfboards as a component of surfwax.
- The primary component of glide wax, used on skis and snowboards.
- As a food additive, a glazing agent with E number E905
- The paraffin test is used in forensics to detect granules of gunpowder in the hand of a shooting suspect.

Food-grade paraffin wax is used in some candies to make them look shiny. Although edible, it is nondigestible; it passes right through the body without being broken down. Non-food grade paraffin wax can contain oils and other impurities which may be toxic or harmful.

Impure mixtures of mostly paraffin wax are used in wax baths for beauty and therapy purposes.

Paraffin wax is not used much to make original models for casting, as it is relatively brittle at room temperature and usually cannot be cold-carved without excessive chipping and breaking. Soft, pliable waxes such as beeswax are preferred for modelling.

Soaps and Detergents

The term *detergent (L.,* detergere-to clean) is rather a general one and is used to denote any cleansing agent. This broad definition includes soaps as well. However, according to present-day popular terminology, the word Detergent generally refers to *synthetic detergents,* also called Syndets.

Composition of Soaps

As we have already discussed, soaps are the sodium or potassium salts of higher fatty acids containing from 12 to 18 carbon atoms. They are generally obtained by the hydrolysis of fats and oils with sodium hydroxide. The mixture of sodium salts of higher fatty acids so produced are called Sodium Soaps.

Sodium carboxylates are the common toilet soaps. Potassium carboxylates or Potassium Soaps are obtained when the saponification of a fat or oil is carried with potassium hydroxide. Potassium soaps are softer than sodium soaps and they are used for special purposes when rapid solution is desired e.g.,.in making shaving creams or liquid soaps *(shampoo)*. The composition of sodium or potassium carboxylates constituting soap depends on the percentage of fatty acids bonded to glycerol in the original triglycerides. Solid fats give mixtures with higher proportion of sodium or potassium salts of higher fatty acids (palmitic acid, stearic. acid) and give *hard soaps*. The vegetable oils give mixtures with a greater proportion of unsaturated fatty acids (oleic acid and linoleic acid) and give *soft soaps*.

Soap Manufacture

In India the main source of soap is coconut oil which is available in abundance in the southern states. Palm oil, groundnut oil and cottonseed oil are also used. Therefore, in actual practice, mixtures of solid fats (hardened oils) and oils are' blended to produce a soap having properties best suited for a particular use.

Soaps can be made from fat blends in two ways:

(a) Saponification of fats with alkali solutions;

(b), Direct neutralisation of fatty acids. .

The saponification methods are most commonly used for the manufacture of soaps. Increasing amounts, however, are now produced by direct neutralisation of fatty acids obtained from fat. splitting.

Saponification Methods

The saponification of a fat with alkali solution *(lye)* may be done by boiling *(Boiling Process)* or in cold *(Cold Process)*. In the modern *Continuous Process,* saponification is carried inclosed vessels at high temperature and pressure, and soap separated by centrifugation.

(1) Boiling Process (Hot Process). The manufacture of soap by the older *(Boiling Process)* is carried by the following steps. .

(a) Boiling. The saponification of the fat is done by boiling the fat with sodium hydroxide solution *(roda lye)* in a large cylindrical steel vessel known as *soap pan* or *kettle*. The soap pan is usually open at the top. The lower part of the pan is funnel shaped and contains a system of steam heating coils which can be either 'open' or 'closed.' Molten fat and appropriate quantity of *soda lye* are simultaneously run into the pan. Steam is then admitted through the open 'steam coils' to boil the mixture which is thus kept in a good state of agitation all the time. Alkali is maintained in sufficient excess, more of' it being added if necessary. Boiling is continued unless the greasy nature of the mix has almost. disappeared and the fat is thus saponified to the extent of about 80 per cent.

$$\text{FAT + SODA LYE} \rightarrow \text{GLYCEROL + SOAP}$$

***(b)* Salting Out**. This step involves the separation of soap and glycerol, a process known as 'Salting Out'. Use is made of the fact that soap is insoluble in concentrated salt solution *(Common Ion Effect),* while glycerol is readily soluble. Solid salt' or brine is added to the mixture of soap, glycerol and excess lye resulting from step *(a),* which is then boiled and allowed to settle. The soap is thrown out of solution as a curdy mass which being of lower density than glycerol/brine mixture floats to the surface. The aqueous layer which also contains *spent lye,* salt and dirt is drawn oft' from the bottom of the pan and pumped to the glycerol recovery plant.

The soap left in the pan is dissolved in water and after boiling for a short time is salted out, the lye being removed after settling. This washing operation is repeated so as to reduce the glycerol content of the soap and to remove impurities. The soap which is relatively pure is once again boiled with fresh *soda lye* to complete the saponification. After settling out as before, the spent lye is run off and reused, finally the soap is boiled with water and left to settle in the pan for two to ten days.

***(c)* Finishing**. The upper layer of soap obtained from step (2) is caned 'neat soap'. While it is still liquid, the warm neat soap is pumped away using a *skimmer pipe,* which can be raised or lowered inside the pan on a swivel joint. The molten soap is received in a: steam. Jacketed pan fitted with a mechanical stirrer *(crutcher).* Here soap is mixed with glycerol, colour perfumes, germicides etc till it becomes a homogeneous mass. The crutched soap is then poured into open-topped 'moulds' or 'frames', and after solidification, cut into small bars using steel wire cutters.

For making toilet soaps, traditionally the neat crutched soap is made into thin shreds, dried by hot air and milled with perfumes, colour etc to thin shavings. These are then stamped into cakes.

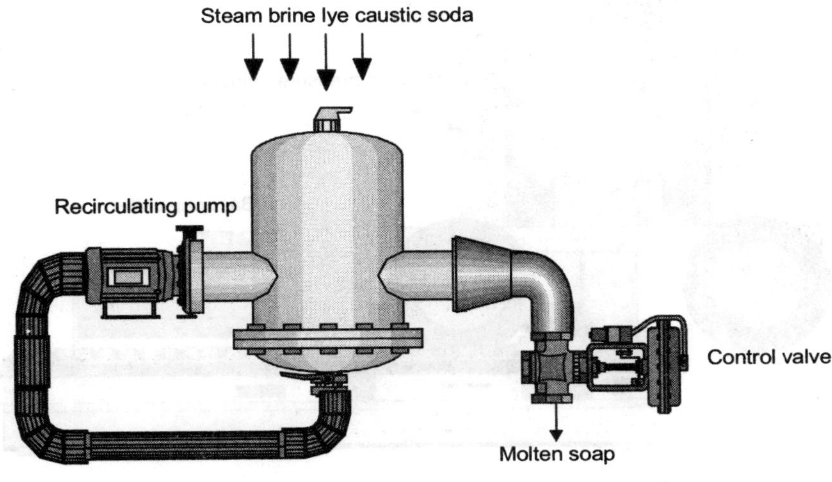

Fig. 19.3 : Showing the Steam Heated Crutcher

(2) Cold Process: The manufacture of soft coconut oil or potassium soaps cannot be carried out by the Boiling Process. This is because of their greater solubility in water which prevents them from being salted out. In this case, the *Cold Process* is used. The fat or oil is mixed with the required amount of *soda lye* in a steam heated vessel called crutcher (See.19.3). The saponification is allowed' to take place in 'cold' with mechanical string. The process is continued till the soap begins to set. At this stage, the hot liquid soap is run into frames where saponification is completed. The by-product glycerol is not recovered and re-mains in the soap.

The Cold Process is also employed in India to prepare *'Washing Soap'* on a small scale for household use.

(3) Modern Continuous Process. In this process saponification can be carried out in about 15 minutes as compared to hours required for the open-pan method. This is achieved by reacting the fat/alkali mixture at elevated temperature and pressure in a closed vessel. This operation not only has the advantage of speed but is economical of space, heat and man power. After cooling, the soap is washed is washed and salted out separation of the soap and *spent lye* layers being effected by centrifugation. After filtering and centrifuging, the molten soap is partially dried and cooled by spraying from the top of a vacuum chamber (Fig. 19.4), The soap is scrapped from the walls of the spray drier and passed under vacuum through two 'plodders' which knead the soap and introduce colour and perfume. The soap comes out of the second 'plodder' in the form of a continuous bar which is chopped into tablets.

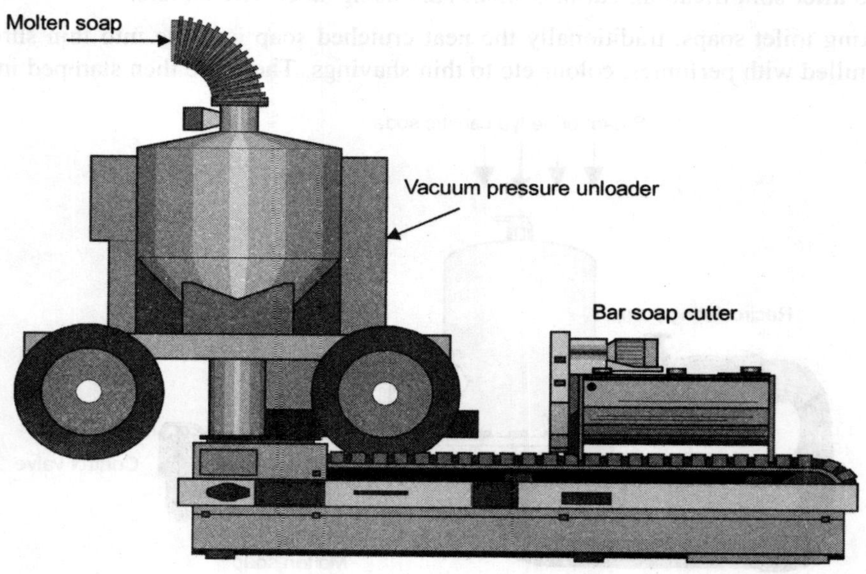

Fig. 19.4 : Showing the Manufacture of Bar Soap

Direct Neutralisation of Fatty Acids. Soap manufacture by direct neutralisation of fatty acids is of recent introduction. The methods developed for the purpose are continuous and hence more economical. The fatty acids required in the process are obtained by hydrolysis of fats in the presence of specific catalysts.

(1) Ittner Process. In this process the hydrolysis of fat is carried out with water under pressure and at elevated temperature in the presence of lime or zinc oxide as catalyst (Fig. 19.5).

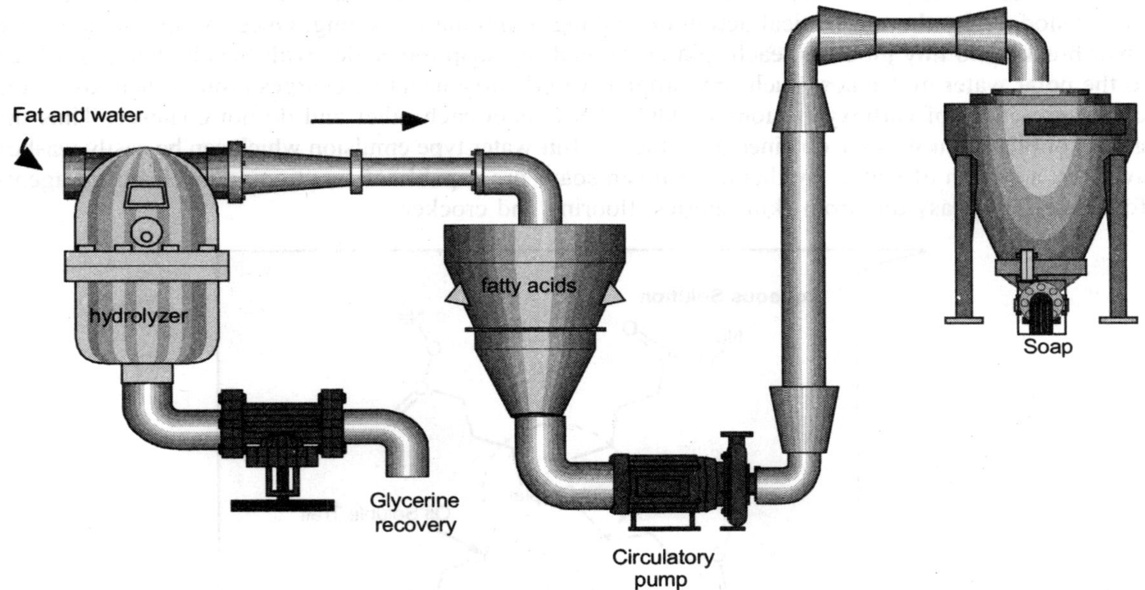

Continuous Soap Manufacturing Process

Fig. 19.5 : Showing the Continuous Soap Making Ittner Process

Hot water is fed into the *hydrolyser* near the top and fat near the bottom. The hydrolysis is rapid and complete. The fatty acids thus produced rise to the surface and are drawn out at the top, while glycerol is removed in water leaving at the bottom. The fatty acids are then pumped to another vessel, called *neutralizer.* Here they are neutralized with sodium hydroxide, or the cheaper sodium carbonate to form soap.

(2) Twitchell Process. In Twitchell process, the hydrolysis of fats is done using a catalyst consisting of dilute sul ph uric acid and aromatic sulphonic acid. All other details are the same as for Ittner Process. The drying and finishing of soaps obtained by the above methods is done exactly as described under the Modern Continuous Centrifugation Process.

Mechanism of Cleansing Action of Soap

The cleansing action of soap depends on the fact that a soap molecule is made of two parts:

(i) a long non-polar hydrocarbon chain which is oil-soluble; and *(ii)* the salt-like polar head which is water-soluble. Thus sodium stearate ($C_{17}H_{35}COONa$), a typical soap, can be represented as

Non polar hydrocarbon tail soluble oil

Polar water soluble head

When soap solution is poured onto a solid clothing or a greasy dish, the hydrocarbon tail of soap molecules dissolves in the oily or greasy layer. The grease layer with soap tails 'pegged' into it is then dislodged by the mechanical action of rubbing, tumbling or boiling. Once loosened, the grease layer breaks into tiny globules, each 'pin-cushioned' by soap molecules with 'heads' being attracted to the polar water molecules. Such tiny droplets of oil carry negative charges around them by virtue of the presences of carboxylate ions (-COO). They repel each other and do not coalesce. Thus the greasy or oily impurities are obtained as stable <:>il-in-water type emulsion which can be easily washed away by a stream of water. By this mechanism soaps are capable of functioning as cleansing agents for removing greasy dirt from skin, fabrics, flooring and crockery.

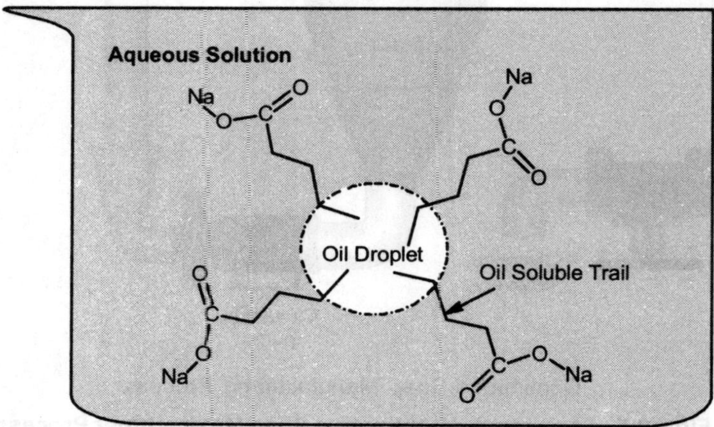

Fig. 19.6 : Emulsified oil droplet from grease layer sticking to clothor any other surface

Limitation of Soaps as Detergents

Under certain circumstances the chemistry of soaps limits their usefulness as detergents. While sodium soaps are water-soluble, calcium and magnesium soaps are insoluble. Since hard water contains Ca^{2+} and Mg^{2+} ions, sodium soaps when added to it are converted to insoluble calcium magnesium or ferric soaps. For example,

$$\underset{\substack{\text{Sodium soap} \\ (\textit{soluble})}}{R-\overset{O}{\overset{\|}{C}}-\overset{-}{O}\overset{+}{Na}} + Ca^{2+} \longrightarrow \underset{\substack{\text{Calcium soap} \\ (\textit{insoluble})}}{\left(R-\overset{O}{\overset{\|}{C}}-\overset{-}{O}\right)_2 Ca^{2+}} + 2Na^{+}$$

$$\underset{\substack{\text{Sodium soap} \\ (\textit{soluble})}}{R-\overset{O}{\overset{\|}{C}}-\overset{-}{O}\overset{+}{Na}} + Mg^{2+} \longrightarrow \underset{\substack{\text{Magnesium soap} \\ (\textit{insoluble})}}{\left(R-\overset{O}{\overset{\|}{C}}-\overset{-}{O}\right)_2 \overset{2+}{Mg}} + 2Na^{+}$$

Thus ordinary soap in hard water produces insoluble precipitates of calcium and magnesium soaps which appear on the surface as insoluble sticky grey *scum*. This not only results in the waste of sodium soap but also discolours and hardens the fabrics being washed. This is the reason that the use of soaps for home laundrying has declined since World War II.

Detergents and Syndets

Soap is the oldest skin cleansing agent. Sumarian clay tablets dating from 2500 B.C. and documenting the early usage of soap have been found between the Tigris and Euphrates. Today, **syn**thetic **det**ergents known as syndets are used in skin-beautifying cleansing.

> The currently popular expression "syndet" comes from the syllables "syn" for synthetic and "det" from detergent. Detergents refer to cleaning agents of all kinds, including soaps and synthetic surfactants. The term "synthetic surfactant" is sometimes used as a synonym for "syndet".

Development of syndets

These disadvantages of soap led the scientific community to search for new body cleansing substances. As a result of these efforts, the end of the 1950's, the era of synthetically produced detergents, the syndets, began. Syndets have distinct advantages over alkaline soaps:

- no alkalizing effort, so an acidic pH can be set (physiological value 5.5)
- therefore also suitable for skin with a reduced alkali reducing capacity
- no formation of insoluble soap scum
- little swelling of the horny layer

As a result of their manifold advantages, the syndets, in solid and liquid form, have gained a firm place in skin cleansing - especially for diseased skin.

Surfactants as active cleansing agents

Surfactants are molecules or ions that are comprised of a very water-soluble (hydrophilic) part, the head, and a fat-soluble (lipophilic) long-chained segment. Surfactants accumulate preferably at interfaces, which the hydrophilic part is orientated towards the water and the lipophilic part towards the oil phase (e.g. lipophilic dirt).

The importance of the surfactant components

Individual surfactants have specific properties, such as the ability to create foam (anionic surfactants), or leave behind a pleasant sensation on the skin (amphoteric surfactants). Therefore, most cleansing products consist of a mixture of surfactants.

> Besides active cleansing agents, the term surfactant is used to describe emulsifiers, foaming agents, solvents and other similar substances.

The skin compatibility of the surfactants is also variable. Even within a class of surfactants, the skin compatibility varies: from the group of anionic surfactants, sodium lauryl sulphate has a marked irritation potential, while other anionic surfactants, for example sodium laureth sulphate and sodium sulphosuccinatex display a much better skin compatibility because of a slightly altered molecular structure.By combining different surfactants, the skin compatibility can be influenced: for example, when combined with an amphoteric surfactant the skin irritating sodium lauryl sulphate achieves a much better skin compatibility.

How Soaps and Detergents Work

Soaps are sodium or potassium fatty acids salts, produced from the hydrolysis of fats in a chemical reaction called saponification. Each soap molecule has a long hydrocarbon chain, sometimes called its 'tail', with a carboxylate 'head'. In water, the sodium or potassium ions float free, leaving a negatively-charged head.

Soap is an excellent cleanser because of its ability to act as an emulsifying agent. An emulsifier is capable of dispersing one liquid into another immiscible liquid. This means that while oil (which attracts dirt) doesn't naturally mix with water, soap can suspend oil/dirt in such a way that it can be removed.

The organic part of a natural soap is a negatively-charged, polar molecule. Its hydrophilic (water-loving) carboxylate group ($-CO_2$) interacts with water molecules via ion-dipole interactions and hydrogen bonding. The hydrophobic (water-fearing) part of a soap molecule, its long, nonpolar hydrocarbon chain, does not interact with water molecules. The hydrocarbon chains are attracted to each other by dispersion forces and cluster together, forming structures called *micelles*. In these micelles, the carboxylate groups form a negatively-charged spherical surface, with the hydrocarbon chains inside the sphere. Because they are negatively charged, soap micelles repel each other and remain dispersed in water.

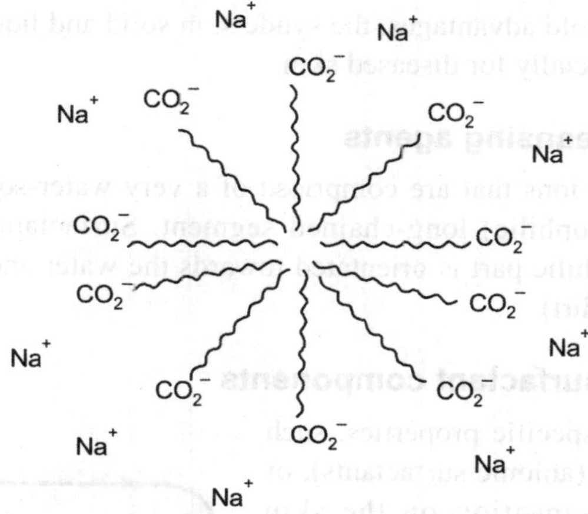

Diagram of Soap Micelle

Grease and oil are nonpolar and insoluble in water. When soap and soiling oils are mixed, the nonpolar hydrocarbon portion of the micelles break up the nonpolar oil molecules. A different type of micelle then forms, with nonpolar soiling molecules in the center. Thus, grease and oil and the 'dirt' attached to them are caught inside the micelle and can be rinsed away. Although soaps are excellent cleansers, they do have disadvantages. As salts of weak acids, they are converted by mineral acids into free fatty acids:

$$CH_3(CH_2)_{16}CO_2^-Na^+ + HCl \longrightarrow CH_3(CH_2)_{16}CO_2H + Na^+ + Cl^-$$

These fatty acids are less soluble than the sodium or potassium salts and form a precipitate or soap scum. Because of this, soaps are ineffective in acidic water. Also, soaps form insoluble salts in hard water, such as water containing magnesium, calcium, or iron.

$$2\,CH_3(CH_2)_{16}CO_2^-Na^+ + Mg^{2+} \longrightarrow [CH_3(CH_2)_{16}CO_2^-]_2Mg^{2+} + 2\,Na^+$$

The insoluble salts form bathtub rings, leave films that reduce hair luster, and gray/roughen textiles after repeated washings. Synthetic detergents, however, may be soluble in both acidic and alkaline solutions and don't form insoluble precipitates in hard water.

Manufacture of Detergents

(I) Sodium Alkyl Sulphates are produced commercially from aliphatic long-chain alcohols (C_{10}-Cu) available from the hydrogenolysis of appropriate fats or oils. The alcohol is first sulphated with sulphuric acid. The resulting alkyl hydrogen sulphate when neutralised gives the sodium salt. For example, the most important detergent of this class *sodium lauryl sulphate* is synthesised from lauryl alcohol obtained by the hydrogenolysis of coconut or palm oil by the following steps.

$$\underset{\text{Lauryl alcohol}}{CH_2(CH_2)_{10}CH_2{-}OH} + \underset{\text{Sulphuric acid}}{HO{-}\overset{O}{\overset{\|}{\underset{\underset{O}{\|}}{S}}}{-}OH} \longrightarrow \underset{\text{Lauryl hydrogen sulphate}}{CH_3(CH_2)_{10}CH_2{-}O{-}\overset{O}{\overset{\|}{\underset{\underset{O}{\|}}{S}}}{-}OH} + H_2O$$

$$\underset{\text{Lauryl hydrogen sulphate}}{CH_2(CH_2)_{10}CH_2{-}O{-}\overset{O}{\overset{\|}{\underset{\underset{O}{\|}}{S}}}{-}OH} + NaOH \longrightarrow \underset{\text{Sodium lauryl sulphate (a detergent)}}{CH_3(CH_2)_{10}CH_2{-}O{-}\overset{O}{\overset{\|}{\underset{\underset{O}{\|}}{S}}}{-}\overset{-}{O}\overset{+}{Na}} + H_2O$$

(2) ***Sodium Alkylbenzenesulphonates. (ABS detergents)*** are manufactured by a Friedel- Crafts alkylation of benzene with long-chain alkenes (C_{10}-C_{12}). The resulting alkylbenzene is then sulphonated to give alkylbenzenesulphonic acid which is converted to its sodium salt.

(*i*)

$$\underset{\text{1-alkene}}{R{-}CH{=}CH_2} + C_6H_6 \xrightarrow{AlCl_3} \underset{\text{alkylbenzene}}{R{-}\overset{CH_3}{\overset{|}{C}H}{-}C_6H_5}$$

(*ii*)

$$R{-}\overset{CH_3}{\overset{|}{C}H}{-}C_6H_5 \xrightarrow{H_2SO_4} \underset{\text{Alkylbenzenesulphonic acid}}{R{-}\overset{CH_3}{\overset{|}{C}H}{-}C_6H_4{-}\overset{O}{\overset{\|}{\underset{\underset{O}{\|}}{S}}}{-}OH} \xrightarrow{NaOH} \underset{\text{Sodium alkylbeozenesulphonate (a detergent)}}{R{-}\overset{CH_3}{\overset{|}{C}H}{-}C_6H_4{-}\overset{O}{\overset{\|}{\underset{\underset{O}{\|}}{S}}}{-}\overset{-}{O}\overset{+}{Na}}$$

For example,

$$CH_3(CH_2)_9CH{=}CH_2 + C_6H_6 \xrightarrow{AlCl_2} CH_3(CH_2)_9\text{-}CH(CH_3)\text{-}C_6H_5 \xrightarrow[2.\ NaOH]{1.\ H_2SO_4} CH_3(CH_2)_9\text{-}CH(CH_3)\text{-}C_6H_4\text{-}\overset{-}{S}O_3\overset{+}{N}a$$

1-dodecene (from petroleum sources) — p-dodecylbenzene — Sodium p-dodecylbenzene sulphonate (a detergent)

For example, alkylbenzene sulphonates are used in powder detergents (*e.g.*, Surf)

How Detergents-cause Water Pollution? Its Remedy

Soaps being sodium salts of long-chain fatty acids, are readily destroyed or degraded by microorganisms in septic tanks and sewage treatment tanks. They are *biodegradable* or "soft" and hence did not cause water pollution. Unfortunately till 1960s the commonest synthetic detergent was Alkyl Benzene Sulphonate(R-C8Hf,-SO.-Na+), or ABS type. It was made from a tetramer of propylene,

$$H_3C\text{—}CH(CH_3)\text{—}CH_2\text{—}CH(CH_3)\text{—}CH_2\text{—}CH(CH_3)\text{—}CH_2\text{—}CH{=}CH_2$$

4,6,8-trimethylnon-1-enetetraamr of propylene

by Friedel-Crafts reaction, followed by sulphonation and neutralisation. The carbon-chain R in the detergent was highly branched. It was "hard" and not *biodegradable.* Microorganisms degrade long carbon chains by first converting the terminal mdhyl group to a carboxyl group. Then they consume the rest of the chain, two carbons at a time, by further oxidation. Branching of the chain blocked this process of degradation and made the ABS detergent "hard" or nonbiodegradable. Such a detergent continued to foam and to make suds, clogging the waster. water disposal plant, and then killing fish and wild life in rivers and streams. The detergent even found way into ground water and thus contaminated the city drinking water. This posed a serious problem. The remedy was found in 1966 when Linear (or long-chain) Alkyl Sulphonate or LAS detergents were introduced in the market. These are "soft" and *biodegradable.* For illustration,

$$H_3C\text{—}CH(CH_3)\text{—}CH_2\text{—}CH(CH_3)\text{—}CH_2\text{—}CH(CH_3)\text{—}CH_2\text{—}CH(CH_3)\text{—}C_6H_4\text{—}O\text{—}S(=O)_2\text{—}Na$$

"hard" or non-biodegradable ABS detergent (prepared from propylene tetramar)

$$H_3C\text{—}CH_2\text{—}CH_2\text{—}CH_2\text{—}CH_2\text{—}CH_2\text{—}CH_2\text{—}CH_2\text{—}CH_2\text{—}CH_2\text{—}CH_2\text{—}CH(CH_3)\text{—}C_6H_4\text{—}O\text{—}S(=O)_2\text{—}Na$$

"soft" biodegradable LAS detergent (prepared from 1-tridence)

LAS detergents are now made from C_{10} to C_{12}. l-alkenes produced by polymerization of ethylene, CH_2.=CH_2, or by cracking of kerosene fraction of petroleum. These detergents, although a bit costly to prepare, have solved the 'water pollution' problem.

The modern LAS detergents naturally will not foam in water. But the housewife does not reconcile with it since she associates the formation of foam with efficiency. For her psychological satisfaction, the manufacturers often add sudsing agents to their products. A packet of detergents contains about 20% of active detergent and an equal amount of sodium sulphate to make up the bulk of the powders. A further 30 to 50% is made up of inorganic phosphates which complex with the calcium and magnesium ions present in hard water and enhance the cleaning efficiency of the detergent. Other ingredients are sodium perborate, a bleaching agent; *fluorescers,* organic compounds that make 'yellow' clothes appear 'white' and foaming agents.

The discharge of detergents with high phosphate content into the rivers and lakes from community sewage system has created a totally different pollution. The excess of phosphates being nutrients promote the growth of algae and weeds which appear as green surface sludge. These plants also deplete the water of available oxygen for fish and other sea animals. It is likely that legislation will have to be introduced to remove, or at least reduce, the phosphate content of detergents to eliminate this source of pollution.

Ionic Detergents

There are four main classes of detergents, anionic, cationic, nonionic, amphoteric.

Cationic Detergents

Another class of detergents have a positive ionic charge and are called "cationic" detergents. In addition to being good cleansing agents, they also possess germicidal properties which makes them useful in hospitals. Most of these detergents are derivatives of ammonia.

A cationic detergent is most likely to be found in a shampoo or clothes "rinse". The purpose is to neutralize the static electrical charges from residual anionic (negative ions) detergent molecules. Since the negative charges repel each other, the positive cationic detergent neutralizes this charge.

Anionic Detergents

Anionic means a negatively charged molecule. In the early days I always remembered this by **an**ionic (**a n**egative). The detergency of the anionic detergent is vested in the anion. The anion is neutralized with an alkaline or basic material, to produce full detergency.

$$CH_3CH_2CH_2CH_2CH_2CH_2CH_2CH_2CH_2CH_2CH_2{-}O{-}\overset{\overset{\displaystyle O}{\|}}{\underset{\underset{\displaystyle O}{\|}}{S}}{-}O{-}Na^+$$

sodium lauryl sulfate

Alkyl Aryl Sulphonates

Linear alkyl benzene sulphonate would be the highest quantity used of any detergent in the world, and the alkyl aryl sulphonates as a group would represent more than 40% of all detergent used.

They are cheap to manufacture, very efficient, and the petroleum industry is a starting point for the base raw material. The most important alkyl aryl condensate is DDB (dodecyl benzene). DDB is sulphonated to DDBSA (dodecyl benzene sulphonic acid), and this in turn is used as a detergent base, where it is neutralised with a base, such as sodium hydroxide, monoethanolamine, triethanolamine, potassium hydroxide, etc.

Long Chain (Fatty) Alcohol Sulphates

Made from fatty alcohol's, and sulphated, these are used extensively in laundry detergents. They can be produced with varying carbon chain lengths, but a C_{12} - C_{18} alcohol sulphate is a good choice.

Other groups

Are the olefine sulphates and sulphonates, alpha olefine sulphates and sulphonates, sulphated monoglycerides, sulphated ethers, sulphosuccinates, alkane sulphonates, phosphate esters, alkyl isethionates, sucrose esters.The anionic detergents are used extensively in most detergent systems, such as dishwash liquids, laundry liquid detergents, laundry powdered detergents, car wash detergents, shampoo's etc.

Cationic Detergents

Another class of detergents have a positive ionic charge and are called "cationic" detergents. In addition to being good cleansing agents, they also possess germicidal properties which makes them useful in hospitals. Most of these detergents are derivatives of ammonia. A cationic detergent is most likely to be found in a shampoo or clothes "rinse". The purpose is to neutralize the static electrical charges from residual anionic (negative ions) detergent molecules. Since the negative charges repel each other, the positive cationic detergent neutralizes this charge.

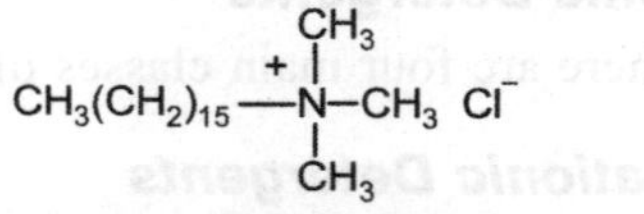

$$CH_3(CH_2)_{15}-\overset{+}{N}(CH_3)_3\ Cl^-$$

Trimethylhexadecylammanium chloride

Neutral or non-ionic detergents

Nonionic detergents are used in dish washing liquids. Since the detergent does not have any ionic groups, it does not react with hard water ions. In addition, nonionic detergents foam less than ionic detergents. The detergent molecules must have some polar parts to provide the necessary water solubility. In the figure on the right the polar part of the molecule consists of three alcohol groups and an ester group. The non-polar part is the usual long hydrocarbon chain.

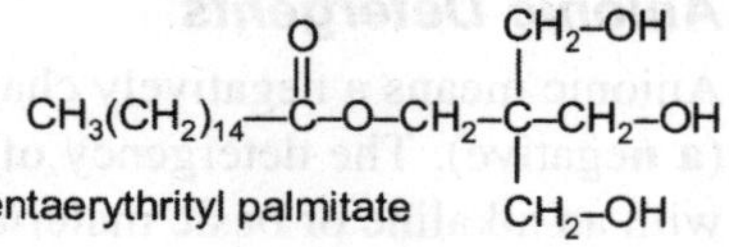

$$CH_3(CH_2)_{14}-\overset{O}{\overset{\|}{C}}-O-CH_2-C(CH_2-OH)_3$$

Pentaerythrityl palmitate

Bile Salts - Intestinal Natural Detergents

Bile acids are produced in the liver and secreted in the intestine via the gall bladder. Bile acids are oxidation products of cholesterol. First the cholesterol is converted to the trihydroxy derivative containing three alcohol groups. The end of the alkane chain at C # 17 is converted into an acid, and finally the amino acid, glycine is bonded through an amide bond. The acid group on the glycine is converted to a salt. The bile salt is called sodiumglycoholate. Another salt can be made with a chemical called taurine. The main function of bile salts is to act as a soap or detergent in the digestive processes. The major action of a bile salt is to emulsify fats and oils into smaller droplets. The various enzymes can then break down the fats and oils.

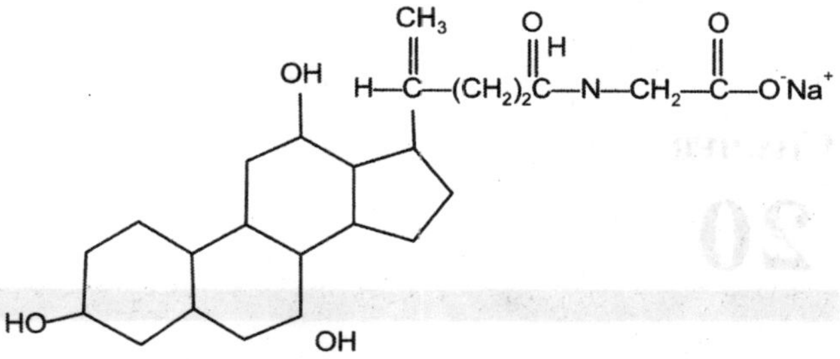

sodium glycoholate

Nonionic Detergents

The vast majority of all nonionic detergents are condensation products or ethylene oxide with a hydrophobe. This group of detergents is enormous, and the permutations endless. They would be the single biggest group of all detergents, and I do not propose going into the endless chemistry of these products.

Amphoterics

These have the characteristics of both anionic detergents and cationic fabric softeners. They tend to work best at neutral pH, and are found in shampoo's, skin cleaners and carpet shampoo. They are very stable in strong acidic conditions and have found favour for use with hydrofluoric acid.

CHAPTER

20

Aromaticity and Aromatic Hydrocarbons

Introduction

The word aromatic comes from the Greek word Aroma meaning scent, as name depicts these compounds possess a typical scent of themselves, sometimes aromatic hydrocarbons are called arenes.

The Huckel Rule for Aromaticity

An aromatic hydrocarbon is a cyclic compound where all the atoms in the ring(s) are sp2 having a free "p" orbital to provide a pipeline or "conduit" in which Pi electrons can travel through being distributed throughout the molecule. This is referred to as a delocalized Pi electron system. A man by the name of Huckel came up with a mathematical rule which assumed a monocyclic planar ring system having 4n + 2 Pi electrons within the cyclic system where n = any integer beginning with the integer zero.

n	4n + 2 Pi electrons
0	$4(0) + 2 = 2$
1	$4(1) + 2 = 6$
2	$4(2) + 2 = 10$
3	$4(3) + 2 = 14$
4	$4(4) + 2 = 18$

There are some cyclic hydrocarbons which have the proper number of Pi electrons according to the 4n + 2 Rule, but they are not aromatic. That is because at least one of the carbons within the ring is something other than sp2. For example, cycloheptatriene has six Pi electrons and seems to follow the 4n + 2 rule but one of the seven carbons is sp3 and therefore the ring is not planar. The ring must be planar in order for the Pi electrons to be delocalized in the ring. This gives the compound extra stability. Benzene is the most commonly known aromatic hydrocarbon having six Pi electrons with all six carbons sp2 and therefore the ring is planar (a necessary condition). There are two conditions that must be present before a cyclic hydrocarbon can be aromatic:

1. The ring must be planar (i.e.: all the atoms in the ring must be sp2 hybridized)
2. There must be 4n + 2 Pi electrons in the ring

Classifications of Aromatic Hydrocarbons

There are several classes of aromatic hydrocarbons.

1. ***Benzenoid aromatics*** - cyclic compounds made up of one or more benzene like rings in which the Pi electrons are conjugated. Examples are benzene, naphthalene, phenanthrene, and Anthracene.

benzene

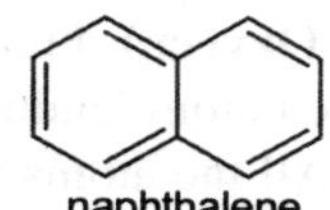
naphthalene

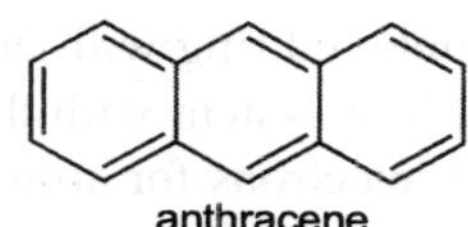
anthracene

2. Non-Benzenoid Aromatics-compounds that have ring systems that are larger or smaller than benzene rings. An example is the cycloheptatriene Cation, $C_7H_7^+$. Earlier we had considered the cycloheptatriene molecule(Fig 1) as having the proper number of Pi electrons but wasn't aromatic because one of the carbon atoms within the ring was sp^3 and not sp^2. If one uses a mild reducing agent like Lithium on cycloheptatriene, a hydride ion can easily be removed from the sp3 carbon to create a carbocation whose positively charged carbon becomes sp2 making the ring planar. This is called the cycloheptatriene Cation. It still has six Pi electrons acceptable according to the Huckel Rule to be aromatic. Another example is the cyclopentadiene anion. The neutral molecule cyclopentadiene has only four Pi electrons and therefore does not conform to the Huckel Rule. However if you removed a proton using a mild base you add two more electrons to the Pi system giving a total of six acceptable to the Huckel Rule. All the carbons become sp2 in the ring and the anion is aromatic. Another example is the aromatic annulenes except for the Annulene(Benzene). Aromatic annulenes are monocyclic compounds where all the Pi electrons are conjugated. Not all annulenes are aromatic. Annulene and annulene are examples of non-aromatics because they have the wrong number of Pi electrons, annulene would not be counted as a non-benzenoid since this is Benzene itself and would be classified as a Benzanoid.

The resonance stabilized conjugated structure of Benzene was first proposed by **Kekule** who, legend has it, had a dream vision of snakes chasing their tails around a circle. When he woke up he proposed the Benzene structure consisting of six carbons in a closed ring with alternating double and single bonds between its carbons. This makes all the carbons sp_2 hybridized and the geometry around each carbon trigonal planar. Therefore, the benzene ring is flat or planar.

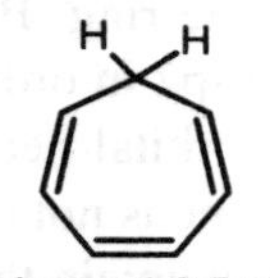

cyclohepta-1,3,5-triene
cycloheptatriene

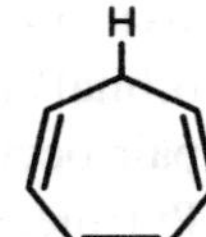

cyclohepta-1,3,5-triene
cycloheptatriene cation

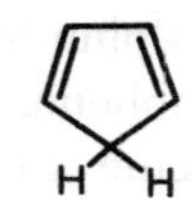

cyclopenta-1,3-diene
cyclopentadiene

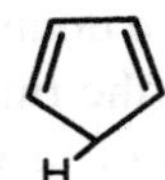

cyclopenta-1,3-diene
cyclopentadiene anion

Kekule did, however, explain the resonance by suggesting two non-equivalent structures that were in equilibium with one another. There has never been any experimental evidence to suggest such an equilibrium mixture exists. A more modern picture based on the more modern molecular orbital approach to bonding and Resonance Theory suggests that there are two equivalent structures that are resonance hybrids of one another. The composite structure that would be closer to reality is a single molecule that has attributes of both resonance structures. All carbon-carbon bond lengths would be equal and lie between what a single bond between two carbons and a double bond between two carbons would be.

3. ***Heterocyclic aromatics:*** These are not actually pure hydrocarbon but ring systems that have at least one atom in the ring that is not carbon. Not all heterocyclic compounds are aromatic only those that follow the Huckel conditions above.

Examples of heterocyclic aromatics include :

- Furan (five membered ring with one atom Oxygen). Furan has two double bonds within the five membered ring system with the Oxygen atom lending one of its lone pair to supply the necessary six Pi electrons for aromaticity. All the atoms within the ring are sp2. Furan has a second lone pair that makes it very basic.
- Pyridine (six membered ring system with one atom a nitrogen.). Pyridine is a six membered ring that resembles a Benzene ring except that one of the atoms is a Nitrogen atom. The ring has three double bonds giving it the necessary six Pi electrons. The Nitrogen atom has an additional lone pair of electrons which gives Pyridine a decidedly basic character.

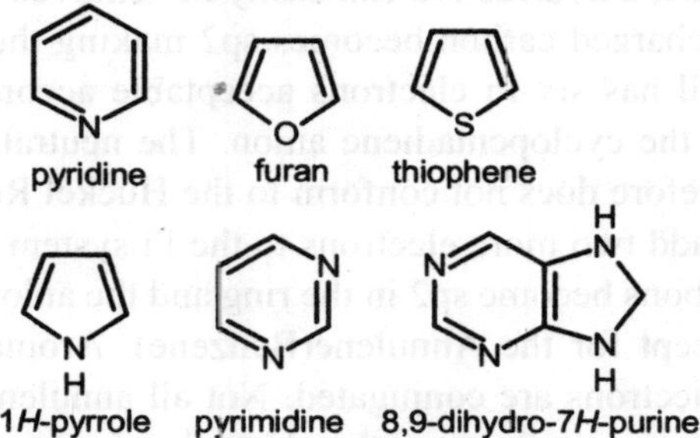

- Pyrrole is a five membered ring with a nitrogen in the ring (Fig 3). The ring system has two double bonds with the Nitrogen lending its lone pair of electrons as the necessary fifth and sixth Pi electron in the ring. Because the Nitrogen atom is bonded to two sp2 carbons, its orbitals must also be sp2 in order to have maximum overlap for maximum stability. The lone pair occupy the "p" orbital necessary to complete the "p" orbital conduit spoken of before. Pyrrole, unlike Pyridine is not basic at all but surprisingly acidic. The lone pair is locked into the ring structure to maintain the more stable aromatic configuration.
- Thiophene is a five membered ring system with sulphur in the ring. Thiophene like Furan has two double bonds and sulphur that lends a lone pair of electrons to establish aromatic character to the ring system. Thiophene, like Furan, has an extra lone pair of electrons on the Sulphur atom to give the molecule a basic character. Indeed, Thiophene is more basic than Furan because Sulphur is a larger atom and the electronegativity of Sulphur is less than that of Oxygen. This allows the lone pair to be donated more easily than Oxygen.

Fullaranes

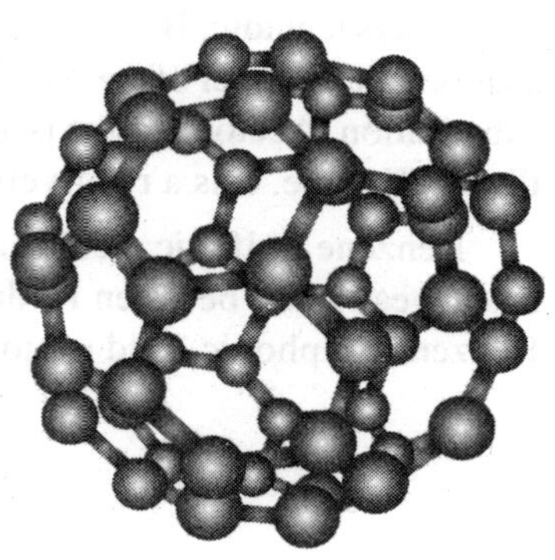

Fullerenes are an exciting new classification of aromatic organic molecules that consist of pure carbon. The formula for Fullerenes are either C_{60} or C_{70}. The shape of these remarkable allotropic forms of pure carbon are shaped like soccer balls or icosahedrons. They are also called Buckministerfullerenes named after an architect, Buckminister Fuller, who developed the geodesic dome like structures as a viable architectural structure. This structure is a network of hexagons and pentagon. Regardless of the number of carbon atoms in the Fullerene molecule, there must always be 12 five membered faces with a varied number of six membered faces. Each carbon atom in the molecule is sp2 hybridized. Since each carbon has four valence electrons, three of them are in the three sp2 hybrid orbitals with the fourth electron in the unhybridized "p" orbital and delocalized. The Fullerenes possess fascinating properties because of their unusually high electron affinity. They readily accept electrons from metals such as Potassium to form fulleride salts like K_3C_{60}. These salts have the Potassium cations attached to the exterior of the Buckey Balls, but recent research has shown that the metal cations can be encapsulated within the geodesic structure. Apparently, these salts become superconductors below 18 Kelvin.

Nomenclature of Aromatic Hydrocarbons

Aromatic hydrocarbons in general are referred to as Arenes. They are further distributed into following classes.

1. Monosubstituted Benzene Derivatives

Benzene is an aromatic monocyclic hydrocarbon of six carbons. It has six Pi electrons that are delocalized. This results in a molecule that is more stable compared to a molecule where the double bonds are localized as in cyclohexatriene. Halobenzenes are monosubstituted benzene where the substituent is a halogen atom. Bromobenzene is shown in. If a Chlorine atom was the substituent then the serivative would be called chlorobenzene. What would Iodobenzene and Flourobenzene look like? A monosubstituted benzene where the substituent is an alkyl group like methyl shown in as methylbenzene is called an aryl benzene. What do you suppose Ethyl benzene would look like? How about iso-propylbenzene? Methyl benzene is also called toluene.

Br CH_3

benzene bromobenzene toluene(methyl benzene)

O^- N^+ O NH_2 O CH

nitrobenzene aniline (amino benzene) benzaldehyde

Nitrobenzene is a monosubstituted derivative with a nitro (-NO_2) group attached. Aminobenzene also called aniline has the amino group(-NH_2). Aniline is a very reactive derivative that is a weak base. Benzoic acid is an aromatic weak acid with the carboxyl group (-COOH) having an ionizable Hydrogen that makes it acidic. Benzoic Acid is relatively inactive toward electrophilic substitution. Benzonitrile is an aromatic member of the Nitrile Family. The Cyano (-CN) group is highly unreactive toward electrophilic substitution. Acetophenone is an aromatic ketone. Another acceptable name for the compound is Methyl Phenyl Ketone. It is a relatively unreactive molecule toward electrophilic aromatic substitution.

Benzene Sulfonic Acid is a sulphur containing aromatic compound. The Sulphonyl group(-SO_3H) has a weak bond between Hydrogen and one of the Oxygen atoms which makes this compound acidic. Benzene Sulphonic Acid is moderately unreactive to electrophilic substitution.

CH_3 O OH O CH_3

methoxybenzene benzoic acid acetophenone

HO S O O N HO

benzenesulfonic acid benzonitrile phenol

Phenol has the Hydroxyl group (-OH) attached to a Benzene ring. The bond between the Oxygen and the Hydrogen is very weak and the resulting Phenoxide ion is unusually stable because of the resonance stabilization of the anion. Phenols are extremely reactive toward electrophilic substitution. Methoxybenzene also called Anisole is an aromatic member of the ether family. It is moderately reactive toward electrophilic substitution.

2. Di-substituted Benzene Derivatives

There are two ways that disubstituted Benzene derivatives can be named:

1. Ortho-Meta-Para Designation
2. Numbering Approach

Ortho-Meta-Para Designation of Di-Substituted Benzene Derivatives

Two substituents attached to a Benzene ring can be positioned in respect to one another:

1. ***Ortho Isomer***-The two substituents are attached to adjacent carbon atoms within the Benzene ring. If we used the numbering approach, they would be attached to carbons #1 and #2. We could name the first molecule as a Bromobenzene by hyphenating the first letter of the word ortho to the word Chloro followed by the monosubstituted benzene derivative corresponding to the group attached to the #1 carbon, o-Chlorobromobenzene. An alternative name would be

to hyphenate the numeral 2 to the word chloro following by the bromobenzene, 2-Chlorobromobenzene.

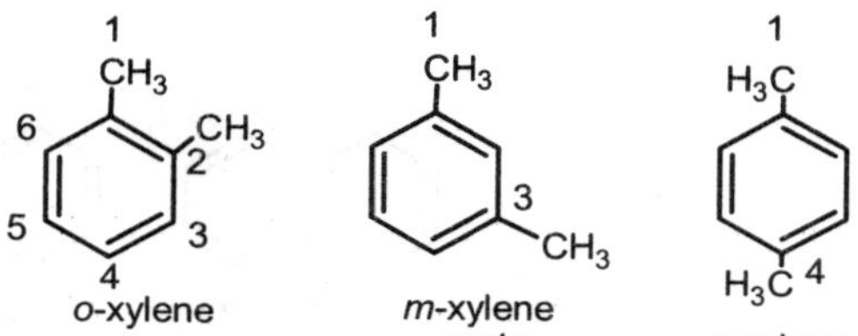

2. ***Meta Isomer***-The two substituents are attached to the 1 and 3 carbon. The two are said to be meta to one another, and the name would be m-Iodotoluene or m-Iodomethylbenzene An alternative name would be to locate the Iodo group by hyphenating the carbon number it is attached to. Here the Iodo group is attached to the #3 carbon so the name would be 3-Iodotoluene or 3-Iodomethylbenzene.

o-xylene ortho

m-xylene meta

p-xylene para

3. ***Para Isomer***-The two substituents are attached to the #4 carbon and the #1 carbon respectively. The two are said to be para to one another, and the name would be p-cyanophenol. An alternative name would locate the cyano group on the #4 carbon so the name would be 4-cyanophenol.

3. Polysubstituted Benzene Derivatives

Polysubstituted Benzene derivatives of three substituents or more must be named by locating all but the group attached to carbon #1 by hyphenating the group name to the carbon number it is attached to.

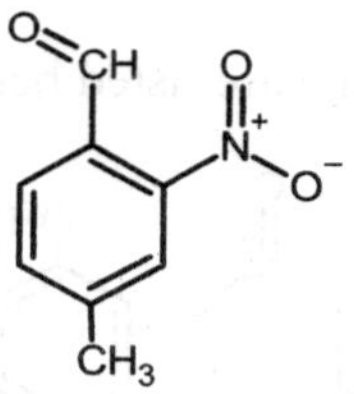

4-methyl-2-nitrobenzaldehyde

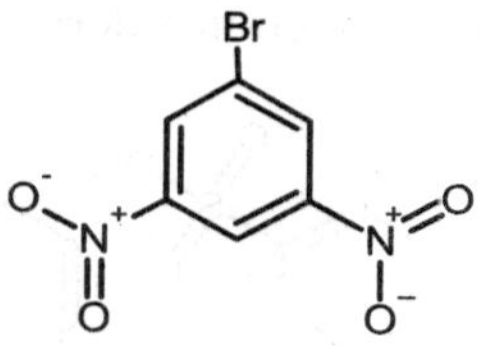

1-bromo-3,5-dinitrobenzene

The first example above in shows three substituents with the CHO group on the number one carbon. A nitro group is attached to carbon #2 and a methyl group attached to carbon #4. These substituents would be named first in alphabetical order followed by the monosubstituted Benzene derivative corresponding to the group attached to carbon #1. The name for this compound would be 4-methyl-2nitrobenzaldehyde. The second example above shows three substituents with two of them of the same kind. The Bromine is attached to the #1 carbon thus establishing the monosubstituted benzene derivative part of the name. There are two nitro groups attached to the #3 and #5 carbon so these would be named first using the Greek prefix “di” to represent two of the same type. The name would be 3,5-dinitrobromobenzene.

Fused Polycyclic Arenes

A number of polycylic arenes are known which contain two or more benzene rings fused in ortho positions. Some important member of this class known by their trivial names naphthalene anthracene 1,4-dihydrophenanthrene

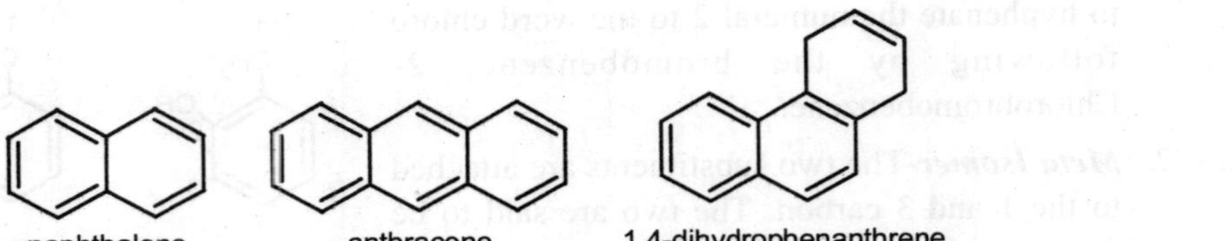

Functional Derivates of Arenes

As already stated arenes form almost all classes of functional groups derivatives as do alkanes. The IUPAC names of these compounds are given either as the substitution products of arenes or by naming the aryl group followed by the name of the function. The groups (or radicals) derived form benzene and toluene by removal of hydrogen atom are.

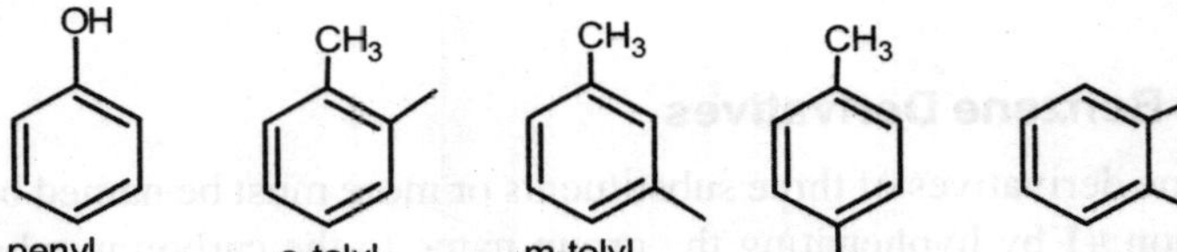

OH CH_3 CH_3 CH_3

penyl o-tolyl m-tolyl p-tolyl o-phenylne

The mono- substitution functional derivatives of benzene are listed below.

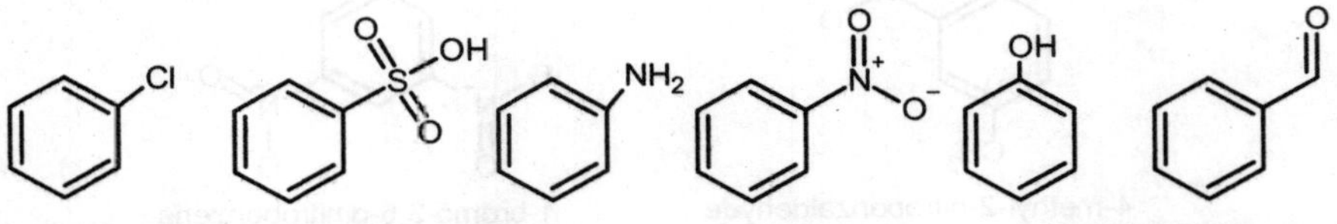

chlorobenzene benzenesulfonic acid aniline nitrobenzene phenol benzaldehyde

In case of monosubstitution products of toluene the systematic names are used, unless the derivative has a recognized trivial name. Sometimes the name of derivatives is given by writing the name of the tolyl group followed by the name of the functional name.

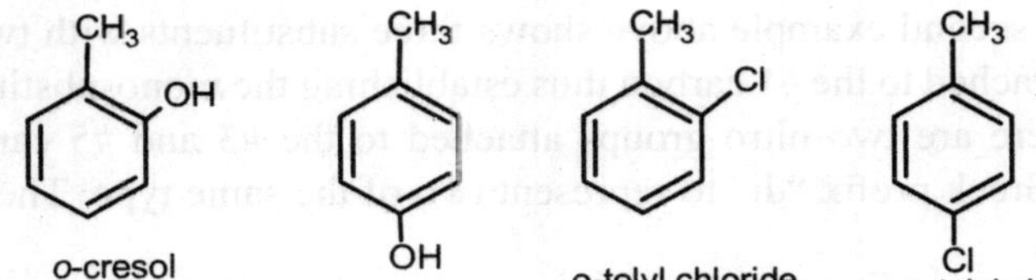

o-cresol (2- hydroxytoluene)

p-cresol (4- hydroxytoluene)

o-tolyl chloride 1-chloro-2-methylbenzene

p-tolyl chloride 1-chloro-4-methylbenzene

In tri or tetra substitution products, the IUPAC rules cited while naming polysubstituted arenes are used, for example.

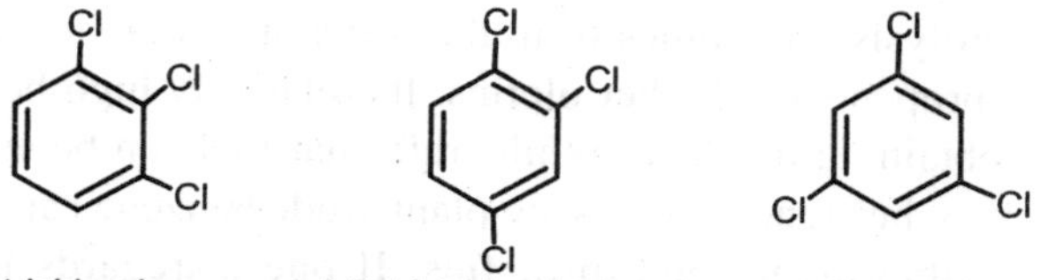

1,2,3-trichlorobenzene 1,2,4-trichlorobenzene 1,3,5-trichlorobenzene

Sources of Aromatic Compounds

Aromatic compounds are almost exclusively taken from the fossil fuels of coal and crude oil. However, the small concentrations of aromatic compounds that are primarily present do not allow for economically viable isolation. Raw materials that are suitable for the profitable yield of aromatic compounds are only formed through thermal or catalytic transformation processes in coking plants or refineries.

There are basically three sources of raw materials available for the technical isolation of aromatic compounds:

1. Products from the coking off hard coal,
2. Reformat fuel from the processing of raw benzine,
3. Pyrolysis fuel from the generation of ethylene/propene.

Tar from hard coal, water and gas from coking plants are historically the most significant sources of aromatic compounds, which are today becoming less and less significant: coke, which with 80% of the weight is the main product of the coking of hard coal, has not only been subject to a drop in its consumption due to the technical improvements in the blast furnace smelting process, but also the change in gas suppliers from town gas coking plants to fission gas and natural gas from crude oil has contributed to the decline. The use of heavy heating oil for the production of steel does, for the time being, still have some significance. Nowadays, crude oil is for practically all industrial countries the necessary basis for the quickly growing need for aromatic compounds. For example, in 1960 in Western Europe there was as yet hardly any taking of aromatic compounds from crude oil, whereas in the USA at the same time 83% was already being isolated from it. Since 1990 there has no longer been a basis of tar for bezene in the USA. In 1992, for example, 93% of the benzene in Western Europe came from crude oil and only 7% was taken from coal. Toluene and xylene are mainly taken from crude oil. Aromatic compounds from reformat and pyrolysis fuel. In the process of the refining of crude oil, both in the refinement in reforming and in the cracking process for the production of olefin, there appear at the same time fractions which are rich in aromatic compounds which, as reformat fuel and as pyrolysis or cracked fuel represent valuable sources of aromatic compounds. In this process, the reformat fuel comes both from the paraffinic crude oils, which consist more than 50% of straight chain and branched chain alkenes and of naphthenic crude oils, which mainly contain cycloalkenes. In the preparation through distilling, both types of crude oil yield fractions which - due to an insufficient octane count for use as petrol for engines – have to be converted in a reforming process. The preparation through distilling of the reformed crude product leads to a fraction that is rich in aromatic compounds which –

due to its content of higher boiling aromatic compounds – is particularly suitable for obtaining toluene and xylene isomers. The pyrolysis fuel comes from the water steam or steam cracking of naphtha for the generation of ethylene, propene and higher olefins. Its relatively high benzene content makes it a leading product for use in obtaining benzene. Whilst reformat fuel can be used directly for obtaining aromatic compounds, cracking benzenes and coking plant crude benzene can only be freed through the hydrogenation of polymerisable mono- and di-olefins. If one disregards further influences on the distribution of aromatic compounds, for example through the different reforming or cracking conditions (two levels) and through boiling levels of the naphtha cuttings, it is possible to obtain more or less the following simplified picture of distribution:

Typical composition of reformat and pyrolysis fuel (in % of weight).

Product Reformat fuel Pyrolysis fuel

Benzene	**3% 40%**
Toluene	13% 20%
Xylene	18% 4-5%
Ethyl benzene	5% 2-3%
Higher aromatic compounds	16% 3%
Non aromatic compounds	**45% 28-31%**

The isolation of aromatic compounds from reformat or pyrolysis fuels consists to a great extent of part cuttings from the separation of non aromatic compounds and the subsequent dividing of the aromatic mixture into the individual components. Therefore, for the separation of aromatic-non aromatic-mixtures, different and very individual separating procedures have come into use. These procedures are very individual because of the different separating problems with the various mixtures for use and because of the increasingly high requirements for cleanliness for the individual components. Special separating process for non aromatic compounds / aromatic compounds and aromatic compound mixtures

Five different procedures have been developed for obtaining aromatic compounds from carbon hydride mixtures. The following table summarizes the problems in separating, the technologies that are used and the necessary conditions for the basic or economical feasibility of the individual procedures:

Procedure for the obtaining of aromatic compounds:

1. Azeotropic Distillation
2. Extractive Distillation
3. Liquid-liquid extraction
4. Crystallization through freezing out
5. Adsorption on solids

Transformation procedures for aromatic compounds

The different production quantities in individual countries of pyrolysis fuel, reformat and hard coal tar as well as their characteristically different composition as aromatic compounds make it necessary to have transformation procedures so as to be able to balance out excess supplies, for example of toluol and the demand for benzene and xylene.

The following are important procedures:

1. Hydrodealkylation of the toluene,
2. Isomerisation of the m-xylene,
3. Disproportioning of the toluene and the transalkylation with trimethylbenzoles.

Important Material and Energy Flows:

Input	Output
Pyrolysis fuel	Benzene
Thermal energy	Xylene (o-xylene; 1.2-dimethyl benzene)
Power	Toluene (methyl benzene)
Reformat fuel	Xylene (p-xylene; 1.3-dimethyl benzene)

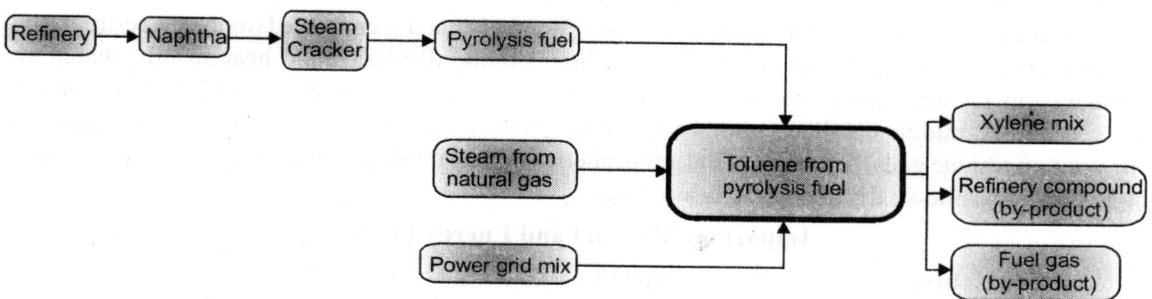

Fig. 20.1 Showing the Flow diagram for the recovery of Aromatics from Naptha

Steam Cracker: Naphta and LNG were splited by a steam cracker for the production of the products ethane (ethylene), propylene, C4-cut, pyrolysis gas, pyrolysis tar, fuel gas and hydrogen. The steamcracking can be subdivided in the following single steps:

1. fission of naphta in tube furnaces
2. quenching
3. fission gas compaction and defecation
4. desiccation, cooling and ow temperature-distillation.

The use of LNG is going to gain strengthened importance, due to the good ethylene yield

Refinery: After desalting, the crude oil is introduced to the distilling column for the atmospheric distillation(fractionation of the crude oil by separation according to density/boiling/condensation areas.Gases go up to the head and are introduced to the liquid gas system. The lighter fractions of the benzines are taken in the upper trays and are led in three different processes.

Firstly, to the reformer for catalytic transformation from aliphatic paraffins to Iso paraffins and from cycloparaffins to aromatic compounds; with a reduction of the thermal value (take care when the energy level is being explicitly balanced).

Secondly, some of the benzine is introduced into the deisopentaniser, in which the light benzine is again distilled in order to obtain iso pentane for the manufacture of super benzene. The super benzene constituent for the mixture of benzene and naphtha leaves the deisopentaniser as a product.

Thirdly, some is taken for the refining of benzine. The outlets for deisopentaniser, reformer and refining lead to the benzene mixing system and super or noraml petrol follow as products. The middle distillates of the atmospheric distillation are introduced directly into the hydrofiner (desulphurization). The sulphur is introduced to a Claus plant and the desulphurised product is introduced to the middle distillate mixture or it is removed directly as kerosine. The residue from the atmospheric distillation is, partly, introduced to the vacuum distillation. Here there is a distillation in gas oil wax distillate and vacuum residue. A further part of the atmospheric residue is introduced into the visbreaker (mild thermic cracking). Small amounts are introduced directly into the heating oil mixing system and the bitumen system. The gas oil, as a product of the vacuum distillation, is introduced to the hydrofiner, desulphurised, and introduced to the middle distillate mixture. Some of the wax distillate which has been taken from the middle trays of the vacuum distillation is introduced to the basic oil production of wax and lubricants. Most of it is introduced to a catalytic cracker and is converted (molecule restructuring to shorter chains). The products are gases, benzines, middle distillates and heating oils, which are introduced to the appropriate levels. The vacuum residue goes into the coker, which also produces gases, benzines, middle distillates and heating oils as well as petroleum coke, which is then purified. The vacuum residue goes, like some of the atmospheric residue, into the visbreaker, which also produces gases, benzines, middle distillates and heating oils.

Important Material and Energy Flows

Input	Output
Crude oil	Propane
	Sulphur
	Propene
	Butane
	Reformat fuel
	Bitumen
	Gasoline (regular and premium)
	Kerosene
	Wax
	Lubricants
	Heavy fuel oil
	Petroleum coke
	Light fuel oil
	Diesel
	Butene
	Refinery gas
	Naphtha

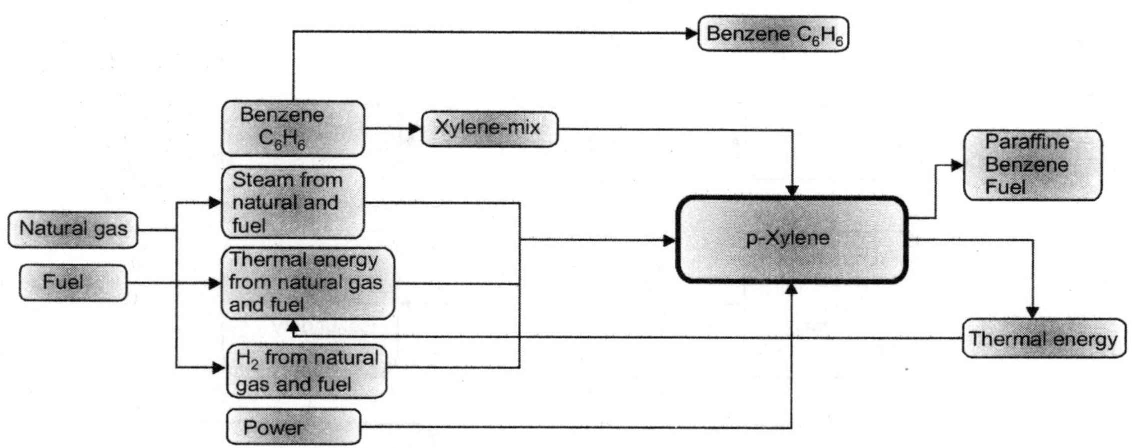

Fig. 20.2. Showing the Flow diagram for obtaining aromatics from Natural Gas

The Crude Oil (Source of Aromatics)

The crude oil deposit are opened up by wells. The crude oil is upgraded in different ways depending on the composition. Salt water, crude oil gas and a part of the sulphur contents are seperated. The energy that is necessary for the upgrading is produced by Disel generators and by burning heavy fuel oil. The upgraded crude oil is transferred in pipelines to the consumer countries. The crude oil is transortated in tankers on the sea. The production is separated into onshore and offshore production. On shore means the production on land, off shore the production on the sea. On some places crude oil and natural gas are produced together. Depending on the quality of the crude oil state of exploitation, the production is divided into primary, scondary and tertiary production, with an increasing effort. The share of the secondary and tertiary production has decreased over the last years, since the price for the crude oil does not allow a profitable production.

Important Material and Energy Flows

Input	Output
Heavy fuel oil	Crude oil
Diesel	

Aromatics from Coal

Coal can be defined as a sedimentary rock that burns. It was formed by the decomposition of plant matter, and it is a complex substance that can be found in many forms. Coal is divided into four classes: anthracite, bituminous, sub-bituminous, and lignite. Elemental analysis gives empirical formulas such as $C_{137}H_{97}O_9NS$ for bituminous coal and $C_{240}H_{90}O_4NS$ for high-grade anthracite.

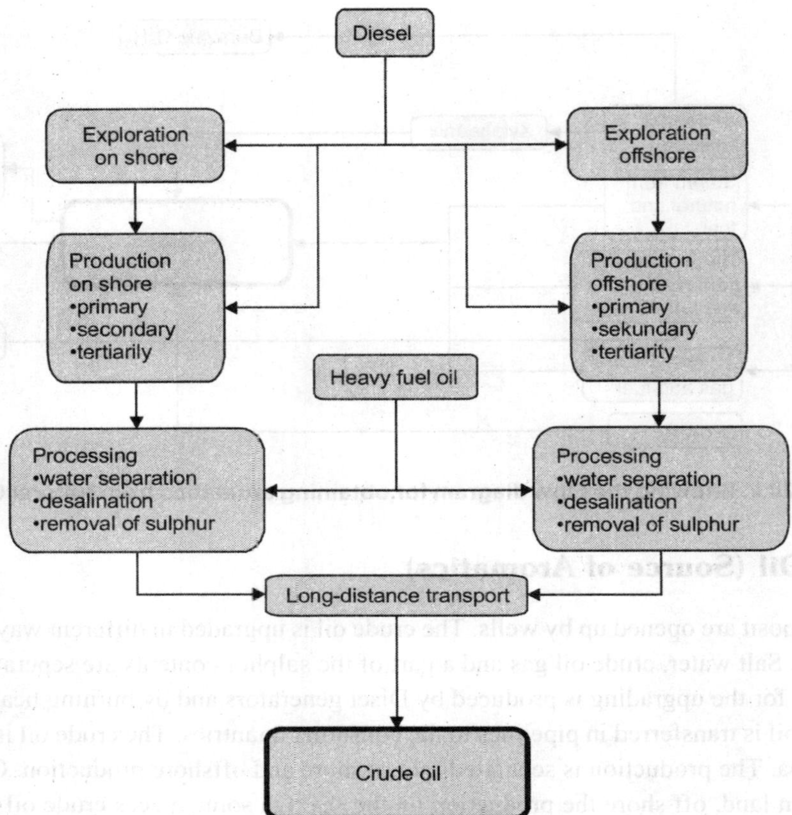

Fig. 20.3. Showing the Flow chart for the recovery of crude oil from diesel for aromatics

Anthracite coal is a dense, hard rock with a jet-black color and a metallic luster. It contains between 86% and 98% carbon by weight, and it burns slowly, with a pale blue flame and very little smoke. *Bituminous coal*, or soft coal, contains between 69% and 86% carbon by weight and is the most abundant form of coal. *Sub-bituminous coal* contains less carbon and more water, and is therefore a less efficient source of heat. *Lignite coal*, or brown coal, is a very soft coal that contains up to 70% water by weight.

Conversion of Different Coal forms Peat to Anthracite

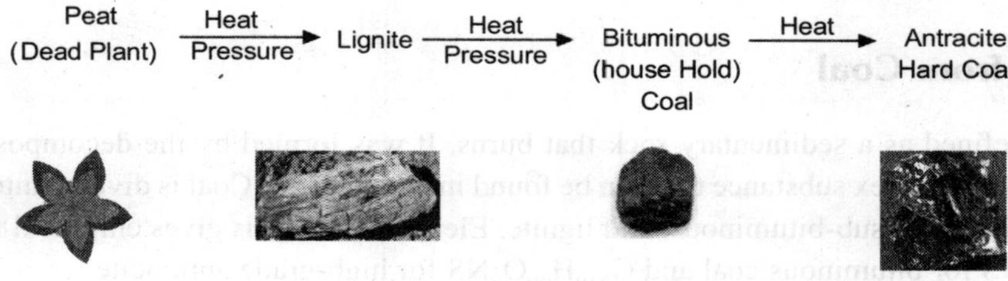

The total energy consumption in the United States for 1990 was 86 x 10^{15} kJ. Of this total, 41% came from oil, 24% from natural gas, and 23% from coal. Coal is unique as a source of energy in the United States, however, because none of the 2118 billion pounds used in 1990 was imported. Furthermore, the proven reserves are so large we can continue using coal at this level of consumption for at least 2000 years.

The transformation of plant material to coal by the above steps is due to progressive decomposition by heat and pressure. Therefore the original material (cellulose) lost moisture and the gases, hydrogen and oxygen and become richer in carbon at each stage listed above. The final product coal has the approximate chemical formula $(C_3H_4)_n$ which shows that it is hydrogen deficient hydrocarbon stuff. Thus the basic structure of coal is probably built up of a large number of interlocked benzene rings, upto thirty (in high ranking coal). Hydrogen is present in the aliphatic side chains. Besides carbon and hydrogen, coal may also contain small percentages of other elements, as shown in the table below

Approximate Composition (Percentages of Various Types of Coal)

Elements	Peat	Lignite	Bituminous	Anthracite
Carbon	54.0	64.0	84.0	93.0
Hydrogen	5.5	5.0	4.5	3.3
Oxygen	35.5	26.0	8.0	2.0
Nitrogen	2.0	1.3	1.3	1.3
Sulphur	3.5	3.3	1.5	0.5

Types of Coal

There are several different types of coal, which have different properties usually dependent on their age and the depth to which they have been buried under other rocks. In some parts of the world (e.g. New Zealand), coal development is accelerated by volcanic heat or crustal stresses. The degree of coal development is referred to as a coal's "rank", with peat being the lowest rank coal and anthracite the highest.

Peat

Peat is the layer of vegetable material directly underlying the growing zone of a coal forming environment. The vegetable material shows very little alteration and contains the roots of living plants. Peat is widely used as a domestic fuel in rural parts of Scotland and Ireland.

Lignite

Lignite is geologically very young (upwards of around 40000 years). It is brown and can be soft and fibrous, containing discernible plant material. It also contains large amounts of moisture (around 70%) and so has a low energy content: around 8-10 MJ/kg. This coal is mined extensively in the Latrobe Valley south east of Melbourne, Australia. As the coal develops it loses its fibrous character and darkens in colour.

Black coal

In Australia, black coal ranges from Cretaceous age (65 - 105 million years ago) to mid Permian age (up to 260 million years ago). They are all black; some are sooty and still quite high in moisture (sub-bituminous coal), including the coal mined at Collie, which is sometimes termed a "black lignite". Coals which get more deeply buried by other rocks lose more moisture and start to lose their oxygen and hydrogen; they are harder and shinier (bituminous coal). These are typical of most of the coals mined in NSW and Queensland, which have energy contents around 24 to 28 MJ/kg. These coals generally have less than 3% moisture, but some power stations in NSW and Queensland burn coal at up to 30% ash.

Anthracite

Anthracite is a hard, black, shiny form of coal which contains virtually no moisture and very low volatile content. Because of this, it burns with little or no smoke and is sold as a "smokeless fuel". In Australia, coals only approach anthracite composition where bituminous coal seams have been compressed further by local crustal movements (at Yarrabee and Baralaba, both in Queensland). Anthracites can have energy contents up to about 32 MJ/kg, depending on the ash content.

Carbonization of Coal

This invention relates to processes for the production of gas, tar-oils, and other coke products from coal and other carbonaceous material and is herein illustrated as applied in a plant for obtaining these products from coal. The solid raw material is reduced to a granular or pulverized condition, such as will permit it to be conveyed through a heated tube by a current of steam of other gas or vapor acting as a carrier for the material and as an atmosphere surrounding it while subjected to heat, suitable for yielding the desired products. The process may be used in treating the fine coal composing parts of shipments received at power plants, the coarser sizes or dust being subjected to other treatments of gasification which inventions form divisions of the present invention. This development provides for economical use of steam as the principle heat-transferring agent and permits the grades of coal as received to be subjected to the type of treatment for which they are best adapted. To carry out economically the features of this process, the fine coal may be treated with superheated steam to yield tar-oils, resins, waxes, gas and fine low-temperature coke. The condensation of these products and the steam gives a large supply of heat which may be utilized for the production of a further supply of steam at a lower pressure. This further supply of steam may be superheated in the form of the invention herein disclosed in distilling a further amount of coal or it may be disposed of advantageously in other processes. According to this invention I overcome the operating difficulties of other processes in obtaining rapid distillation of coals. The difficulties are due principally to the fusing properties of the coals and their poor heat conductivity and I overcome these controlling factors to such an extent that coals can be distilled very quickly to any degree of volatilization regardless of their fusing properties or conductivity of heat, or the grade of coal and amount of mineral matter therein. The present invention provides for the successful use of culm or slack coal as well as for many other types of carbonaceous materials, even sawdust being utilizable. In the form of the invention herein illustrated the finely divided carbonaceous material, such as coal which has preferably been preheated, is introduced into an externally heated pipe carrying a stream of steam. When the carbonaceous material is of the fusing type, air also may be introduced advantageously, or the coal dust may be subjected to a pretreatment with air or combustion gases in the same type of apparatus. To distill the tar-oils and obtain only part of the total possible gas, from the coal, the stream of steam and air, if added, is heated to a temperature well within the temperature range of low temperature carbonization which will prevent excessive cracking of the tar-oils. Too high a temperature must not be used because the coke formed in the process will react with the steam to form water gas and the tar-oils will be decomposed largely into gases and heavy oils of poor quality. I have used temperatures from 700 F to 1200 F satisfactorily. The finely divided carbonaceous material presents an immense amount of surface to react with the steam and to take up the heat thereof, making it possible to obtain substantially complete gasification of the carbonaceous materials present with great rapidity. Coal carried in steam of high velocity, say 100 to 600 feet per second, in a one-inch pipe 100 feet long has been found sufficiently reactive under the above conditions to complete the gas-forming reactions. Under the conditions described below the turbulent flow of the steam and its contact with the heat-transmitting walls of the apparatus have accomplished very efficient transfer of heat from the combustion gases surrounding the pipe to the rapidly moving steam and to the fine carbonaceous material inside the pipe. Under these conditions the reaction appears to be practically a surface one between the steam and the carbonaceous material. It also appears that either the reaction is almost instantaneous by which the particles are consumed, or, there is rapid disintegration of the carbonaceous particles by the wear to which they are subjected in the pipe and the carbon dust is then reacted upon more efficaciously because of the much greater reacting surface provided. I have used one form of this process for distilling finely divided coal at low temperatures for the production of coal gas, and obtained a large yield of rich gas and tar-oils and a low temperature coke dust which was easily ignitable and suited for power plant use. In carrying out the process, the preheated coal was placed in a pressure chamber or magazine, hereinafter described, and then was led by a suitable screw device into the upper end of a one-inch pipe coiled around a vertical axis and carrying steam. The steam carried the coal down through the coil and hot combustion gases surrounding the coil, heated the stream of steam and coal so as to distill the combustible volatiles from the coal. To effect this result the temperature within the pipe was kept above 700 F and below 1200 F. The products passed out of the lower, hottest part of the coil and entered a hot walled or heat-insulated cyclone dust collector which precipitated most of the coke and ash and permitted the volatile products to pass on to the condenser. Clogging of the pipes by any adhering coal was found to be entirely prevented by dropping pieces of iron such as small metal punchings into the pipe ahead of the coal injector. These effectively removed any adhering coal and also prevented scale from forming. The pieces of iron were easily separated from the volatilized coke residues and could be returned by mechanical devices for reuse. The latent heat released by condensing the volatiles from the above process is shown in the present disclosure as conserved or reused in producing a second supply of steam which is used for the complete gasification of carbonaceous particles so that the products of this second steam treatment may be used for distilling coarse coal. The new steam from the first unit may, of course, be used in a second similar unit for distilling coal. One of the economies resulting from the present invention is obtained by operating the above described carbonization and gasification process under suitable pressure so that the volatiles produced, together with any undecomposed steam, and oil vapors may be made to give up their sensible and latent heats to produce fresh steam at a lower pressure to be used in a succeeding unit operating at a lower pressure. The accompanying drawing which is an elevation with parts in vertical section shows diagrammatically an apparatus for carrying out the process.

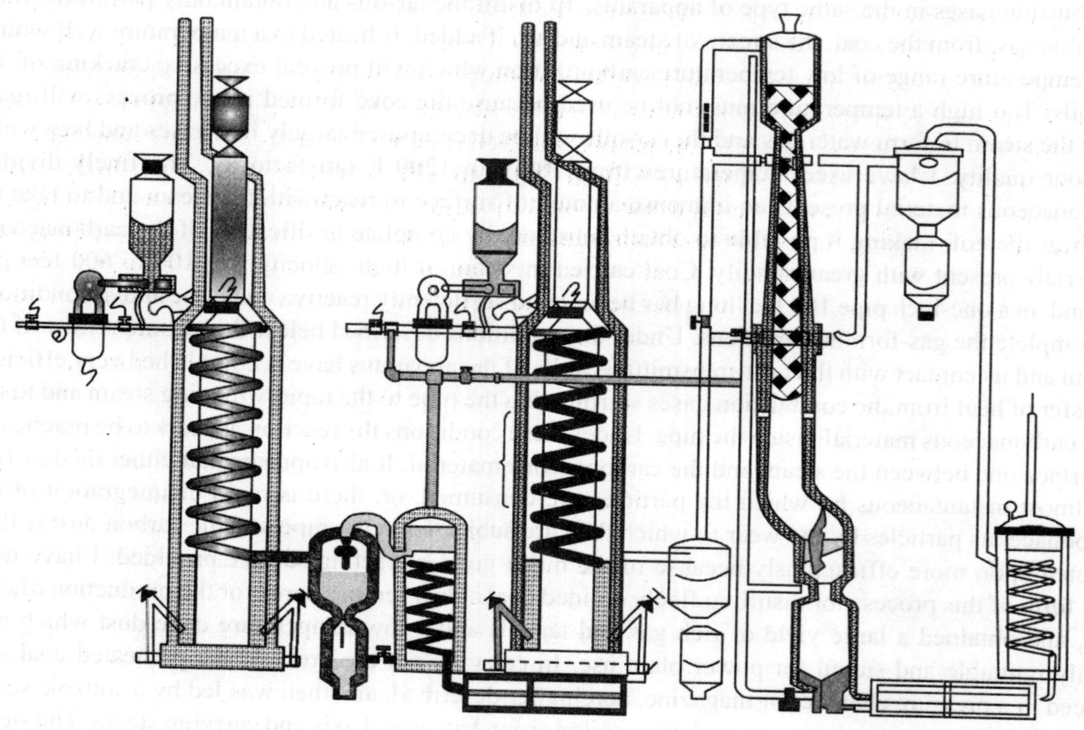

Fig. 20.4. Showing the Carbonization of Coal.

In the form of the invention here shown steam is used for distillation of coal in one retort, shown as a coil, the resulting vapors and gases are cooled in a heat exchange device and the fresh steam at lower pressure thus obtained is used with further additional heat to completely gasify other coal or coke in a similar retort. The hot gaseous products from this second lot of coal are shown as heating a retort through which sized coke is fed to produce low-temperature coke, a rich gas, and to obtain the condensable volatiles from it. The coiled pipe device for gasification was found to work satisfactorily with one-inch pipe of uniform diameter, so that the speed of the steam, owing to its increased volume on high temperatures, is increased as it flows through the pipe. Using dry steam at about 15 pounds pressure at a temperature of about 265 F the volume would be double in rising to 1000 F and treble in rising to 1700 F, The apparatus was found to work satisfactorily under these and other conditions. The pulverized fuel gasifier furnace 1 shown, consists of a heat-insulated outer wall 2, inside of which is a cylindrical center wall or core 3 having a pipe coil 4 supported in the annular space 5 between the wall and the core. Steam from a boiler or other source such as an evaporator enters the system 6 and passes through the steam trap 7 and valve 8 where the steam is throttled down to any desired pressure before passing into the coil 4. The steam passes down through the coil 4 to an exit 9 moving countercurrent to the combustion gases which ascend rapidly in the annular space 5. The required heat is supplied by the combustion of gases in the annular checkerwork 10, the gas and air being introduced by burners 11 at the bottom of the gasifier 1. The combustion gases pass out of the furnace 1 through the stack 12 which also serves as a drier and preheater for the pulverized material such as coal which is to be distilled. The pulverized material is charged into the feed bin 13 in the center of the stack 12 through the magazine 14 which with its top an bottom valves 15 and 16 serves as a lock or means of keeping the bin full at all times while preventing steam from lowing up through the bin 13. The bottom of the feed bin contains a power driven screw feed 17 having a speed regulating device 18 by which the dry and preheated material can be charged continuously at an rate in the steam line 19 entering the top of the gasifier 1. Pieces of iron which are used to scour the inside walls of the coil 4 may be fed at intervals into the steam line 19 from the valve-closed hopper 20 by means of a suitable feeding device such as a power driven crank and slide vlave 21 which contains an aprture to receive one device or more at a stroke. The exit gases carrying whatever there is of ash and the coke pass through the exit 9 and through the conduit 22 to the dust collector 23, which may be a suitable type of cyclone dust collector and is preferably well protected by heat insulation so that there is no separation of condensable vapors here. The solids may be withdrawn from the collector into a double-valved bottom bin 24 without material loss of gas or pressure.

The cleaned gases and vapors pass on through a conduit 25 to a coil or other cooling element 26 forming part of a heat exchange device which also serves as an evaporator. The pressure and therefore the temperature of the vapors in the conduit 25 will be so high that the water in the device 26 yields the steam used in a second coal heating device while the condenser vapors of water and tar, oils, resins, and so forth, together with uncondensed gases flow out of the coil 27 by a main 28 into a separator 29 without material reduction in pressure. In this it is found that four products separate when a coal such as Utah coal is used. The heaviest of these products is a layer of resins carrying wax, which is a sticky viscous mixture when cold, and which can be drawn off by a bottom valve 30. Above this lies a much deeper layer of water carrying principally ammonia and some tar acids which can be drawn off by a valve 31. Above the water floats a layer of oil and wax which when cold is of the consistency of cup grease and can be drawn off by a valve 32. Above the oil layer is a space which is occupied by the uncondensible gases carrying some light oil which may be removed by a scrubber and which can be drawn of through a valve 33. The steam generated in the heat exchange device 26 from the latent and sensible heats of condensed vapors, gases and water may be used for distilling a further quantity of coal in a second unit similar to the one already described. The invention may also be used for the carbonization of lump coal by the sensible heat of hot gaseous products obtained from the treatment of finely divided coal in one of the units mentioned above. I thus apply the invention by gasification of coal with the utilization of the sensible heat of the vapors and gases thus produced for the low temperature carbonization of a second quantity of coal which is preferably lump coal, thus providing for the complete utilization of all forms of the coal received by an ordinary commercial plant. For this purpose the steam generated in the heat exchange device 26 is carried by a valved heat-insulated main 34 through a steam trap 35 and past a valve 37 to gasifying coil 38 into which pulverized coal is introduced as into the coil 4. The coil 38 like the coil 4 lies in an annular space 39 between the heat-insulating outer wall 40 and the core 41. The coil 38 like the coil 4 is heated by burners 42 at the bottom so as to heat the lower part of the coil to the highest temperature while the products of combustion rise around the upper part of the coil and pass off through a stack 43 which, like the stack 12, surrounds a feed bin 44 fed by a magazine 45 similar to the magazine 14. There may also be provided a variable coal feeding device 46, and a variable feeding device 47 for feeding pieces of iron or other solid materials for maintaining the coil 38 free from carbon and scale on its inner surface. The exit end 48 of the coil passes into a cyclone heat-insulated dust collector 49, similar to the dust collector 23, which removes the hot solids from the gas and vapor stream. There is provided a double-valved bottom bin 50 for removing the solids collected in the dust collector 49. The heat insulated outlet pipe 51 of the dust collector 49 conducts the still very hot products of gasification into a retort diagrammatically sown at 52 where the gases are used to carbonize coal or other material. The retort may be 30 or more feet high and steam may be fed into it from a number of coils 38. The properly sized coal preferably free from fines, is fed into the tapered upper part of the retort by a valved opening 54. To enable the coal to be preheated it comes from a valve-closed bin 55 provided with heating means described below. For heating the coal in the retort by such gases as come form the main 51 two procedures have been found useful. According to one procedure the descending coal is heated by a counter-current up-flow of steam and gas. The principal heat may be supplied by the hot gases coming from the main 51. To effect this an annular manifold 56 surrounds the retort 52 at the large bottom end of its tapered portion 53 and through this the hot fluid gas and vapors coming from the main 51 are led by a valved connection 57. The entering gases will be about 1200 F to 1850 F and will distill off the volatiles in the coal with great rapidity while the steam reacts to a considerable extent with the resulting coke and by increasing the rate of flow of the coal so that it carries volatile oil ingredients into steam hot enough to crack them, the steam reacts with the carbon particles released by the cracking and forms water gas and gaseous and condensable light hydrocarbons. The volatile products and gases pass upwardly through the coal and pass out of a manifold 58 below the opening 54. Preferably there is a column of coal between the manifold 58 and the opening 54 high enough to provide a continuous supply of coal to the distilling zone while the valve 54 is closed and the bin 55 is being recharged. The coal may be preheated by passing part of the highly heated gases from the main 51 through a valved insulated pipe 59 into the bin 55. The gases pass down through the material I the bin 55 and out of the system at the manifold 58 thereby drying and preheating the coal and simultaneously serving the valuable functions of preventing tar and oil vapors from the distilling coal from entering the bin 55 and condensing there. The gases and vapors leaving by the manifold 58 enter a heat insulated and valved main 60 which convey them to a heat insulated cyclone dust collector 61 where the dust is removed. From this a heat insulated min 62 carries the vapors and gases almost at issuing temperature to an evaporator or heat exchange device 63 which may be like the device 26. Here the latent and sensible heat of condensable materials are recovered and the four products, resin, water, oil and gas, are separated in a separator 64 like the separator 29 and removed as desired by valves 65, 66, 67 and 68. In addition to the heat provided at the manifold 56 the contents of the retort are further heated by steam rising from the residues or form the coke formed. The lower part 70 of the retort may be cylindrical or straight instead of tapering. It may terminate in one of the customary forms having an elbow 71 forming a nearly horizontal support for the load of the charge in the retort. The feed of the contents past the elbow 71 may be controlled by any suitable means, and extend toward pivoted hanging arms 74 which detain the upper portion of the charge at the nearly horizontal section 71. After passing the elbow or horizontal portion 71 the residues or coke drop into a bottom bin 75 and rest on the closed bottom valve 76 thereof. A valve 77 between the bin 75 and lower part of the retort may be closed when it is desired to remove the contents of the bin 75 which required the opening of valve 76. The residues of coke lying upon the valves 76 and 77 can be used to heat gas, water or steam which may be provided by introducing the necessary gas, vapor, water or steam through valved pipes 78 and 79 which are connected by a main 80 to a suitable source of steam 81. The steam or other material introduced at 78 and 79 may be laden with admixed solutions of light-giving salts or odoriferous materials, thus enabling any desired properties to be given to the coke discharged. The heat of the coke or residues lying upon the valves 76 or 77 superheats the steam or vapors so that they aid in heating the charge in the retort thus recovering a very large proportion of the heat usually lost in carbonizing processes. The second procedure which has been found useful in heating the charge in the upper part 53 of the retort, is to close the valved connection 57 and open a valve 82 at the upper manifold 58 thus admitting steam from the main 51 to the upper manifold. The valve 33 in the main 50 is also closed, and a valve 34 in a branch 85 of the main 60 and extending to the lower manifold 56 is also opened with the result that heating steam and vapors from the main 51 enter the top of the retort and flow down with the charge of coal to the manifold 56 and then pass off to the branch 85 in the dust collector 61. It is found that rather more oil is obtained from a charge heating according to the latter method than is obtained when the charge and the retort is heated by a counter-current flow of steam and vapor. In order to preheat the charge in the bin 55 steam may be drawn either from a steam main 81 or from the steam and vapor main 51. There is provided in the main 80 leading from the main 31 a valve 86 which may be open to allow steam to flow up an extension 87 into the bin 55. As an alternative method of heating the bin 55 the valve 33 of the extension 59 of the main 51 may be opened. Either of these methods of heating the bin 55 serves to keep vapors and steam from condensing in the bin 55 as they arise from the upper part of the retort 53. The evaporators 26 and 63 are provided with valved cold water inlets 89 at their bottoms. The oils condensed by the cold water are easily separated into portions which contain a large proportion of various products. The amount and character of these products varies with the material treated. To obtain the above described separation without any separate operations for separating the materials the separators may consist of long, narrow, closed chambers in which the condensed material moves toward the outlet very slowly, with the result that stratification is effected as the streams approach the decanting valves. The gases leaving the separators are usually scrubbed to remove the uncondensed light oils and ammonia. It will be seen that the apparatus shown is one in which successive gasifiers work efficiently at successively lower pressures, thus one utilizes the latent heat and the sensible heat of steam, gases and superheated vapors from the preceding one, generating from 50% to 75% by weight as much steam as the weight of material so condensed. It has been found that when complete gasification temperatures are used very little coke collects in the dust collectors 23 and 49. For the convenience of operating the apparatus, steam gases are placed on each evaporator and at the outlet to each coil. In one low pressure test run substantially saturated steam was used at a pressure approximating 6 psi and into this was fed coal, all passing 16-mech and averaging rather smaller than 40-mesh at the rate of about a half pound of coal per pound of steam, the steam flowing at the rate of about 0.85 lb/minute. The coil of ne-inch pipe about 100 feet long was heated at the lower end by gases having a temperature of approximately 1830 F and the steam as it passed through the coil increased in temperature from the admission end where it was 230 F to about 1600 F. Gas was produced in excess of the capacity of the meter to measure and substantially all the coal was gasified. It was found that powdered low temperature coke from Utah coal instead of coal, gave a flow of gas which could be controlled instantly by varying the temperature or flow of steam or amount of coke dust supplied, so that the process was well adapted to produce gas for a rapidly varying demand. The ash did not clinker and the small amount of coal too coarse to completely gasify was swept out of the coil by steam. In other runs saturated steam was successfully used at 10 to 25 pounds pressure delivering about 1.5 pounds and 2.5 pounds of steam per minute respectively. Varying amounts of coal and coke were used.

Where substantial steam pressures are used they accelerate the water-gas reaction and also permit steam to be regenerated from the heat evolved in the condensation of undecomposed steam and from the vapors from one charge as well as from the sensible heat of the gases formed, and thereby another charge may be distilled by the fresh steam. The gas formed is rich in ammonia since the best conditions exist for the conversion of nitrogen compounds of the coal into ammonia. When steam was used at 25 pounds pressure and the flow was 2.5 pounds per minute, the steam at the exit end of the coil was 1300 F and the combustion gases in contact with the heated coil were at 1750 F the steam reached the exit end at 1480 F and when the combustion gases were at 2000 F the steam issued at 1800 F.

Carbonization of Bituminous Coal

The products of carbonization of bituminous coal are shown in the table below:

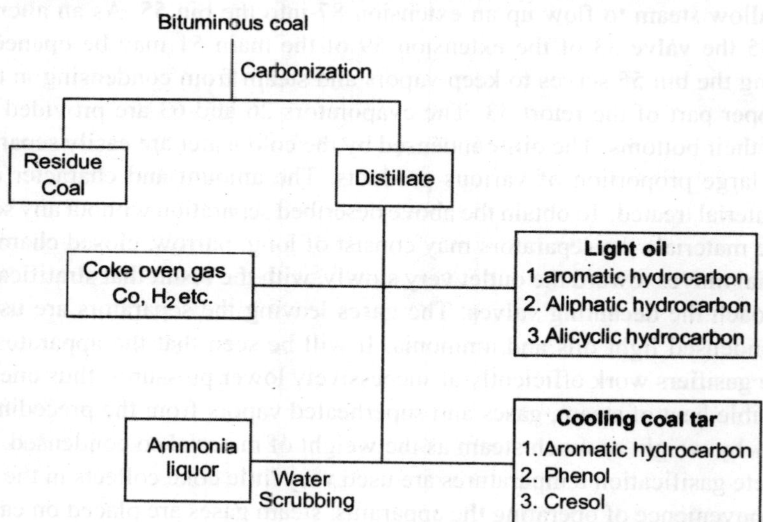

The exact nature and properties of the by products obtained by carbonization of coal are determined by the temperature conditions employed for the process. Actually, carbonization is carried in two ways, depending upon whether coal or the non-volatile matter are required as the main product.

- Low Temperature Carbonization
- High Temperature Carbonization

The by products from the either process are of same type, are indicated in the chart given above although their properties will vary with the process and the type of coal carbonized. For example one ton of bituminous coal yields various products in the following proportions

	I	II	III	IV	V
Low temp carbonization	Coal tar 17 – 90 gal	Ammonia 2.5 lb	Light oil 2.5 – 3.5 gal	Coke oven gas 4000 scf	Coke 0.75 tons
High temp carbonization	6 – 7 gal	3 – 4 IB	3 gal	10, 000 scf	0.7 – 0.8 tons

Coal tar: It is obtained in much greater properties from low carbonization which is excellent source of naphtha and small amounts of numerous aromatics from high temperature carbonization coal tar is obtained in poor yield and small amounts of aromatics can be recovered from it.

Ammonia: High temperature carbonization yields less of ammonia gas ammonia gas is converted into the fertilizer ammonium sulphate by reaction with sulphuric acid and subsequent separation of the solid substance in the centrifuge.

Light Oil: The light oil obtained from high temperature carbonization is a rich source of benzene (72%) toluene (13%) and xylene (4%) BTX is the name of the mixture of these hydrocarbons.

Light oil derived from low temperature carbonization on the other hand is rich in alkanes (46% vol) alkanes (16%) cycloalkanes (8%) cycloalkenes (9%) and also contains some aromatics (16%)

Coke Oven Gas: low temperature carbonization gas contains mainly methane and higher alkanes (65%) although hydrogen (10%) carbon monoxide (5%) carbon dioxide (9%) etc. It can also be used for supply as coal gas for domestic connection. High temperature carbonization gas is made of hydrogen (50%) alkanes (34%) carbon monoxide (8%) carbon dioxide (3%) etc. This can be used as reducing fuel gas in metallurgical operations.

The composition of coal tar and light oil produced by low temperature and high temperature carbonization of one ton bituminous coal

Product	Low temp carbonization		High temp carbonization	
Coal tar	Aromatic hydrocarbon (BTX)	<0.5%	Aromatic hydrocarbon(BTX)	0.6%
	Cresols	3.5	Cresols	1.0
	Xylenols	6.5	Xylenols	0.5
	Other phenols	13.0	Other phenols	1.5
	Naphtha	36.0	Naphthalene	8.9
	Other aromatics	3.0	Other aromatics	10.0
Light oil	Alkanes	46 vol%	Benzene	72%
	Alkanes	16	Toluene	13
	Cyclo alkanes	9	Alicycles	5
	Aromatics	16	Aliphatics	6
	Others	5		

Coke: Low temperature carbonization coke contains 8 – 20% volatile matter and is used as smokeless domestic fuel. The coke produced by the high temperature carbonization process is highly used for metallurgical purpose.

As clear from the above coal tar and light oil are product of coal carbonization which are infact the effective source of aromatic hydrocarbons and their derivatives. It will be worthwhile to have knowledge of the relative amounts of aromatics available in the low and high temperature carbonization process.

Coal Gas Manufacture and Recovery of Aromatics

As stated earlier coal gas can be produced by low temperature carbonization of bituminous coal. It was long used as a domestic fuel. Since the availability of bottled 'natural' or refining gas the carbonization as has because obsolete in advanced petroleum countries like USA etc. However the present petroleum crisis, particularly in India has once again derived the attention to coal gas. At present only the cities of Bombay and Kolkata have gas plants that supply gas as a domestic fuel as a very limited scale to have plants supply coal gas to families in Calcutta. The plant used for manufacture of coal gas is shown in the Fig. 20.5.

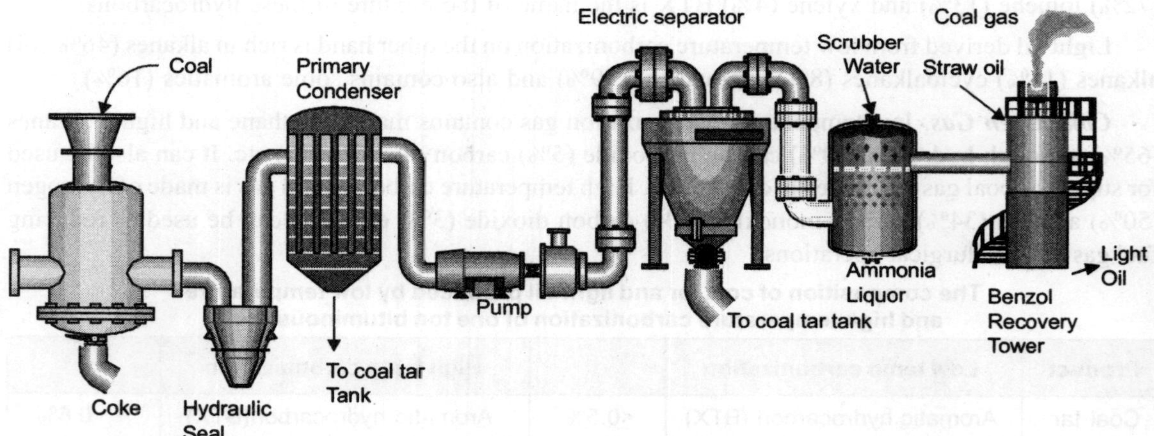

Fig. 20.5. Showing the separation of aromatics and light oil from coal far

The coal is treated in a large vertical iron retorts at about 600°C by burning producer gas. The gas (Hydrogen, methane, carbon monoxide, carbon dioxide acetylene, benzene ammonia and hydrogen sulphide) water vapours of aromatics coal dust etc pass out of the retort. The residue in the retort is coke used as domestic fuel.

The gases vapours and dust first passes through the HYDRAULIC SEAL where they are practically cooled and tar separated to some extent. The purpose of the seal is to act as valve and therefore not to allow the gases and vapours to return to the retort when it is opened for cleaning. The gases the volatile organic matter and coke dust are then passed through PRIMARY CONDENSER fitted with cold water pipes. Tar is condensed here and collects at the bottom and led to the tar tank. The remaining gases vapours and smoke are made to pass through a PUMP which helps them circulating. The ELECTRICAL SEPARATOR precipitates smoke and particles of organic matter which settle down and are taken to coal tar tank. The uncondensed gases that escape.

The uncondensed gases that escape from the electrical separator are scrubbed with water by a ROTARY SCRUBBER remaining ammonia liquor. This is latter reacted with sulphuric acid to produce ammonia sulphate.

The cooled gases free from ammonia are then scrubbed with straw oil in BENZOL RECOVERY TOWER. The light oil collecting at the base of the tower is fractionated to recover benzene, while coal gas leaves near the top through pipes. Before distribution for domestic use, the coal gas is passed over iron oxide to remove hydrogen sulphide gas and produces sulphur dioxide which is poisonous.

Fractional Distillation of Coal Tar

Practically all tar in USA and in most other countries is produced by high temperature carbonization in coke ovens by the tar is burnt as fuel in furnaces, while the remaining 84% is processed for the recovery of aromatics.

Coal gas is made free from sulphur by passing over hot iron oxide it produces sulphur dioxide while burning.

The dark brown sticky liquid called coal tar is an aromatic hydrocarbon, phenol, bases etc. The first step in the separation of coal into components is distillation in fractionating column four main usually collected leaving behind a residue of pitch, which is mainly used for road surfing.

Table: Fractions obtained in Coal Tar Distillation

Fraction	Temperature	Percentage above by vol	Chief constituents
1. Light oil	<70°C	5	Benzene, toluene
2. Middle oil	170-230°C	7.5	Xylene
3. Heavy oil	230-270°C	10	Phenols, cresols, naphtha
4. Anthracene oil	270-400°C	20	Cresols, naphthalene
5. Pitch	residue left	57.5	Anthracene

The above fractions collected at different temperature worked up for the recovery of their aromatic components.

Fraction I – Light Oil

The fraction upto about 170°C is called light oil, because it is lighter than water. Besides the hydrocarbons, benzenes toluene and xylene (BTX) light oil also contains traces of acidic substance phenol and cresols and basic substances pyridine and methyl pyridines.

The main constituent of light oil are stated below: -

To isolate the above components light oil is washed conc. Sulphuric acid which removes alkanes basic substances such as pyridine and Thiophene. This is followed by treatment of the oil with dilute

Neutral

benzene toluene p-xylene 1H-pyrrole

Acidic

phenol m-cresol

Basic

pyridine 2-methylpyridine

sodium hydroxide which removes acidic compounds phenols. In fact the sequence of operations is as follows:

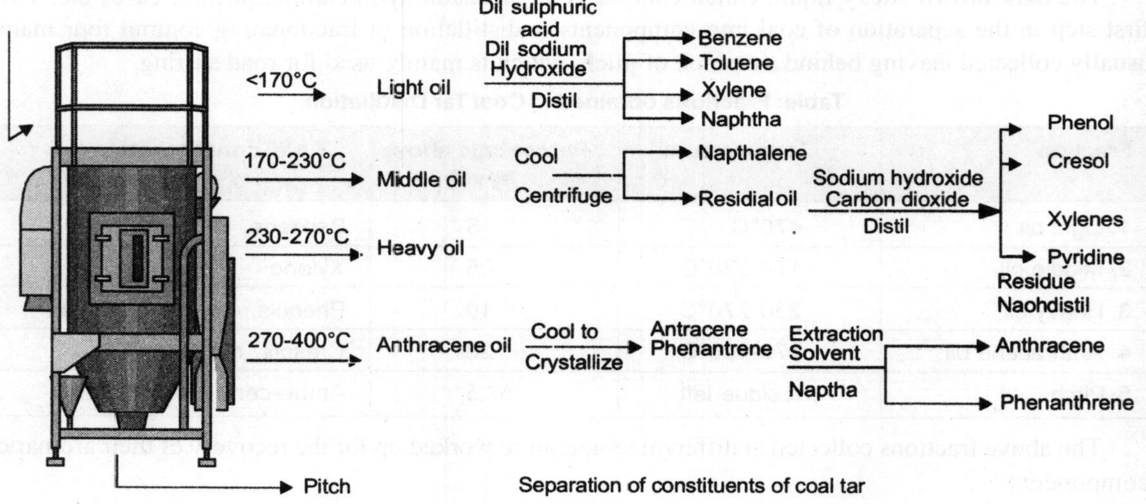

Separation of constituents of coal tar

Fig. 20.6. Showing the separation of constitution of Coal Tar.

The purified oil is then desired and subjected to further fractionation to get the following products.

(i) benzene (ii) toluene (iii) a mixture of o, m and p xylenes and (iv) a residue solvent naphtha. The solvent naphtha consisting of cumene and higher benzene homologous e.g. mesitylene is used, almost exclusively as solvent for paints resin rubber etc.

Fraction – II Middle Oil

The fraction collected between 170°C - 270°C is called middle oil (being in the middle of the light oil and the heavy oil fractions) or carbolic oil. It consists of naphthalene, phenol or carbolic oil, cresols pyridine and methyl pyridine.

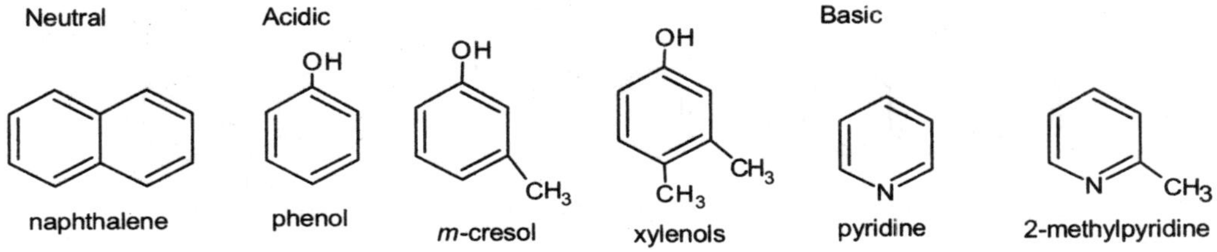

Neutral: naphthalene
Acidic: phenol, *m*-cresol, xylenols
Basic: pyridine, 2-methylpyridine

On cooling the middle oil, crystals of crude naphthalene are obtained as deposits which are removed by centrifugation. Naphthalene thus obtained by purified by treating, while molten with aqueous NaOH aqueous H_2SO_4 and water. Thereafter it is dried and finely sublimed to get naphthalene.

The oil from which naphthalene is separated is treated with worm aqueous sodium hydroxide to remove phenols (acidic substances). The resulting solution is saturated with carbon dioxide which sets free the phenols. The mixture of phenols is wasted with water and fractionally distilled to yield phenol a mixture of isomeric cresols and xylenols. All these products are valuable disinfectants and are also important intermediates for the manufacture of industrial chemicals e.g. salicylic acid, aspirin, phenacetin many dyes and explosives.

After the extraction of the phenols the remaining oil is washed with dilute sulphuric acid which removes tar bases as these as their salts. The resending salt solution is treated with NaOH and distilled to obtain pyridine. The residual oil is mixed with heavy oil fraction.

Fraction – III Heavy Oil

This fraction of coal tar is also called creosote oil obtained between 230 - 270°C. Since it is between 230 - 270°C and also it is also heavier than water it is named as heavy oil. The mail constituent in this heavy oil are:

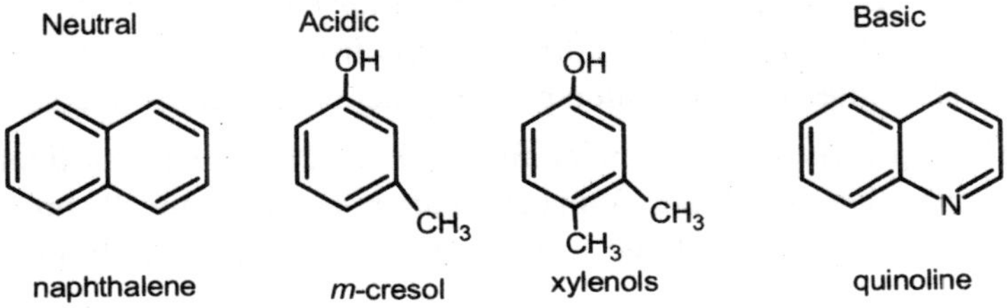

Neutral: naphthalene
Acidic: *m*-cresol, xylenols
Basic: quinoline

Heavy oil may be treated in such the same way as the Middle oil yield naphthalene, cresols, qunolines. Since these compounds can be conveniently produced by simpler synthetic methods, this fraction is not worked for their recovery. Heavy oils find a principal use as a wood preservative under the name creosote oil.

Fraction–IV Anthracene Oil

This fraction distilling between 270 - 400°C gets its name from its chief ingredient Anthracene phenthracene and carbazole are the two other neutral components of anthracene oil.

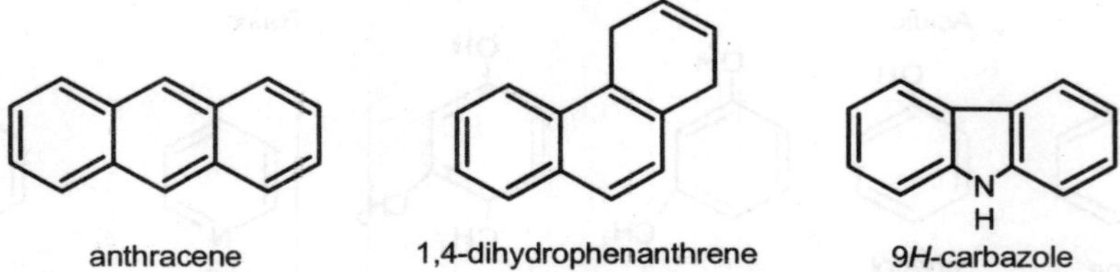

Since the oil shows green florescence, it is also called ***green oil***, for isolating anthracene, the oil is run into tanks and allowed to cool. The crystals of anthracene are formed. Which are separated from the oil by filtration under pressure. On digesting there with solvent naphtha phenanthrene dissolves in preference living behind anthracene and carbazole. The solid mass is then extracted with pyridine to remove carbazole and the residue is submitted to yield 85 – 95% pure anthracene. This hydrocarbon is the starting material for alizarin dyes.

Fraction – V Pitch

The residue left is called pitch. Although most of it is carbon (42 – 94%), It contains a number of five and six membered fused ring compounds. It also contains some percentage for tar oils which determines the softening temperature rages of the pitch. Pitch is used for (i) making varnishes and water proofing for roofs (ii) for making acid resistant shine ware (iii) as pulverized fuel

Aromatics from Petrolium

As mentioned earlier, benzene, toluene, xylene and several other aromatic hydrocarbons, were originally obtained as a by product in the manufacture of coke for steel. This was the only source of all aromatic compounds until 1940. In the past 30 years the demand for aromatic compounds has far out shipped the aromatic hydrocarbon is now petroleum industry. For illustration in 1969 the petrochemical benzene as was in 1061 gallons as compared to 93 million gallons from coal tar. Petronaphthalene production was 353 million pounds while only 45 million pounds where obtained from coal tar.

Aromatics are also present in the naphtha about 10 to 12% in naphtha fraction (40 - 100°C) of petroleum. Straight separation of aromatic from petroleum is not economically profitable. The following process is generally employed for the large scale production of aromatics from petroleum.

(i) Catalytic Reforming: it is the process of converting $C_6 - C_8$ aliphatic hydrocarbon present in petroleum naphtha into aromatic hydrocarbons. The $C_6 - C_8$ fraction of light naphtha at 500°C and 25 – 35 atmospheric pressure over a platinum alumina catalyst gives a 45 – 55% yield of aromatics such in benzene toluene and xylene.

$$\underset{\text{hexane}}{CH_3CH_2CH_2CH_2CH_2CH_3} \xrightarrow{\text{cyclisation}} \underset{\text{cyclohexane}}{C_6H_{12}} \xrightarrow{\text{aromatization}} \underset{\text{benzene}}{C_6H_6}$$

$$\underset{\text{2-methylhexane}}{C_7H_{16}} \xrightarrow{\text{cyclisation}} \underset{\text{methylcyclohexane}}{C_6H_{11}CH_3} \xrightarrow{\text{aromatization}} \underset{\text{toluene}}{C_6H_5CH_3}$$

Middle distillates from petroleum may be similarly reformed to naphthalene. Resulting a new catalyst platinum – rhenium – alumina has been introduced with functions. Satisfactorily at 10 to 20 atmospheric pressure and increases the yield of aromatics 25%. An illustrative diagram of benzene – toluene – xylene (BTX) is given in the fig. 20.7.

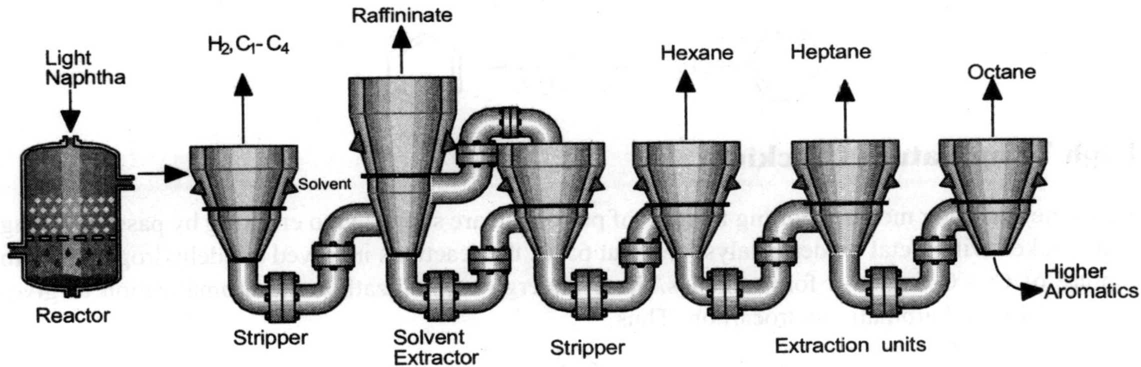

Fig. 20.7. Showing the Catalytic Reforming Process

Light naphtha fraction ($C_6 - C_8$) of petroleum is first passed through a reformer having a fixed bed of platinum – alumina catalyst.

Here the higher alkanes present in naphtha undergo reactions (cracking, crystallization, aromatization etc.) to reform benzene and its homologous ($C_6 - C_8$). The product mixture them undergoes to the STRIPPER, where the gaseous components (Hydrogen, Carbon 1 – carbon 4) are removed. The mixture freed from gaseous is treated with a suitable solvent to extract the aromatics with which it forms less volatile azeotropic mixture. The relatively volatile non-aromatics (reffinate C_5 alkanes) escape at the head of the SOLVENT EXTRACTOR). The most commonly used solvent for the extraction is diethylene glycol water mixture (under process). The extract containing the aromatics is sent to the stripper for the recovery of the solvent which is reused. The extract freed from the solvent is finally fractionated in the SEPERATOR UNITS for the recovery of benzene, toluene and xylenes. The separation of isomeric xylene is difficult and is accomplished by the following process.

Isomeric Xylenes —Distil→ Distilate → O - xylene; Residue → M - xylene + P - xylene → Solid → M - xylene; Mother Liquor → P - xylene

Benzene and toluene are used as raw materials for the preparation of a large number of derivatives drugs, explosives, plastics etc. The o and p – xylenes are convenient starting points for the manufacture of pthalic and terepthalic acids.

The mixture of BTX obtained by the selective extraction described above yields benzene, toluene and xylene but the yield of benzene is far less than the other two components. Since benzene is the most needed aromatic, toluene and xylenes are converted into benzene by heating with hydrogen obtained from stripper in the presence of chromic oxide catalyst. This reaction can be illustrated as follows:

$$C_6H_5-CH_3 + H_2 \xrightarrow[Cr_2O_3]{\Delta} C_6H_6 + CH_4$$

High Temperature Cracking

Kerosene and other medium boiling fraction of petroleum are subjected to cracking by passing through stills packed with metal oxides catalyst at about 65°C. the reactions involved are dehydrogenation and rupture of C – C bonds to form alkanes.These undergo crystallization and aromatization to give a mixture of liquid aromatic hydrocarbon. Thus,

$$\underset{\text{buta-1,3-diene}}{CH_2=CH-CH=CH_2} + \underset{\text{ethylene}}{CH_2=CH_2} \longrightarrow \underset{\text{cyclohexene}}{C_6H_{10}} \xrightarrow{Cr_2O_3} \underset{\text{benzene}}{C_6H_6} + 2H_2$$

Polymerization

Acetylene a petroleum product, when passed through a red-hot brick chequer work surface, polymerizes to form benzene.

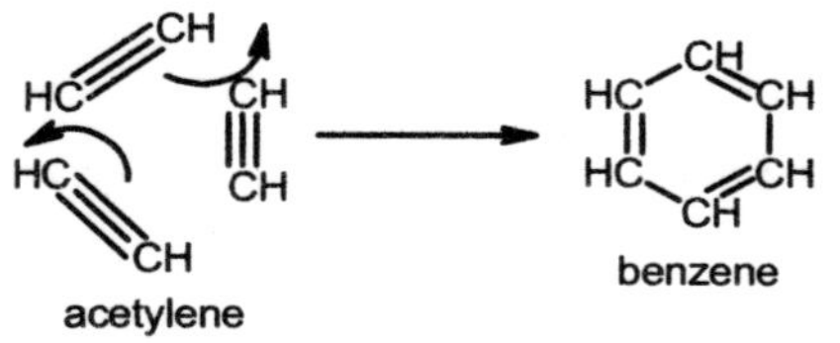

HC≡CH, HC≡CH, HC≡CH → benzene

acetylene benzene

Acetylene in turn is produced by passing rapidly a liquid petroleum fraction through the brick chequer work in the first step of the process.

Alkyl Benzene Syntheis

A number of higher homologous of benzene have been very recently obtained by synthetic reactions from petroleum products, for example.

(a) Ethyl benzene Synthesis : When a mixture of benzene and ethylene derived from petroleum source is passed over liquid $AlCl_3$ – HCl catalyst at 95°C and 5 psi pressure ethyl benzene is formed.

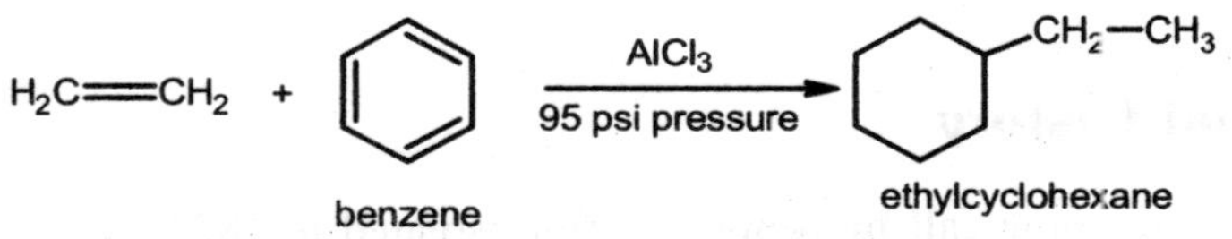

$$H_2C{=}CH_2 + \text{benzene} \xrightarrow[\text{95 psi pressure}]{AlCl_3} \text{ethylcyclohexane } (C_6H_{11}{-}CH_2{-}CH_3)$$

benzene ethylcyclohexane

(b) Cumene Synthesis : Isopropyl benzene or cumene is synthesised from benzene and propylene (obtained from petroleum) by using phosphoric acid as the catalyst at 250°C.

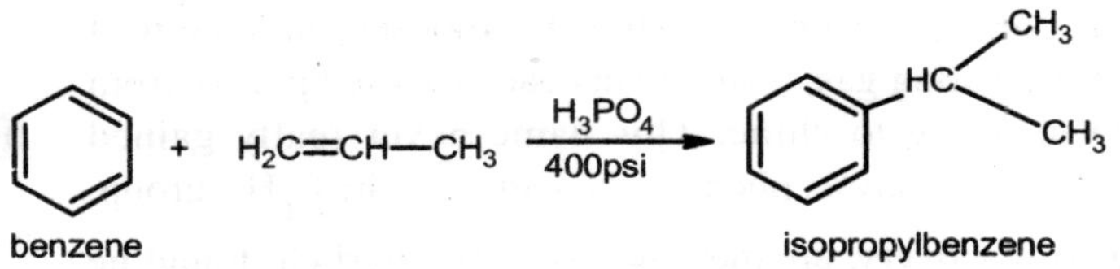

$$\text{benzene} + H_2C{=}CH{-}CH_3 \xrightarrow[\text{400psi}]{H_3PO_4} C_6H_5{-}CH(CH_3)_2$$

benzene isopropylbenzene

CHAPTER

21

Benzene

Introduction and History

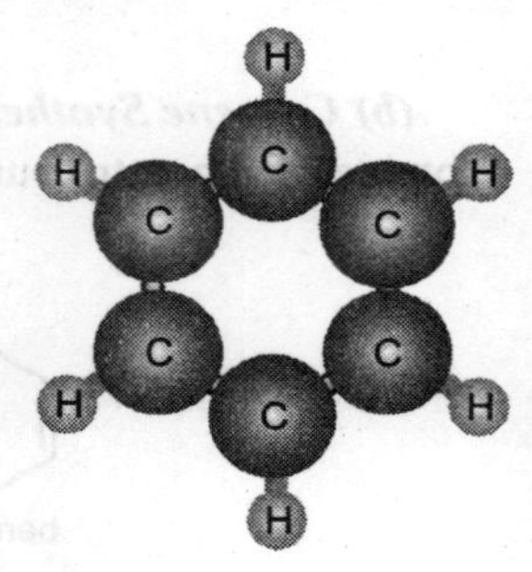

The hydrocarbon that we now call **benzene** was first isolated in 1825 by **Michael Faraday** from an oily film that deposited from the gas used for lighting. Faraday did some experiments, and discovered that the new compound had equal numbers of carbons and hydrogens, and so named it 'carbureted hydrogen'. Another chemist, **Laurent**, proposed that due to it being discovered in illuminating gas, it should instead be called pheno, from the Greek *pentaetia* meaning to shine. This name never really gained acceptance, but persists to this day as phenyl - the name for the $C_6H_5^-$ group.

Nine years after its discovery, another chemist, Mitscherlich, found he could produce the same substance by heating a chemical that had been isolated from gum benzoin - so he decided to call the compound benzin. Other chemists rejected this name because it implied the compound was similar to alkaloids, such as quinine. Another suggestion was the German name, benzol, from the German öl, meaning oil. But in France and England the name benzene was used instead, to avoid the -ol ending confusing it with an alcohol.

Preparation of Benzene

(I) ***From Alkene***: Benzene can be prepared by polymerization of approximate alkanes acetyls will polymerize at high temperature to yield arenes. Thus

$$3H—C\equiv C—H \xrightarrow{500°C} \text{benzene}$$

acetylene benzene

Recently catalysts for e.g. cobalt carbonyls and other metal complexes have been found to catalyze the trimerization of acetylene in solution at low temperature. Thus hexamethyl benzene has been prepared from 2-butyne in the presence of dimethyl cobalt.

H_3C CH_3 dimethyl alcohol → hexamethylbenzene

2 butyne (3 mols)

very recently this method has been used for the large scale production of the benzene from acetylene derived from series.

(2) Decarbooxylation of Aromatic Acids: benzene can be prepared by heating aromatic acids or their sodium salts with soda lime.

(benzoato-κO) sodium sodium benzoite + NaOH → benzene + Na_2CO_3

(3) Cyclisization and Aromatization of Long chain alkanes: Alkanes are now synthesized on a large scale by passing the vapour of normal alkanes containing six to nine carbons over a metal catalyst (platinum spattered on aluminium oxide) at 500°C. The reaction first involves cyclization and is followed by aromatization by loss of hydrogen.

hexane $\xrightarrow{pt\,Al_2O_3}$ cyclohexane $\xrightarrow{pt\,Al_2O_3}$ benzene

Recovery of Light Oil from Coke-oyen Gas

Light oil is obtained from coke-oven gas by scrubbing with wash oil; see Figure 21.1. The petroleum wash oil, or" "straw oil" as it is often called. absorbs the light oils from the gas more or less completely as requirementl dictate. Straw oil is a fraction obtained from crude petroleum and commonly has a specific gravity between 0.830 and 0.875 at 15.5°C. Typical distillation specifications call for an initial boiling point of approximately 285°C.; and 95% distilled to 350°C. In European practice, where petroleum distillates are more costly, various coal-tar distillates have been used as wash oils, for example, creosote oil. Ctesylic acid has also been used for the same purpose. Solid adsorbents such as silica gel and activated carbon have been used commercially instead of wash oil. In the U.S. petroleum distillates are most frequently used;

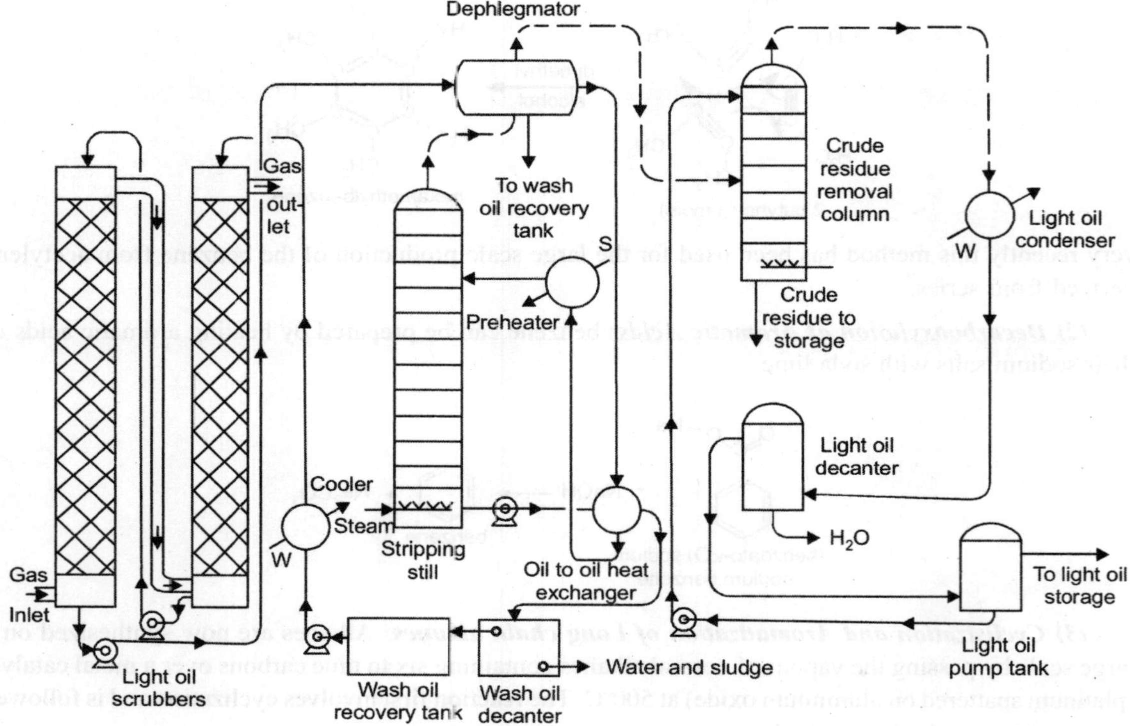

Fig. 21.1. Showing the Removal of the light oil from coke-oven gas

The scrubbing is done in a series of steel tower-packed with wood grids. The tower is equipped at the bottom with a gat-; inlet and a wash-oil outlet, at the top with a gas outlet and a wash-oil inlet. They are 8-12 ft. in diameter and range from 60 to 100ft. in height. Two or more towers arc usually used in series with countercurrent flow of gas to wash oil. Thus maximum opportunity is, afforded for the wash oil to absorb the light oil of the coke-oven gas to the desired degree.

These towers or "scrubbers"' operate at their greatest efficiency when gas and oil temperatures are at about 20°C. At higher temperatures it is necessary to increase circulation of the wash oil to maintain efficiency. There are other conditions as well that affect recovery of the light oil, often adversely, such as sludge formation. Sludge, which deposits on heat exchangers, may result when high-boiling compounds and tar are removed from the gas or it may be due to polymerization. Many procedures have been proposed for preventing or removing sludge, including use of antioxidants, electrostatic precipitators, centrifuges, filters, and settling tanks. The methods used are generally characteristic of individual plants.

The benzolized wash oil" leaving the last tower is pumped through a heat exchanger, or otherwise heated by steam or recovered heat. It is then charged into a stripping still at one of the upper plates and allowed to flow downward from plate to plate. Columns for this purpose vary in diameter from 6 to 8 ft. and may contain 1020 plates. Low-pressure steam (10-15 p.s.i.g.) enters the bottom of the column and flows upward, effecting a steam distillation by which the light oil is stripped from the benzolized wash oil. Light-oil vapors and steam are passed through a dephlegmator and a crude-residue removal column and are then condensed together, and the water layer is separated. The recovered light oil is pumped to storage. Hot wash oil leaving the bottom of the column is partially cooled in the heat exchanger by the benzolized wash oil from the scrubbing towers, and then further cooled in atmospheric coolers, before being recycled in the scrubbing system.

Coke-oven light oil is further processed to refined benzene and other aromatic compounds. In the U.S., this is usually done at the coking plant.

The composition of the light oil is governed by the characteristics of the coal coked and the coking temperature. Fast coking times result in higher flue temperatures and in little or no paraffin content in the light oil. Light oil produced from coking the same coal at lower flue temperatures or longer coking times has higher paraffin content. A difference of from 0 to 7% paraffin content may result. Most coke ovens in the U.S. operate at high temperatures. The boiling range of crude light oil reaches a maximum of approximately 200 to 215 degrees Centigrade. Although the composition varies from plant to plant, Table III may be considered to indicate a typical analysis showing the major constituents.

Table III. Analysis of Coke-Oven Light Oil

Fraction	Per cent by volume
Forerunnings (cyclopentadiene, other low-boiling hydrocarbons, carbon disulfide, etc.)	3
Benzene	65
Toluene	14
Xylenes, cumene, etc.	12
Cumarone, indene, naphthalene, heavy oils, etc.	6

Table IV. Analysis of Carbureted Water-Gas Light Oil

Fraction	Per cent by volume
Pre-benzene (low-boiling)	1.5
Benzene	32.0
Toluene	25.0
Xylenes, cumene, etc.	21.0
High-boiling (including unsaturates)	20.5

It will be noted that the major component of light oil is benzene.

The potential yield of light oil per ton of bituminous coal varies widely, depending upon the quality of the coal, design and conditions of the ovens, oven temperatures, coking time, and type of scrubbing equipment. A representative figure would be 3 gal of light oil per ton, which means a potential, on the average, of 1.95 gal. of benzene per ton of coal charged. For a discourse on light oil processing in the largest plant of its kind in the world,

Carbureted water-gas light oil contains a higher percentage of unsaturated compounds than coke-oven light oil. These consist of styrene, methylstyrene, indene, arid other unsaturated. The composition depends on the method of carbureting and the character of the petroleum oil used, and varies considerably at different plants. A typical analysis of carbureted water-gas light oil is shown in Table IV.

Recovery and Refining of Benzene from Light Oil

Although special grades of benzene may be produced for particular uses, the major proportion of crude benzene in the U.S. is usually processed into 5 commercial grades. In decreasing order of benzene content these are: (1) thiophene-free benzene, (2) nitration-grade benzene, (3) industrial-grade benzene, (4) industrial 90% benzene, and (5) motor benzol. Specifications for these grades are shown in Table V. Thiophene-free benzene, having' the same characteristics as nitration-grade, is in addition treated to remove thiophene and carbon disulfide. Motor benzol is a semi-refined mixture of benzene. and its homologs. The specifications for motor benzol, used for blending with gasoline, would permit the use of anyone of the other grades.

The quality of the product desired determines the method and extent of processing of the crude benzene-containing oils. Crude benzene contains compounds whose distillation temperatures lie close to that of benzene, so that separation by fractionation alone would be expensive, if not impossible. Some of these compounds are unsaturated hydrocarbons easily polymerizable by 66°Be. sulphuric acid (93%), others are compounds so affected by the strong acid that they no longer resist separation by fractionation, and some are soluble in and removable by the sulphuric acid. Hence the refining process consists of two distinct operations: (1) treatment of crude benzene with concentrated sulphuric acid, and separation of the acid; and (2) fractional distillation of the acid-treated crude. The operations may be either batch or continuous, depending upon economic factors such as the quantity of crude available and the number of products regularly produced.

Acid Washing. At most of the smaller plants, the light oil is acid-washed before fractionation, or else after a simple distillation. that removes the residue. At some plants the light oil is fractionated into several crude cuts or fractions: fore runnings, which are not acid-washed; crude benzene, which is washed; motor benzol, which usually requires no washing; crude benzene-toluene-xylene, which is washed; .crude light solvent; which may be washed in certain cases; and a residue, which is not - washed.

Sulphuric acid (66°Bé.) is the usual washing medium because of its economy and ease of handling. The washers may be large, mechanically agitated, cast-iron or steel tanks having conical bottoms.

Lead-lined agitation tanks may also be used. In most plants washers are equipped with cooling coils in order .to control the temperature rise during washing operations.

The washers are charged to capacity (for example 5000 gal.) and 20 gal. of. 66°Bé sulphuric acid is added from the acid .measuring tank. The mass is agitated 30 minutes, and then allowed to settle for 30 minutes and the acid sludge drawn off through a small discharge valve at the bottom of the cone. This light-acid wash removes substantially all water from the charge; Approximately 150 gal. of 66°Be. sulphuric acid is added and the mass agitated for 40 minutes, then allowed to settle 40 minutes before drawing off the acid sludge. At this point the acid- washed oil may be tested in the laboratory to

Table V. Benzene Specifications.[a]

Grade	$d^{15.5}_{15.5}$	Color	Distillation range, 760 mm. pressure	Solidifying point	Acid-wash color, not darker than	Acidity	Sulfur compounds	Copper corrosion
Thiophene-free benzene	0.882-0.886	Not darker than a solution of 0.0030 gram of K2Cr2O7 in 1 litre of water	Total not more than 1°C, (anhy-cluding 80.1°C.	Not lower than 5.0°C. (anhy-drous basis)	No. 2 color standard	No free acic	Free of H_2S, SO_2, thio-phene, and CS_2	Copper strip shall not show iridescence, nor a gray or black deposit or discoloration
Nitration-grade benzene	0.8820-0.8860	"	Total not more than 1°C, in-cluding 80.1°C	Not lower than 5.0°C. (anhy-drous basis)	No. 2 color standard	"	Free of H_2S and SO	"
Industrial-grade benzene	0.875-0.886	"	Total not more than 2°C, in-cluding 80.1°C	—	No. 3 color standard	"	"	"
Industrial 90 benzene	0.870-0.886	"	First drop not below 78.2°C. Percentage re-covered to 100°C, at least 90. Dry point, not above 120°C,	—	No. 6 color standard	"	"	"
Motor benzol	Not lower than 0.870	"	First drop at 76-82°C Percentage recovered to 100°C, at least 60. Per-centage re-covered to 120°C, at least 90. Dry point, not above 170°C.	—	No. 12 color standrd	"	Total sulfur con-tent, not more than 0.40%	Copper strip may show iri-descent or "peacock" discoloration, discoloration, but not gray or black dis-coloration or deposit

[a] For reference to methods of testing see page 438.

determine whether another acid wash is necessary. When tests show that the acid treatment is completed, 100-150 gal. of caustic soda solution (6-10% NaOH) is added in order to neutralize residual acid. Neutralization is usually complete after 20-30 minutes of agitation. The soda sludge is drawn off after settling for 30-45 minutes. The oil is finally washed by agitation with water, which is drawn off after a settling period. The washed and neutralized light oil is pumped to storage. Approximately 0.5 lb. of sulphuric acid is required to wash one gal of light oil. In addition to the action of the acid on the unsaturates, some of the ardmatics are sulfonated, especially if the temperature becomes high (55-60°C.). Best yields are obtained at temperatures below 50°C. A good overall yield would be approximately 95% based on the charge to the washer. The crude benzene and benzene-toluene- xylene fractions are acid-washed in substantially the same manner as described above before fractionating into refined products.

Although at many plants sludge is either burned or dumped in an isolated location, some plants are equipped for the recovery of sulphuric acid from the sludge. One process for such acid recovery is the **Ufer process**. In this process the light oil is washed with larger quantities of sulphuric acid of special density, and before withdrawing the sludge, a definite quantity of water is added and the mass is agitated. The resinous components contained in the strong acid sludge are liberated and dissolve in the oil. The diluted acid (approximately 40oBe.) is drawn off after settling and may be used without further treatment in the ammonium sulfate recovery plant. High acid recovery, low caustic consumption, and high oil yields are claimed for the process. Resins dissolved in the washed light oil may be recovered from the final still residues. The dilution of the strong sulphuric acid necessitates the use of lead-lined equipment.

Distillation. Because of the relatively small quantity of light oil handled at the average plant (7000-15,000 gal. per day) and the wide variety of benzene products, most plants are equipped with batch distillation units. The, characteristics of a typical distillation unit follow: still material, steel; capacity, 11,000 gal team coil area, 200 sq.ft.; *column:* material, cast iron; diameter, 5-0 ft., number of plates, 25-30; plate spacing, 16-18 in.; *condenser:* type, submerged coil; area, 300 sq.ft.

These units operate at atmospheric pressure with a linear vapor velocity in the column of 1.5 ft. per second. The boil-up rate for a 5-ft.-diameter column is equivalent to about 2400 gal. per hour of benzene, with the pressure drop across the column 4-6 in. Hg.

At the start of a distillation cycle for nitration-grade benzene the valves are set to return all condensate to the column. The still is then charged with 11,000 gal. of light oil, including any intermediate fractions produced during the previous cycle. The steam may be turned on in the steam coils as soon as they are covered by the charge. From 2 to 3 hours is required to heat the charge and maintain vaporization at capacity boil-up rate. At total reflux, the columns will be well filled with light-boiling materials and non-condensable will escape at the condenser vent. The temperature at the top of the column will be below 80°C.

The forward-flow valve is now partially opened so that part of the condensate is run by gravity to the receivers at a rate of 400 gal per hour. After taking off approximately 400 gal. the temperature at the top of the column will be nearly 80°C. The forward-flow valve is then closed and more light-boiling material is allowed to accumulate in the upper portion of the column, causing lowering of the temperature.

After 12-15 minutes at total reflux the forward-flow valve is again opened and about 150 gal. of light boiling material is run to the receiver at a rate of 400 gal. per hour. At this point the temperature at, the top of the column will again have risen to nearly 80°C. and the distillate will be benzene, contaminated with small quantities of low-boiling compounds.

Forward flow is continued at the above rate the entire distilled product usually about. 400gal. is run to an intermediate tank to he held for redistilling in a subsequent. charge. The distillate is tested periodically, by boiling and by gravity, when tests show that the distillate is nitration-grade benzene, the forward flow is increased to 600 gal per hour and the distilled product is cut to the nitration-grade storage tank. When tests show that the distillate is becoming contaminated by toluene, as indicated by a high dry-point of the distillate, forward flow is reduced to 400 gal. per hour. Reduction in forward flow is continued until the distillate, produced 30t a rate of 300 gal. per hour, no longer meets the specification for nitration-grade benzene.

Forward flow is now increased to 500 gal. per hour and the distillate is run to the intermediate tank to be held for redistillation in a subsequent charge. The distillate is principally benzene contaminated with toluene.

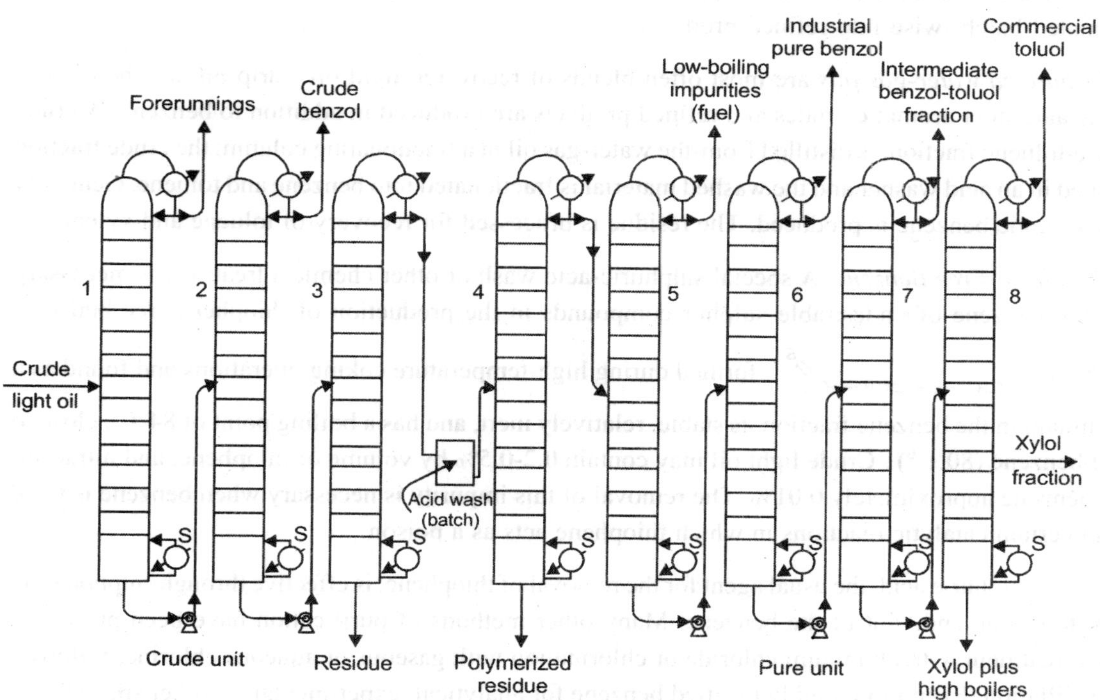

Fig. 21.2. Showing the Continuous plant for refining of coke-oven light oil

The temperature at the top of the column will remain at 80°C. during the period in which nitration-grade benzene is produced, and will gradually rise to 105-110°C. during production of the intermediate fraction, which follows the benzene. The temperature in the still will then be approximately 130°C. The residue in the still may be processed for production of toluene.

With the exception of Thiophene-free benzene, other grades may be produced in a manner similar to the above with adjustments in the forward-flow rate to produce the desired grade of benzene. High reflux and low forward flow correspond to the highest-grade products.

Figure 4 shows a continuous plant in which the crude light oil, recovered from steam distillation of the benzolized wash oil, is separated into crude fore running, crude benzol, and a broad benzol-toluol-xylol fraction. The latter fraction is acid washed by, batch operation and then distilled to rid it of the polymerized residue. The distillate from this still (unit No.4 in Figure) is fractionated by continuous operation into commercial products.

In the semicontinuous process, forerunning and a motor benzol fraction are produced as distillates from a train of two continuous fractionating units. The residue-from the second unit, containing benzene, toluene, xylene, and unsaturated come pounds, is acid-washed in the usual manner and subsequently fractionated batch'-wise into refined products.

Carbureted water-gas oils are most often blends of recovered light oils, drip oil, and holder oil. Toluene and other aromatic crudes and refined products are produced in addition to benzene. A crude benzene-toluene fraction, is distilled from-the water-gas oil in a fractionating column; the crude fraction is washed in an acid washer and the washed material is fractionated into benzene and toluene. Generally nitration-grade benzene is produced. The residue is processed for recovery of toluene and xylene.

Thiophene-Free Benzene. A special sulphuric-acid wash or other chemical treatment is necessary to rid the benzene of undesirable sulphur compounds in the production of thiophene-free benzene.

Thiophene, S\CH—CH=CH—CH//S , formed during high-temperature coking operations and found as a contaminant in the benzene fraction, is stable, relatively inert, and has a boiling point of 84°C., close to that of benzene (80.1 °) . Crude light oil may contain 0.2-0.5% by volume of thiophene, and nitration-grade benzene approximately 0.01%. The removal of this impurity is necessary when benzene is to be used in certain catalytic reactions in which thiophene acts as a poison.

Strong sulphuric acid, the usual agent for the removal of thiophene, is effective through sulphonation of the thiophene and "not of the benzene. Many other methods of purification have been proposed, such as treatment with aluminum chloride or chlorination with gaseous or aqueous chlorine, followed by redistillation. Even more highly purified benzene for analytical, experimental, or other special uses may be prepared by recrystallization of a highly refined commercial grade such as thiophene-free benzene.

Physical Properties

Density and phase	0.8786 g/cm³, liquid
Solubility in water	1.79 g/l (25°C)
Melting point	5.5°C (278.6 K)
Boiling point	80.1°C (353.2 K)
Viscosity	0.652 cP at 20°C
Structure	
Molecular shape	Planar
Symmetry group	D6h
Dipole moment	0 D
Flash point	−11°C

Structure

The formula of benzene (C_6H_6) caused a mystery for some time after its discovery, as no explanation had been found that could account for all the bonds — carbon usually forms four single bonds and hydrogen one.

The chemist Friedrich August Kekulé von Stradonitz was the first to deduce the ring structure of benzene. An often-repeated story claims that after years of studying carbon bonding, benzene and related molecules, he dreamt one night of the Ouroboros, a snake eating its own tail, and that upon waking he was inspired to deduce the ring structure of benzene. However, the story first appeared in the *Berichte der Durstigen Chemischen Gesellschaft* (Journal of the Thirsty Chemical Society), a parody of the *Berichte der Deutschen Chemischen Gesellschaft*, which appeared annually in the late-19th century on the occasion of the congress of German chemists; as such, it is probably to be treated with circumspection.

While his (more formal) claims were well-publicized and accepted, by the early-1920s Kekulé's biographer came to the conclusion that Kekulé's understanding of the tetravalent nature of carbon bonding depended on the previous research of Archibald Scott Couper (1831-1892); further, the Austrian chemist Josef Loschmidt (1821-1895) had earlier posited a cyclic structure for benzene as early as 1862. The cyclic nature of benzene was finally confirmed by the eminent crystallographer Kathleen Lonsdale.

Benzene presents a special problem in that, to account for all the bonds, there must be alternating double carbon bonds:

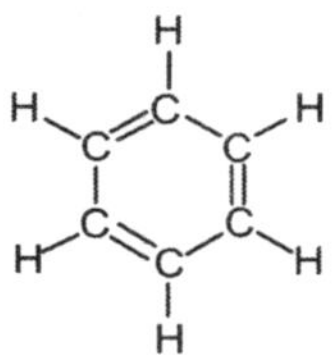

Using X-ray diffraction, researchers discovered that all of the carbon-carbon bonds in benzene are of the same length, and it is known that a single bond is longer than a double bond. In addition, the bond length, the distance between the two bonded atoms in benzene is greater than a double bond, but shorter than a single bond. There seemed to be in effect, a bond and a half between each carbon.

This is explained by electron delocalization. In order to visualise this, one should consider the position of electrons in the bonds of benzene.

One representation is that the structure exists as a superposition of the forms below, rather than either form individually. This type of structure is called a *resonance hybrid*.

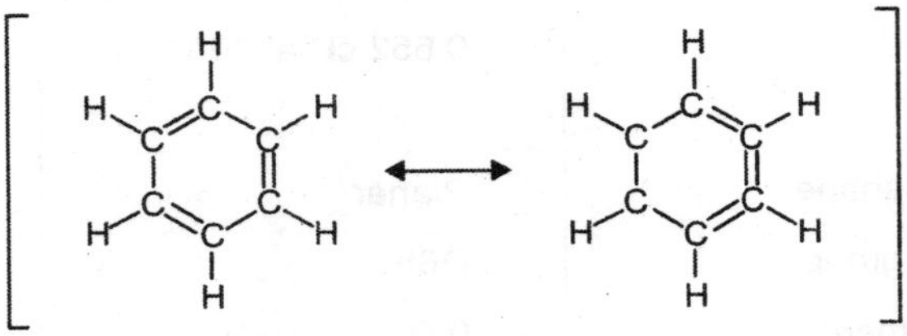

In reality, neither form really exists. Delocalisation must be explained using a higher level of theory than single and double bonds. The single bonds are formed with electrons in line between the carbon atoms - this is called ó (sigma) symmetry. Double bonds consist of a sigma bond and another, ð bond. This second bond has electrons orbiting in paths above and below the plane of the ring at each bonded carbon atom. The π-bonds are formed from atomic p-orbitals above and below the plane of ring. The following diagram shows the positions of these p-orbitals:

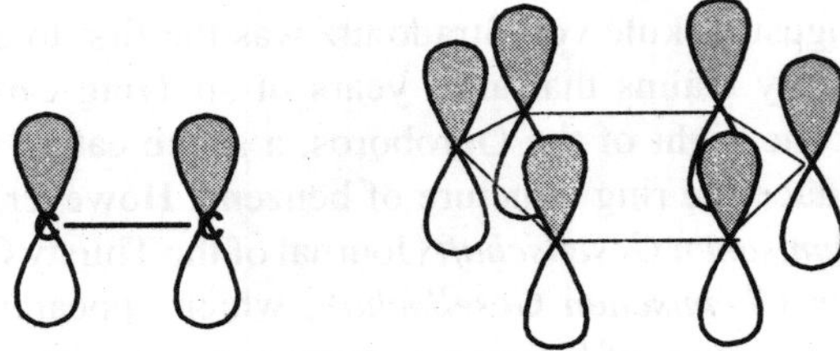

Since they are out of the plane of the atoms, these orbitals can interact with each other freely, and become delocalised. This means that, instead of being tied to one atom of carbon, each electron is shared by all six in the ring. Thus, there are not enough electrons to form double bonds on all the carbon atoms, but the "extra" electrons strengthen all of the bonds on the ring equally. The resulting molecular orbital has π symmetry.

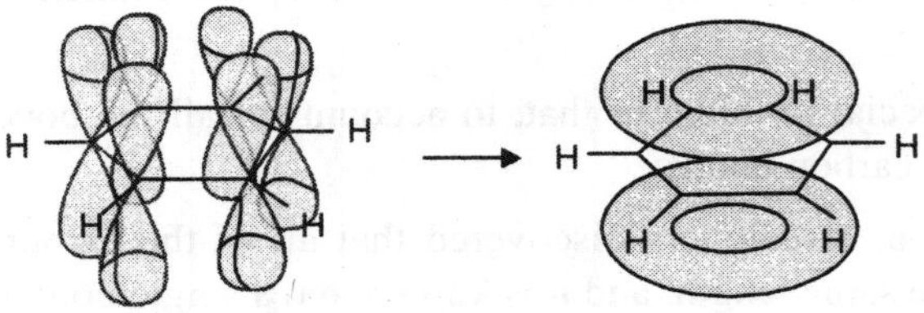

6 p-orbitals

delocalised orbital clouds

This delocalisation of electrons is known as *aromaticity*, and gives benzene great stability. This is the fundamental property of aromatic chemicals that differentiates them from non-aromatics.

To reflect the delocalised nature of the bonding, benzene may be depicted as a circle inside a hexagon in chemical structure diagrams:

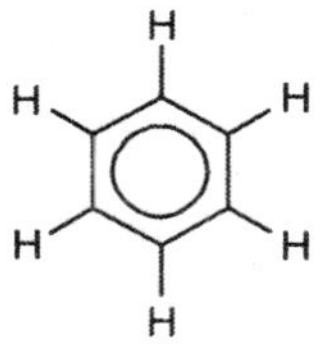

As is common in diagrams of organic structures, the carbon atoms in the diagram above have been left unlabelled.Benzene occurs sufficiently often as a component of organic molecules that there is a Unicode symbol with the code 232C to represent it:

H H H H H H

Bonding in Benzene

The Kekulé structure for benzene, C_6H_6

What is the Kekulé structure?

Kekule was to say later that he must have dozed off at this point. In his **dream** the black balls of carbon turned into black imps with forked tails that began racing around the room and would soon be upsetting the apparatus of the laboratory. He was ready to run the rascals out. Then, almost suddenly, the confusion died away as each imp grabbed the tail of the one ahead of him, the six forming a whirling circle. One hand of each imp held a tail, the other a white handkerchief—and they waved to him as the group whirled by. He said that he came awake with a start, realizing that the imps were acting out the formula for benzene. As his hand grabbed the sketching pencil, the imps were back to black balls again and the handkerchiefs had changed to hydrogen atoms. How simple the *arrangement turned out to be. "The carbon atoms of benzene form a ring."*, he immediately suggested that the carbons are arranged in a hexagon, and he suggested alternating double and single bonds between them. Each carbon atom has a hydrogen attached to it. This diagram is often simplified by leaving out all the carbon and hydrogen atoms! In diagrams of this sort, there is a carbon atom at each corner. You have to count the bonds leaving each carbon to work out how many hydrogens there are attached to it. In this case, each carbon has three bonds leaving it. Because carbon atoms form four bonds, that means you are a bond missing - and that must be attached to a hydrogen atom.

Problems with the Kekulé structure

Although the Kekulé structure was a good attempt in its time, there are serious problems with it . . .

Problems with the chemistry

Because of the three double bonds, you might expect benzene to have reactions like ethene - only more so!

Ethene undergoes addition reactions in which one of the two bonds joining the carbon atoms breaks, and the electrons are used to bond with additional atoms.

Benzene rarely does this. Instead, it usually undergoes substitution reactions in which one of the hydrogen atoms is replaced by something new.

Problems with the shape

Benzene is a planar molecule (all the atoms lie in one plane), and that would also be true of the Kekulé structure. The problem is that C-C single and double bonds are different lengths.

C–C	0.154 nm
C=C	0.134 mm

"nm" means "nanometre", which is 10^{-9} metre.

That would mean that the hexagon would be irregular if it had the Kekulé structure, with alternating shorter and longer sides. In real benzene all the bonds are exactly the same - intermediate in length between C-C and C=C at 0.139 nm. Real benzene is a perfectly regular hexagon.

Problems with the stability of benzene

Real benzene is a lot more stable than the Kekulé structure would give it credit for. Every time you do a thermochemistry calculation based on the Kekulé structure, you get an answer which is wrong by about 150 kJ mol^{-1}. This is most easily shown using enthalpy changes of hydrogenation.

It doesn't matter whether you've done any thermochemistry sums recently or not. This is all so simple that you could understand it even if you had never done any!

Hydrogenation is the addition of hydrogen to something. If, for example, you hydrogenate ethene you get ethane:

$$CH_2{=}CH_2 + H_2 \longrightarrow CH_3CH_3$$

In order to do a fair comparison with benzene (a ring structure) we're going to compare it with cyclohexene. Cyclohexene, C_6H_{10}, is a ring of six carbon atoms containing just one C=C.

When hydrogen is added to this, cyclohexane, C_6H_{12}, is formed. The "CH" groups become CH_2 and the double bond is replaced by a single one.

cyclohexane: six carbons in a ring, but the ***ane*** ending means NO C=C bond.

If you are a bit shaky on names: ***cyclohexene***: ***hex*** means *six carbons*, **cyclo** means *in a ring*, ***ene*** means *with a C=C bond.*

The structures of cyclohexene and cyclohexane are usually simplified in the same way that the Kekulé structure for benzene is simplified - by leaving out all the carbons and hydrogens.

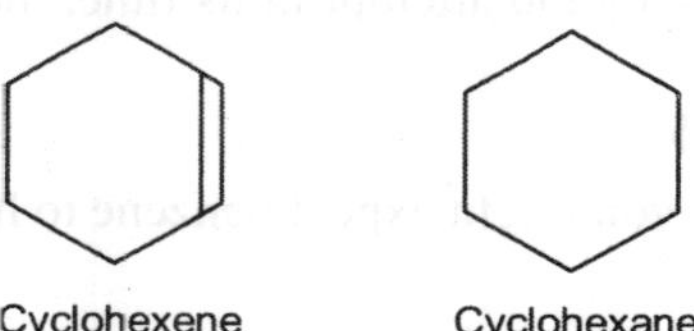

In the cyclohexane case, for example, there is a carbon atom at each corner, and enough hydrogens to make the total bonds on each carbon atom up to four. In this case, then, each corner represents CH_2.

The hydrogenation equation could be written:

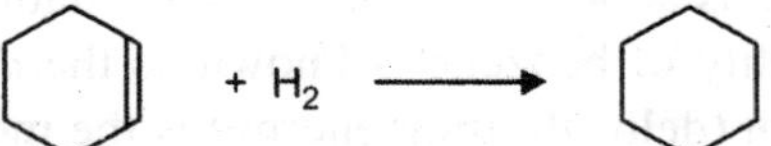

The enthalpy change during this reaction is -120 kJ mol^{-1}. In other words, when 1 mole of cyclohexene reacts, 120 kJ of heat energy is evolved.

Where does this heat energy come from? When the reaction happens, bonds are broken (C=C and H-H) and this costs energy. Other bonds have to be made, and this releases energy.

> "Enthalpy change" can be translated as "heat evolved or absorbed". The negative sign shows that heat is evolved.

Because the bonds made are stronger than those broken, more energy is released than was used to break the original bonds and so there is a net evolution of heat energy.

If the ring had *two* double bonds in it initially (*cyclohexa-1,3-diene*), exactly twice as many bonds would have to be broken and exactly twice as many made. In other words, you would expect the enthalpy change of hydrogenation of cyclohexa-1,3-diene to be exactly twice that of cyclohexene - that is, -240 kJ mol^{-1}.

+ $2H_2$ ⟶

In fact, the enthalpy change is ***-232 kJ mol⁻¹*** - which isn't far off what we are predicting.

Applying the same argument to the Kekulé structure for benzene (what might be called *cyclohexa-1,3,5-triene*), you would expect an enthalpy change of -360 kJ mol^{-1}, because there are exactly three times as many bonds being broken and made as in the cyclohexene case.

+ $3H_2$ ⟶

In fact what you get is ***-208 kJ mol⁻¹*** - not even within distance of the predicted value!

This is very much easier to see on an enthalpy diagram. Notice that in each case heat energy is released, and in each case the product is the same (cyclohexane). That means that all the reactions "fall down" to the same end point.

Heavy lines, solid arrows and bold numbers represent real changes. Predicted changes are shown by dotted lines and italics.

The most important point to notice is that real benzene is much lower down the diagram than the Kekulé form predicts. The lower down a substance is, the more energetically stable it is.

> If you look at the diagram closely, you will see that cyclohexa-1,3-diene is also a shade more stable than expected. There is a tiny amount of delocalisation energy involved here as well.

This means that real benzene is about 150 kJ mol^{-1} more stable than the Kekulé structure gives it credit for. This increase in stability of benzene is known as the ***delocalisation energy*** or ***resonance energy*** of benzene. The first term (delocalisation energy) is the more commonly used.

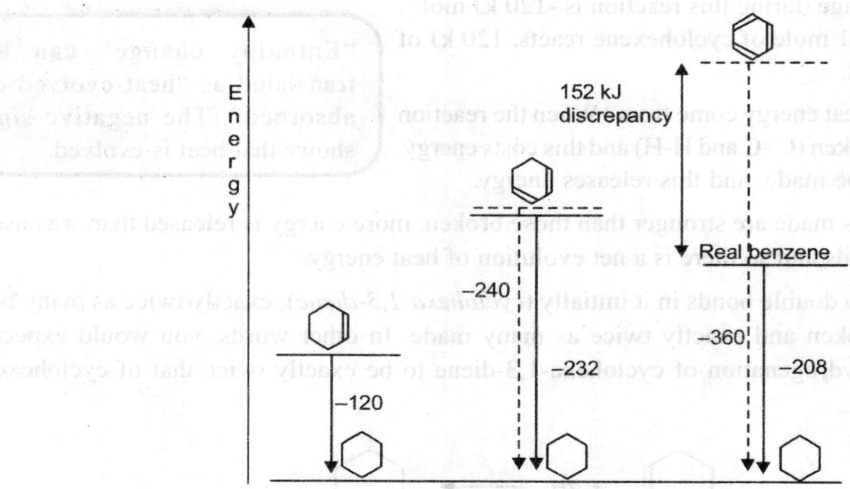

Benzene and Resonance

In 1858, August Kekule proposed a rapid oscillation (I & II) between the 3 C=C and the 3 alternating C-C of hexagonal benzene (C_6H_6).

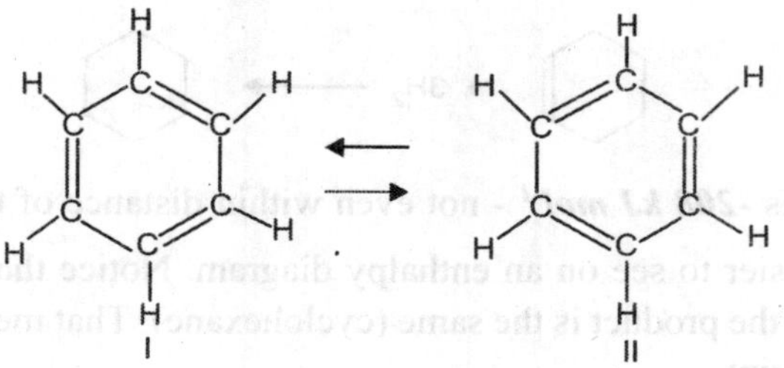

I ⇄ II

However, Kekule's representation of benzene had serious flaws.

1. Heat of Hydrogenation is the heat liberated when hydrogen adds to C=C. Benzene is compared to cyclohexene because both form cyclohexane upon complete hydrogenation. Since benzene supposedly contains 3 C=C, it should release three times the heat of hydrogenation reported for cyclohexene with only 1 C=C. Heats of hydrogenation for cyclohexene and benzene are reported as:

Since cyclohexane releases 28.6 kcal/mol, benzene is expected to release (3x28.6) or 85.8 kcal/mol.

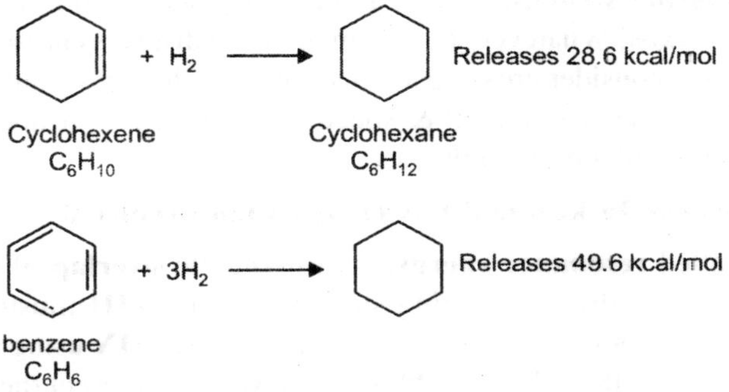

Cyclohexene C_6H_{10} + H_2 → Cyclohexane C_6H_{12} Releases 28.6 kcal/mol

benzene C_6H_6 + $3H_2$ → Releases 49.6 kcal/mol

Why does benzene possess 36 kcal/mol less energy than predicted?

2. Addition versus Substitution: Compounds with C=C (alkenes) typically undergo addition reactions such as bromination. Shown below are reactions of ethene/bromine and benzene/bromine:

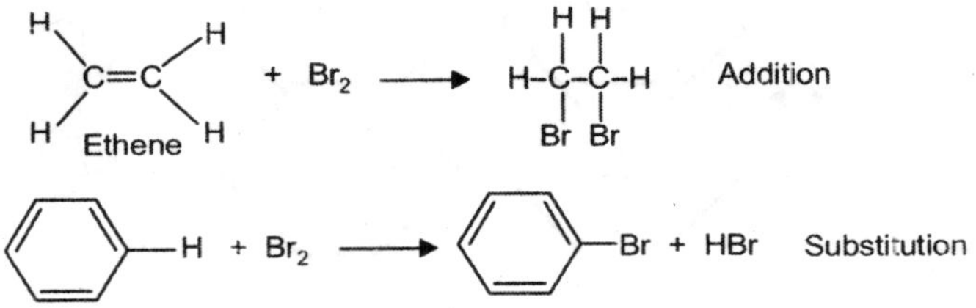

Ethene + Br_2 → H-C(H)(Br)-C(H)(Br)-H Addition

benzene-H + Br_2 → benzene-Br + HBr Substitution

Why does benzene with 3 C=C resist addition and prefer substitution?

3. X-Ray studies reveal the following bond lengths: 1.53A for typical C-C, 1.34A for typical C=C, and 1.39A for the 6 carbon-carbon bonds in benzene.

Why does benzene possess 6 bonds intermediate between C=C and C-C?

Resonance

To help explain molecules like benzene, **Linus Pauling** proposed resonance theory in 1931 (Pauling also gave us hybrid orbitals, electronegativity, and valence bond theory). As a result of Pauling's resonance, benzene is viewed as a hybrid of **III& IV**and represented by**V**.

III ↔ IV = V

The six carbons are arranged in a hexagon with one hydrogen atom attached to each carbon. The 12 atoms of benzene are planar with carbon in the sp2 hybrid state and 120Â° bond angles. Because the six bond lengths are equivalent, Kekule's rapid equilibrium must be ruled out. To describe benzene we need to use **contributing structures III& IV**or one structure with a circle in the middle (**V**). Structures **III &IV** do not exist! Whatever benzene might be, it displays characteristics of its contributing structures. For an analogy, consider crossing a bloodhound with a greyhound: Although the offspring is unique, it displays characteristics (smell & speed) of its parents—it does not flip flop between a greyhound one instant and a bloodhound the next.

Why does benzene possess 36 kcal/mol less energy than predicted?

This missing energy, called **resonance energy,** is attributed to **overlap** of p-orbitals. Structures **VI** and **VII** show pi bonding for the contributing structures while **VIII** illustrates **delocalization** of pi electrons over the 6 carbon atoms. Benzene can be represented as **IX** using molecular models with p-orbitals. The circle in the middle of **V** is an abbreviated way to represent the delocalization of the 6 pi electrons. Resonance energy is the difference in energies between **III** (or **IV**) and **V** (**V** has lower energy). Due to resonance, benzene is 36 kcal/mol more stable than calculations would predict.

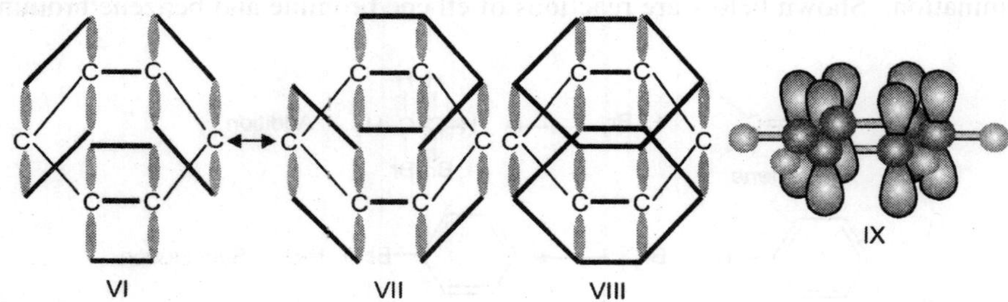

Why does benzene resist addition and prefer substitution?

If benzene were to undergo addition,resonance would be disrupted and this would render the structure less stable. Therefore benzene prefers to substitute (Br for H) and maintain resonance.

Why does benzene possess 6 bonds intermediate between C=C and C-C?

According to resonance, the bonds are not C-C or C=C but a hybrid of the two. X-ray studies confirm this with the intermediate bond lengths.

Bonding in Benzene

An orbital model for the benzene structure

Building the orbital model

Benzene is built from hydrogen atoms ($1s^1$) and carbon atoms ($1s^2 2s^2 2p_x^1 2p_y^1$).

Each carbon atom has to join to three other atoms (one hydrogen and two carbons) and doesn't have enough unpaired electrons to form the required number of bonds, so it needs to promote one of the $2s^2$ pair into the empty $2p_z$ orbital. This is exactly the same as happens whenever carbon forms bonds - whatever else it ends up joined to.

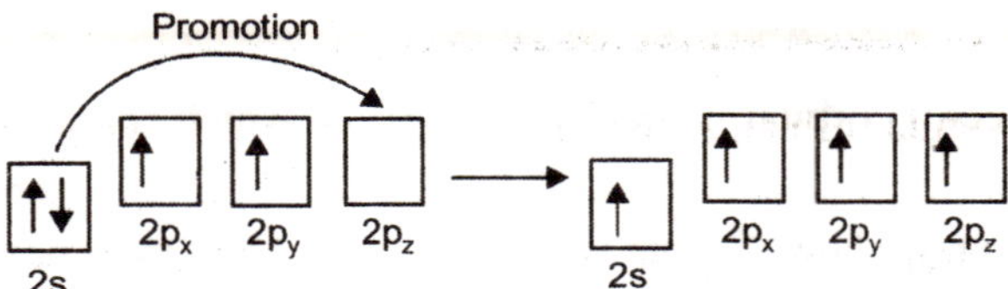

Because each carbon is only joining to three other atoms, when the carbon atoms hybridise their outer orbitals before forming bonds, they only need to hybridise *three* of the orbitals rather than all four. They use the 2s electron and two of the 2p electrons, but leave the other 2p electron unchanged.

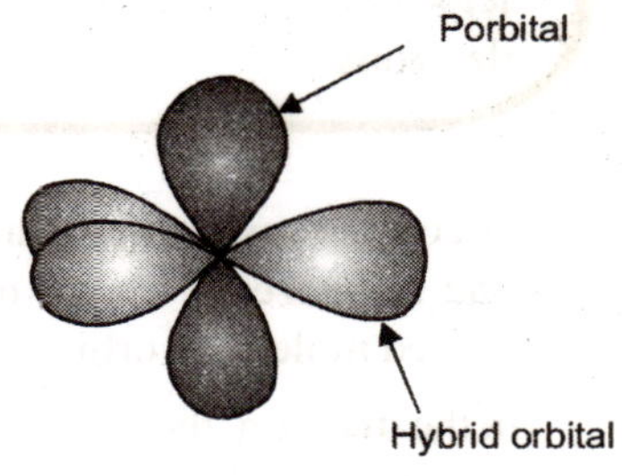

The new orbitals formed are called ***sp² hybrids***, because they are made by an s orbital and two p orbitals reorganising themselves. The three sp^2 hybrid orbitals arrange themselves as far apart as possible - which is at 120° to each other in a plane. The remaining p orbital is at right angles to them.

Each carbon atom now looks like the diagram on the right. This is all exactly the same as happens in ethene. The difference in benzene is that each carbon atom is joined to two other similar carbon atoms instead of just one. Each carbon atom uses the sp^2 hybrids to form sigma bonds with two other carbons and one hydrogen atom. The next diagram shows the sigma bonds formed, but for the moment leaves the p orbitals alone.

> ***Remember*:** A sigma bond is formed by the end-to-end overlap between atomic orbitals.

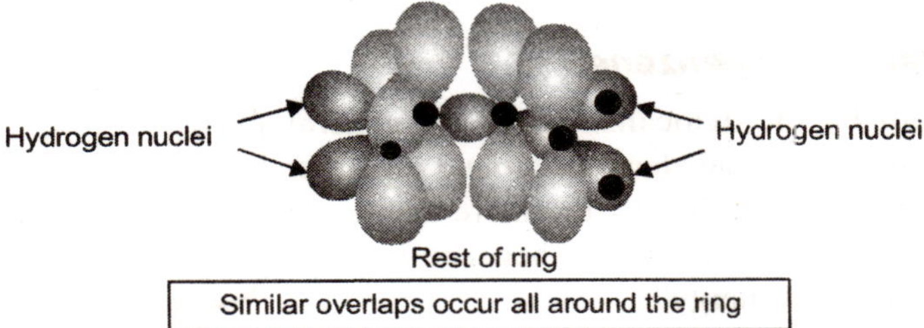

Similar overlaps occur all around the ring

Only a part of the ring is shown because the diagram gets extremely cluttered if you try to draw any more.

Notice that the p electron on each carbon atom is overlapping with those on both sides of it. This extensive sideways overlap produces a system of pi bonds which are spread out over the whole carbon ring. Because the electrons are no longer held between just two carbon atoms, but are spread over the whole ring, the electrons are said to be ***delocalised.*** The six delocalised electrons go into three molecular orbitals - two in each.

Remember: A molecular orbital is the region of space which contains a bonding pair of electrons.

Warning! Be very careful how you phrase this in exams. You must never talk about the p orbitals on the carbons overlapping sideways to produce **a** delocalised pi bond. This upsets examiners because **a** pi bond can only hold 2 electrons - whereas in benzene there are 6 delocalised electrons. Talk instead about a "pi system" - or just about the delocalised electrons.

In common with the great majority of descriptions of the bonding in benzene, we are only going to show one of these delocalised molecular orbitals for simplicity.

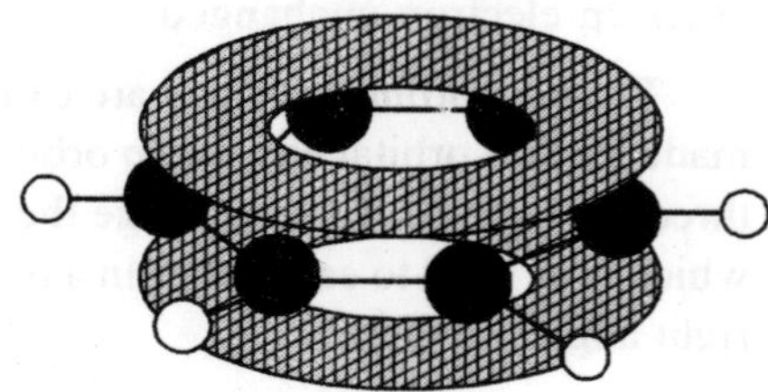

In the diagram, the sigma bonds have been shown as simple lines to make the diagram less confusing. The two rings above and below the plane of the molecule represent *one* molecular orbital. The two delocalised electrons can be found anywhere within those rings. The other four delocalised electrons live in two similar (but not identical) molecular orbitals.

Relating the orbital model to the properties of benzene

The shape of benzene

This is easily explained. Benzene is a regular hexagon because all the bonds are identical. The delocalisation of the electrons means that there aren't alternating double and single bonds.

The energetic stability of benzene

This is accounted for by the delocalisation. As a general principle, the more you can spread electrons around - in other words, the more they are delocalised - the more stable the molecule becomes. The extra stability of benzene is often referred to as "delocalisation energy".

To get the best out of this section you ought to read the article on the Kekulé structure for benzene.

The reluctance of benzene to undergo addition reactions

With the delocalised electrons in place, benzene is about 150 kJ mol^{-1} more stable than it would otherwise be. If you added other atoms to a benzene ring you would have to use some of the delocalised electrons to join the new atoms to the ring. That would disrupt the delocalisation and the system would become less stable. Since about 150 kJ per mole of benzene would have to be supplied to break up the delocalisation, this isn't going to be an easy thing to do.

Other Structure of Benzene

Other structures that were proposed at the time were Armstrong's centroid structure and Ladenburg's prismane.

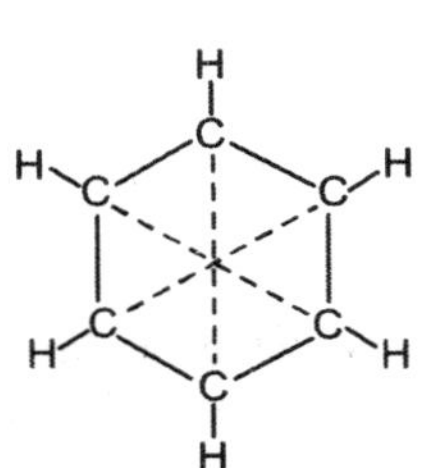

Armstrong's 'centroid'

Ladenburg's prismane

We now know that the best representation for the structure of benzene is indeed, hexagonal, with each C-C bond distance being identical, and intermediate between those for a single and double bond. The *p*-orbitals from each neighbouring C overlap to form a delocalised molecular orbital which extends around the ring, giving added stability and with it, decreased reactivity.

Below are several other attempts that were made to explain the equivalence of all six positions in the benzene ring with the concept of the tetravalent carbon atom.

- Adolf Carl Ludwig Clause (1840 - 1900) proposed long bonds.
- Lothar Meyer (1830 – 1895) suggested free affinities, which was not really compatible with the lack of reactivity of benzene.

- Henry Edward Armstrong (1848 – 1937) imagined that a carbon atom influenced each the other five through centrally directed bonds.

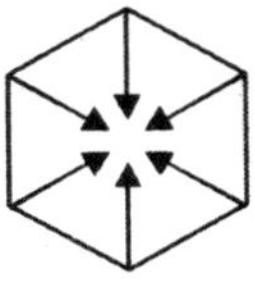

- Adolf von Bayer (1835 – 1917) put forward a similar idea.

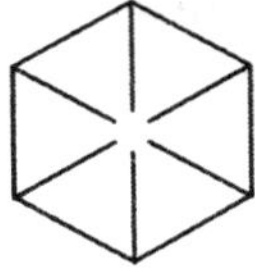

- In 1867 James Dewar (1842 – 1923) displayed some models he designed to show other possibilities, including the structure we now call "Dewar benzene".

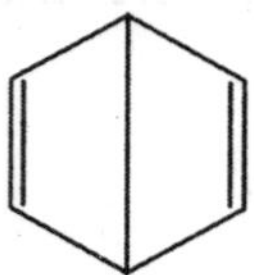

Stability

The extra stability that is conferred by the resonance energy makes benzene by far the most stable compared to some of the valence isomers of C_6H_6 shown below. Their relative stabilities are in the order benzene > benzvalene > Dewar benzene > prismane > bicyclo-propenyl, and their energies from highest to lowest differ by almost 546 kJ/mol.

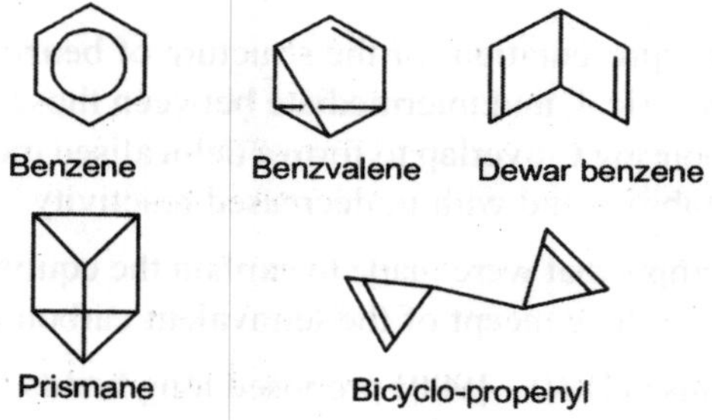

Substituted Benzene

Monosubstitution of Benzene

Benzene is an aromatic monocyclic hydrocarbon of six carbons (Fig 1). It has six Pi electrons that are delocalized. This results in a molecule that is more stable compared to a molecule where the double bonds are localized as in cyclohexatriene. Halobenzenes are monosubstituted benzene where the substituent is a halogen atom. Bromobenzene is shown in Fig 1. If a Chlorine atom was the substituent then the serivative would be called chlorobenzene. What would Iodobenzene and Flourobenzene look like? A monosubstituted benzene where the substituent is an alkyl group like methyl shown in Fig. 1 as

benzene bromobenzene toluene nitrobenzene

aniline benzaldehyde

methylbenzene is called an aryl benzene. What do you suppose Ethyl benzene would look like? How about iso-propylbenzene? Methyl benzene is also called toluene.

Nitrobenzene (Fig 1) is a monosubstituted derivative with a nitro (-NO_2) group attached.

Aminobenzene also called aniline (Fig 1) has the amino group(-NH_2). Aniline is a very reactive derivative that is a weak base. Benzoic acid (Fig 2) is an aromatic weak acid with the carboxyl group (-COOH) having an ionizable Hydrogen that makes it acidic. Benzoic Acid is relatively inactive toward electrophilic substitution. Benzonitrile (Fig 2) is an aromatic member of the Nitrile Family. The Cyano (-CN) group is highly unreactive toward electrophilic substitution.

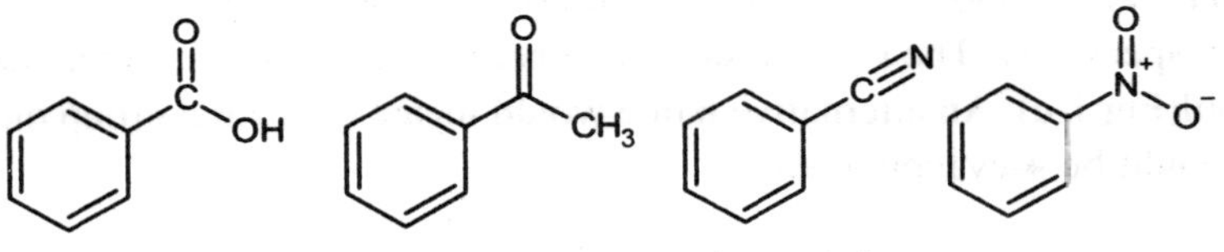

benzoic acid | 1-phenylethanone acetophenone | benzonitrile

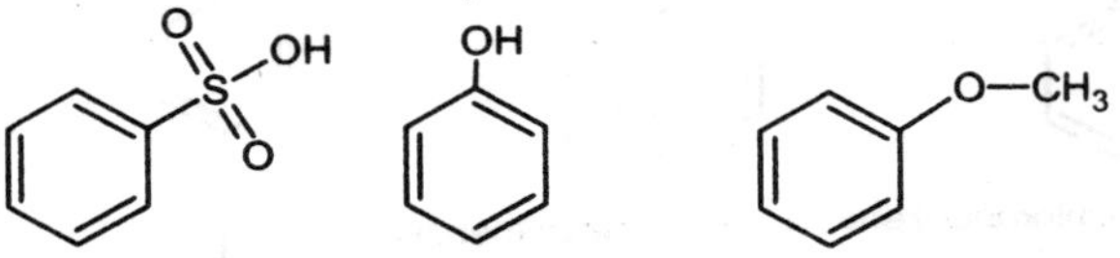

benzenesulfonic acid | phenol | methoxybenzeneanizole

Acetophenone (Fig 2) is an aromatic ketone. Another acceptable name for the compound is Methyl Phenyl Ketone. It is a relatively unreactive molecule toward electrophilic aromatic substitution.

Benzene Sulfonic Acid (Fig 2) is a sulphur containing aromatic compound. The Sulfonyl group(-SO_3H) has a weak bond between Hydrogen and one of the Oxygen atoms which makes this compound acidic. Benzene Sulfonic Acid is moderately unreactive to electrophilic substitution.

Phenol (Fig 2) has the Hydroxyl group (-OH) attached to a Benzene ring. The bond between the Oxygen and the Hydrogen is very weak and the resulting Phenoxide ion is unusually stable because of the resonance stabilization of the anion. Phenols are extremely reactive toward electrophilic substitution.

Methoxybenzene also called Anisole (Fig 2) is an aromatic member of the ether family. It is moderately reactive toward electrophilic substitution.

Disubstitution of Benzene

Ortho-Meta-Para Designation of Di-Substituted Benzene Derivatives

Two substituents attached to a Benzene ring can be positioned in respect to one another:

1. ***Ortho Isomer*** : The two substituents are attached to adjacent carbon atoms within the Benzene ring. If we used the numbering approach, they would be attached to carbons #1 and #2. (Fig 1-a). We could name the first molecule in Fig 1 as a Bromobenzene by hyphenating the first letter of the word ortho to the word Chloro followed by the monosubstituted benzene derivative

corresponding to the group attached to the #1 carbon, o-Chlorobromobenzene. An alternative name would be to hyphenate the numeral 2 to the word chloro following by the bromobenzene, 2-Chlorobromobenzene.

2. ***Meta Isomer*** : The two substituents are attached to the 1 and 3 carbon. The two are said to be meta to one another, and the name would be m-Iodotoluene or m-Iodomethylbenzene (Fig 1-b). An alternative name would be to locate the Iodo group by hyphenating the carbon number it is attached to. Here the Iodo group is attached to the #3 carbon so the name would be 3-Iodotoluene or 3-Iodomethylbenzene.

3. ***Para Isomer*** : The cyano group and the Hydroxyl groups are attached to the #4 carbon and the #1 carbon respectively. The two are said to be para to one another, and the name would be p-cyanophenol(Fig 1-c). An alternative name would locate the cyano group on the #4 carbon so the name would be 4-cyanophenol.

1-bromo-2-chlorobenzene 1-iodo-3-methylbenzene

4-hydroxybenzonitrile

Chemcial Properties of Benzene

It burns with a very luminous flame and smoky flame. Benzene is very stable compound and remains practically unaffected by the ordinary oxidizing and reducing agents or by double acids or boiling aqueous alkalies and also it does not decolorize bromine water. It differs exactly from the aliphatic hydrocarbons and shows the following important reactions which may be taken as typical of the family of aromatic hydrocarbons.

1. Nitration of Benzene or one of its derivatives involves a rather unusual nitrating mixture. Sulphuric Acid and Nitric Acid are mixed where the stroger Sulphuric Acid forces a proton on Nitric Acid.

$$\text{benzene} + HNO_3 \xrightarrow{H_2SO_4} \text{nitrobenzene} + H_2O$$

The first two steps in the Reaction Mechanism (Fig 2) shows how the nitro electrophile was generated.

1 HNO_3 (H—O—N⁺(=O)—O⁻) + H_2SO_4 (H—O—S(=O)(=O)—O—H) ⟶ H—O(H)—N⁺(=O)—O⁻ + HSO_4

2 H—O(H)—N⁺(=O)—O⁻ ⟶ NO_2 + H_2O

3 benzene + O=N⁺—O⁻ ⟶ nitrobenzene + H^+

4 H^+ + HSO_4 ⟶ H_2SO_4

Notice that the Sulphuric Acid is regenerated and therefore serves as a catalyst in this reaction. The generation of a positively charged Oxygen atom (oxonium ion) in step 1 produces an unstable intermediate which spontaneously decomposes losing water and generating the electrophile in step 2.

Mechanism

The electrophilic substitution reaction between benzene and nitric acid

The facts

Benzene is treated with a mixture of concentrated nitric acid and concentrated sulphuric acid at a temperature not exceeding 50°C. As temperature increases there is a greater chance of getting more than one nitro group, $-NO_2$, substituted onto the ring.

Nitrobenzene is formed.

$$C_6H_6 + HNO_3 \longrightarrow C_6H_5NO_2 + H_2O$$

or:

benzene + HNO_3 ⟶ (C_6H_5—NO_2) + H_2O

The concentrated sulphuric acid is acting as a catalyst.

The formation of the electrophile

The electrophile is the "nitronium ion" or the "nitryl cation", NO_2^+. This is formed by reaction between the nitric acid and the sulphuric acid.

$$HNO_3 + 2H_2SO_4 \longrightarrow NO_2^+ + 2HSO_4^- + H_3O^+$$

The electrophilic substitution mechanism

Stage one

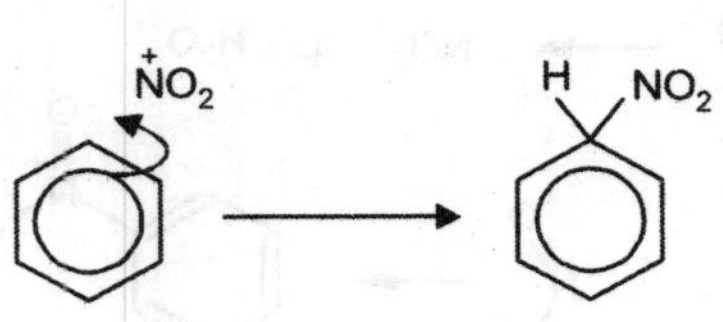

Stage two

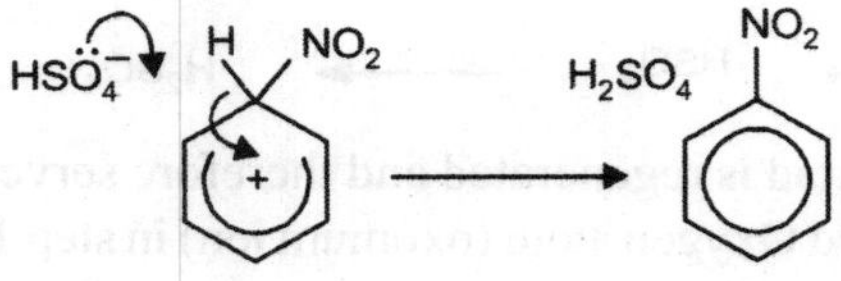

2. Sulphation of Benzene

Aromatic Sulfonation places a Sulfonyl group ($-SO_3H$) onto the Benzene ring. It accomplishes this by the use of Sulphur Trioxide in the presence of Sulphuric Acid. (Fig 1-a)

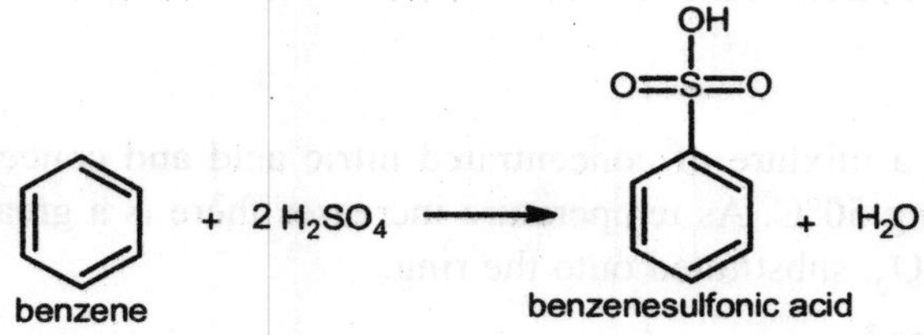

benzene + $2\,H_2SO_4$ → benzenesulfonic acid + H_2O

Sulphur Trioxide gas is dissolved into concentrated Sulphuric Acid at 20% to give what is called "fuming Sulphuric Acid" which earns its name by billowing out white plumes of smokey SO_3 fumes when the bottle is opened. The Sulphur Trioxide (SO_3) serves two purposes. It acts first as the sulfonating agent, and it prevents the unfavourable reversable of this equilibrium by reacting with and effectively removing the water product thus driving this equilibrium to the right.

$$SO_3 + H_2O \longrightarrow H_2SO_4$$

The Sulfonation can be accomplished without the Sulphur Trioxide by using excess Sulphuric Acid (Fig. 1b). The extra Sulphuric Acid will act as a dehydrating agent absorbing the water and preventing the reversal of the equilibrium. This is not as efficient as the Sulphur Trioxide so the reaction is much slower.

The reaction mechanism in the presence of Sulphur Trioxide differs by that of excess concentrated Sulphuric Acid in that step 1 in the mechanism shown in Fig 2 below is not necessary, and the mechanism

begins essentially with step 2. In both environments the SO_3 is apparently the sulfonating agent (step 2) below:

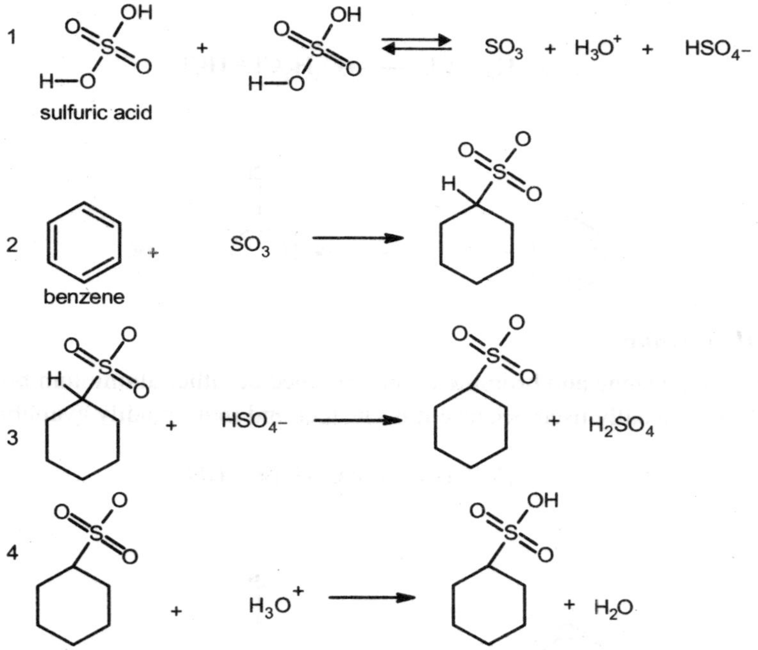

1 sulfuric acid + sulfuric acid ⇌ SO_3 + H_3O^+ + HSO_4^-

2 benzene + SO_3 ⟶ (H, SO_3 adduct)

3 (H, SO_3 adduct) + HSO_4^- ⟶ (benzenesulfonate) + H_2SO_4

4 (benzenesulfonate) + H_3O^+ ⟶ (benzenesulfonic acid) + H_2O

This problem with the water product in this reaction can be used to our advantage if we wish to remove a sulfonyl group by using dilute Sulphuric Acid. The excess water in the diluted acid will drive this equilibrium to the left (Fig 1) according to Le Chatlier's Principle.

4. Halogenation of Benzene

The electrophilic substitution reaction between benzene and chlorine or bromine

The facts

Benzene reacts with chlorine or bromine in an electrophilic substitution reaction, but only in the presence of a catalyst. The catalyst is either aluminium chloride (or aluminium bromide if you are reacting benzene with bromine) or iron.Strictly speaking iron isn't a catalyst, because it gets permanently changed during the reaction. It reacts with some of the chlorine or bromine to form iron(III) chloride, $FeCl_3$, or iron(III) bromide, $FeBr_3$.

$$2Fe + 3Cl_2 \longrightarrow 2FeCl_3$$

$$2Fe + 3Br_2 \longrightarrow 2FeBr_3$$

These compounds act as the catalyst and behave exactly like aluminium chloride in these reactions.

The reaction with chlorine

The reaction between benzene and chlorine in the presence of either aluminium chloride or iron gives chlorobenzene.

$$C_6H_6 + Cl_2 \longrightarrow C_6H_5Cl + HCl$$

or:

$$C_6H_6 + Cl_2 \longrightarrow C_6H_5Cl + HCl$$

The reaction with bromine

The reaction between benzene and bromine in the presence of either aluminium bromide or iron gives bromobenzene. Iron is usually used because it is cheaper and more readily available.

$$C_6H_6 + Br_2 \longrightarrow C_6H_5Br + HBr$$

or:

$$C_6H_6 + Br_2 \longrightarrow C_6H_5Br + HBr$$

The formation of the electrophile

We are going to explore the reaction using chlorine and aluminium chloride. If you want one of the other combinations, all you have to do is to replace each Cl by Br, or each Al by Fe.

As a chlorine molecule approaches the benzene ring, the delocalised electrons in the ring repel electrons in the chlorine-chlorine bond.

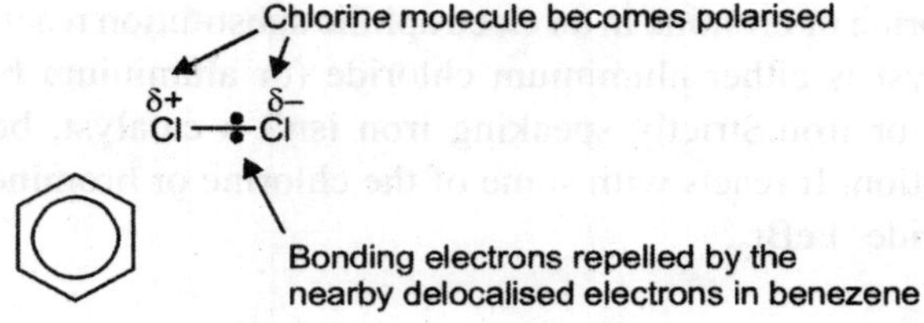

It is the slightly positive end of the chlorine molecule which acts as the electrophile. The presence of the aluminium chloride helps this polarisation.

The electrophilic substitution mechanism

Stage one

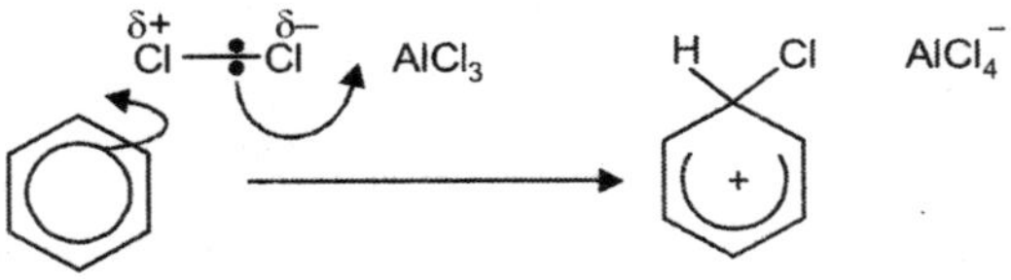

Stage two

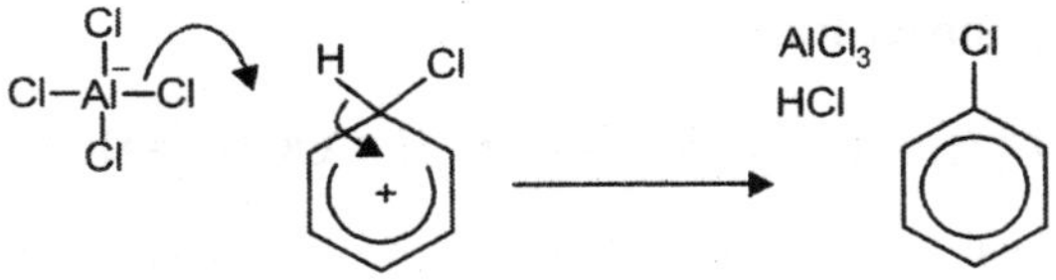

The hydrogen is removed by the $AlCl_4^-$ ion which was formed in the first stage. The aluminium chloride catalyst is re-generated in this second stage.

Friedel-crafts Reactions of Benzene and Methylbenzene

Friedel-Crafts acylation

Friedel-Crafts acylation of benzene

What is acylation?

An acyl group is an alkyl group attached to a carbon-oxygen double bond. If "R" represents any alkyl group, then an acyl group has the formula RCO-. Acylation means substituting an acyl group into something - in this case, into a benzene ring.

The most commonly used acyl group is CH_3CO-. This is called the ethanoyl group, and in this case the reaction is sometimes called "ethanoylation". In the example which follows we are substituting a CH_3CO- group into the ring, but you could equally well use any other acyl group.

Note: Don't worry too much about the name "phenylethanone" - all that matters is that you can draw the structure.

You may find slight variations on the conditions for this reaction. Various recipes I have found vary the temperature and time - for example by having a slightly lower temperature for a longer time.

The facts

The most reactive substance containing an acyl group is an acyl chloride (also known as an acid chloride). These have the general formula RCOCl.

Benzene is treated with a mixture of ethanoyl chloride, CH_3COCl, and aluminium chloride as the catalyst. The mixture is heated to about 60°C for about 30 minutes.A ketone called phenylethanone (old name: acetophenone) is formed.

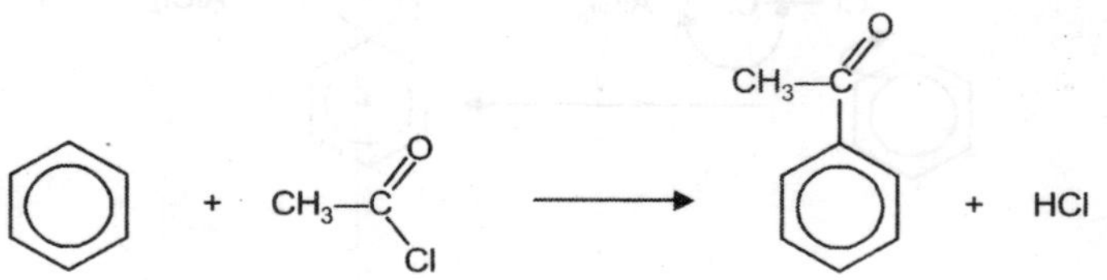

or, if you want a more compact form:

$$C_6H_6 + CH_3COCl \longrightarrow C_6H_5COCH_3 + HCL$$

The aluminium chloride isn't written into these equations because it is acting as a catalyst. If you wanted to include it, you could write $AlCl_3$ over the top of the arrow (see below).

Aromaticity and Hückel's Rule

Introduction

In comparison to the pi electrons of simple alkenes, the pi electrons of benzene are extremely non-reactive towards electrophiles. Such low reactivity suggests that the pi electrons of benzene are unusually stable. Valence bond theory attempts to rationalize this unusual stability by invoking a resonance argument. According to this line of thinking neither valence bond notation 1 nor 2, Figure 1, can adequatley represent the true structure of benzene. Rather, the true structure is closer to 3, which is described as a hybrid of structures 1 and 2.

Rationalizing Benzene

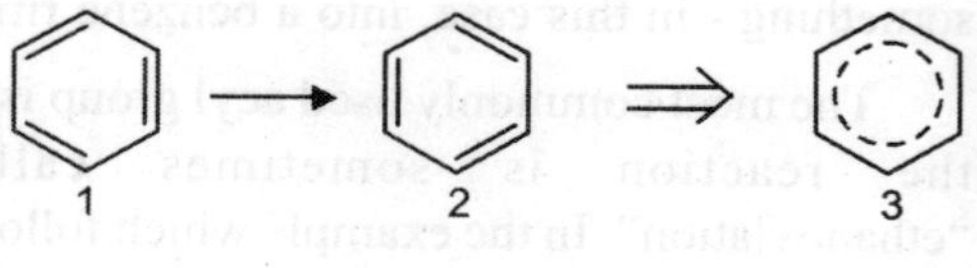

The implication of valence bond structure 1 is that each pi bond is localized between two carbon atoms, which means that the pi electron pair in each of these bonds experiences the nuclear attraction of two carbon atoms. The same goes for the pi electrons in structure 2. In structure 3 however, the pi electrons are no longer confined to regions between two specific carbon atoms. Rather they are free to roam over the entire 6-carbon atom framework. In essence, each electron experiences the nuclear attraction of 6 nuclei. Valence bond theory ascribes the added stability of the pi electrons in benzene to this increased nuclear attraction. Let's be clear- This is nonsense. Valence bond theory fails to adequately describe the physical and chemical properties of benzene. Modifying valence bond theory by incorporating the concept of resonance brings us closer to reality, but it's still a sham. We have seen that molecular orbital theory (MO theory) offers an alternative description of the bonding in molecules where a single Lewis structure does not adequately describe the structure of a compound. We are now going to take another look at MO theory.

MO Theory Revisited

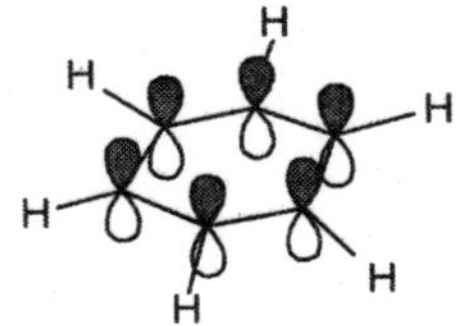

In our introduction to **molecular orbital theory** we saw the molecular orbitals are formed by taking linear combinations of atomic orbitals. To describe the pi system in benzene we will combine one AO from each of the six carbons to form six MOs. Restricting our attention to those AOs that are involved in the formation of the pi system of benzene is justified because the pi system is orthogonal (perpendicular) to the sigma bonded framework of the molecule.

Figure dissects the bonding in a molecule of benzene from a molecular orbital point of view.

Note the following points:

1. This model assumes that each carbon is sp^2 hybridized.
2. The s-bonded system is treated independently of the p-bonded system. This is justified by the assumption that the sp^2 hybridized orbitals are orthogonal to the p orbital on each carbon.

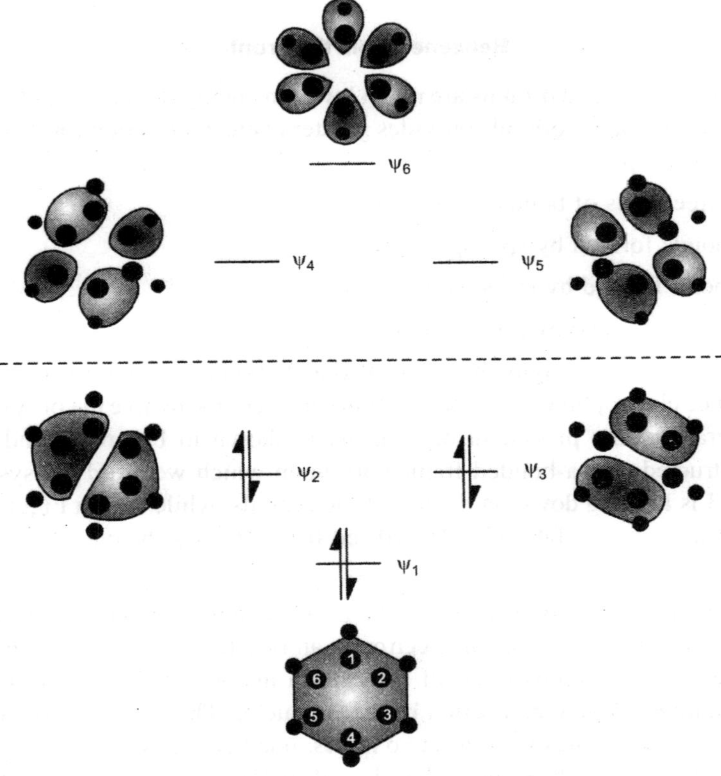

Benzene (From The Top)

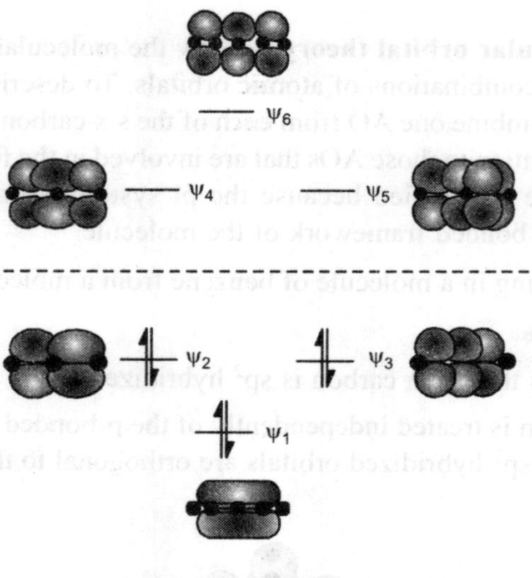

Benzene (From the Front)

3. Electrons in the s-bonded orbitals are much lower in energy than those in the p-bonded system. Head-to-head overlap of orbitals provides greater nuclear attraction for the electrons than does side-to-side overlap.
4. There are three types of bonds in benzene
   - 6 C-C s-bonds formed by sp^2- sp^2 overlap
   - 6 C-H s-bond formed by sp^2-s overlap
   - 3 C-C p-bonds formed by p-p overlap

While MO theory deals with sigma bonds as well as pi bonds, we will restrict our attention to the pi-bonded system since electrophilic aromatic substitution reactions involve the pi system of the aromatic ring. The MO diagram of the pi system of benzene is shown in Figures 3 and 4. This treatment assumes a pre-constructed sigma-bonded framework upon which we build the system of pi orbitals. The view in Figure 3 is looking down on the top of the orbitals, while that in Figure 4 is edge-on. The C and H nuclei appear as black dots. The dashed red line indicates the energy level of an isolated p orbital.

According to MO theory, the pi electrons of benzene occupy three molecular orbitals, y_1, y_2, and y_3, all of which are lower in energy than an electron in an isolated p orbital. The linear combination of p orbitals that generates y_1 extends over all six carbon nuclei. The two electrons that occupy y_1 experience the Coulombic attractions exerted by all six nuclei. They are very stable. A second linear combination of p orbitals generates y_2. It has two nodes, one between C-1 and C-2, the other between C-4 and C-5. A node is an area where the probability of finding an electron equals 0.

Students often ask the question "How do the electrons get from the red zone to the blue zone in y_2 if the probability of finding an electron between these two zones is zero?" This question stems from the *fact that these students are thinking about electrons as discrete particles, like tiny BBs. MO theory does not regard electrons as particles. Rather, it describes their behaviour as waves. Einstein pointed out that at the molecular and sub-molecular level matter exhibits behaviours characteristic of both particles and waves; the so-called "wave-particle duality". In MO theory orbitals are pictures of three-dimensional standing waves, similar to a two-dimensional standing wave such as a sine wave. In a sine wave, e.g. $y = \sin x$, a node simply represents that point at which the value of y changes from positive to negative. It has similar meaning in MO theory.*

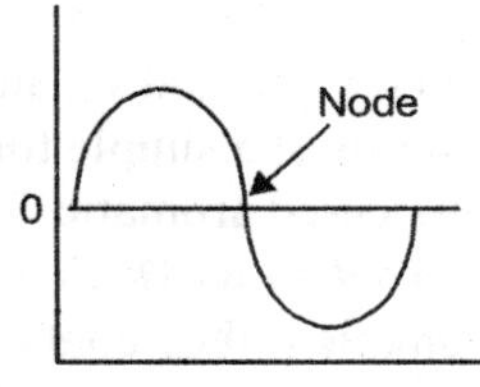

As the number of nodes in an orbital increases, so does the energy of the electrons that occupy that orbital. Since y_2, and y_3 each have two nodes, they have the same energy. They are said to be degenerate.

The most important feature of Figures 3 and 4 is the arrangement of the orbitals. Examine the pattern of MOs shown in Figure 5 for compounds containing 4, 6, 8, and 10-membered rings.

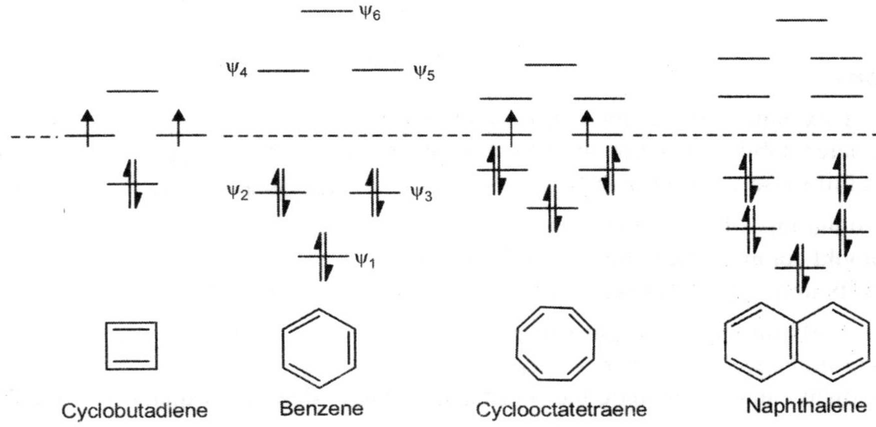

The orbitals are arrayed symmetrically about a center line (———) which represents the energy level of an isolated p orbital. Orbitals that are lower in energy than an isolated p orbital are classified as bonding orbitals. Those that are higher are called antibonding orbitals. Those at the same level as an isolated p orbital are called non-bonding orbitals.

Aromaticity, Stability, Filled Shell Rules, and Hückel's Rule

There is a parallel here with the filled shell rules in that the most stable electronic arrangement occurs when all of the bonding molecular orbitals are filled such as in benzene and naphthalene. These systems are considered to be aromatic. The connotation of the word is that the pi electrons of aromatic molecules possess unusual stability in comparison to pi electrons in simple alkenes.

If you consider the MO diagrams of benzene, naphthalene, and anthracene, you will notice that all contain a single low-energy orbital, y_1, which holds 2 electrons. Benzene also contains 1 pair of degenerate bonding orbitals, y_2 and y_3 which hold 4 electrons. Naphthalene contains 2 pairs of degenerate orbitals,

y_2 and y_3 and y_4 and y_5 which hold 4 electrons each. Anthracene contains 3 pairs of degenerate orbitals, y_2 and y_3, y_4 and y_5, and y_6 and y_7 each of which holds 4 electrons. It is possible to generalize this idea in terms of a simple formula that indicates the number of pi electrons required for a cyclic system to be considered aromatic. This formula was devised by a physicist named Erich Hückel, and it is commonly referred to as Hückel's 4n+2 rule: cyclic planar molecules in which each atom has a p orbital are aromatic if they contain 4n + 2 pi electrons. Here n is an integer, 0,1,2,3,...., that represents the number of sets of degenerate bonding orbitals in the MO diagram of the molecule. It takes 4 electrons to fill each set. The number 2 refers to the two electrons in y_1. In other words, pi systems containing 2, 6, 10, 14,...... electrons display unusual stability.

Anti-aromaticity

Molecules in which the highest occupied molecular orbitals are not completely filled, like those shown for cyclobutadiene and cyclooctatetraene are regarded as anti-aromatic. Here the connotation is that the pi electrons in anti-aromatic molecules are less stable than their counterparts in simple alkenes. In the most common situations anti-aromatic molecules contain 4n pi electrons, i.e. 0, 4, 8, 12, ... electrons.

Aromatic Ions

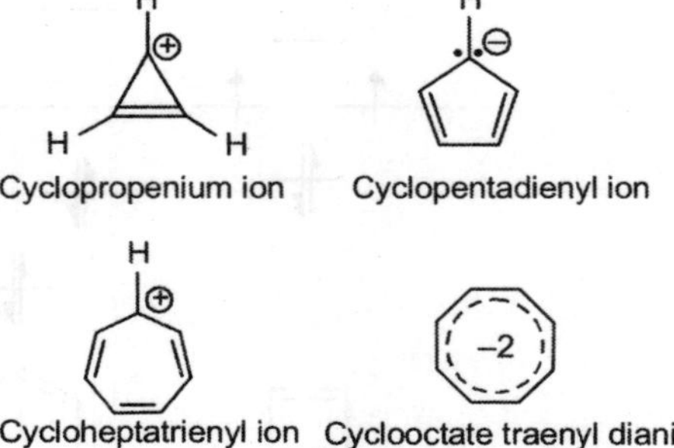

Cyclopropenium ion Cyclopentadienyl ion Cycloheptatrienyl ion Cyclooctate traenyl dianion

There are many examples of organic cations and anions that display greater stability than you might expect at first glance. Several of these are shown in Figure 6.

The MO diagrams of these four ions are shown in Figure You should compare the patterns of orbitals in this figure with the patterns shown in fig below

While each of the ions in Figure 6 makes for an interesting case study, we will consider only one, namely the cyclopentadienyl anion. Compare the acid-base reactions shown in Equations 1 and 2.

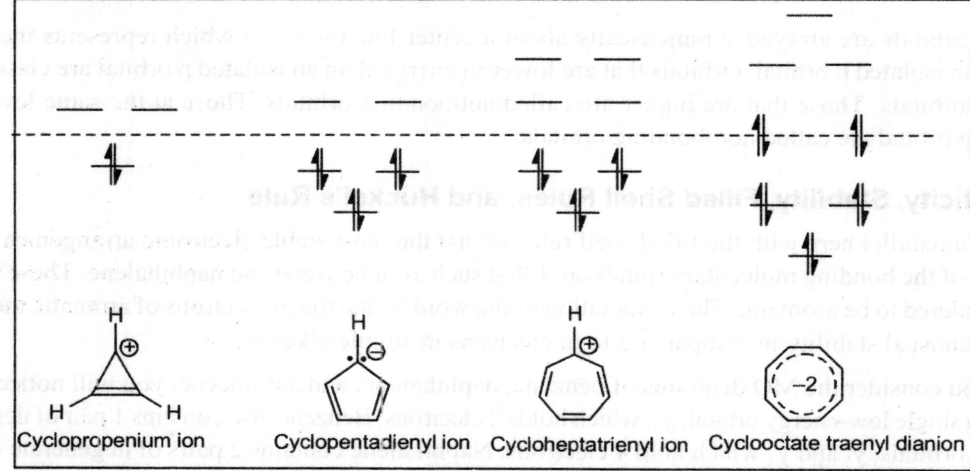

Cyclopropenium ion Cyclopentadienyl ion Cycloheptatrienyl ion Cyclooctate traenyl dianion

The pK_a of cyclopentadiene is 16!!! The equilibrium constant for reaction 1 is approximately 1. The pK_a of 1,4-pentadiene is approximately 45. Why is cyclopentadiene so much more acidic than 1,4-pentadiene? More to the point, why is the conjugate base of cyclopentadiene so much more stable than the conjugate base of 1,4-pentadiene? According to MO theory the pi system of the cyclopentadienyl anion has a filled shell electron configuration. The unusual stability associated with this arrangement means is responsible for the ease of deprotonation of cyclopentadiene.

H H NaOH H (1)

H H NaOH H (2)

Detection of Benzene

1. With nickel cyanide (ammonical) it gives white crystalline ppt.
2. With antimony pentachloride it produces a yellow or red ppt

Uses of Benzene

In the 19th and early-20th centuries, benzene was used as an aftershave because of its pleasant smell. Prior to the 1920s, benzene was frequently used as an industrial solvent, especially for degreasing metal. As its toxicity became obvious, other solvents replaced benzene in applications that directly exposed the user to benzene. Benzene was also used to initially decaffeinate coffee by German importer Lugwig Roselius in 1903. This lead to the production of Sanka, -ka for kaffein, but later discontinued the use of benzene.As a gasoline additive, benzene increases the octane rating and reduces knocking. As a result, gasoline often contained several percent benzene before the 1950s, when tetraethyl lead replaced it as the most widely-used antiknock additive. However, with the global phaseout of leaded gasoline, benzene has made a comeback as a gasoline additive in some nations. In the United States, concern over its negative health effects and the possibility of benzene's entering the groundwater have led to stringent regulation of gasoline's benzene content, with values around 1% typical. European gasoline specifications now contain the same 1% limit on benzene content. By far the largest use of benzene is as an intermediate to make other chemicals. The most widely-produced derivatives of benzene are styrene, which is used to make polymers and plastics, phenol for resins and adhesives (via cumene), and cyclohexane, which is used in Nylon manufacture. Smaller amounts of benzene are used to make some types of rubbers, lubricants, dyes, detergents, drugs, explosives and pesticides. In laboratory research, toluene is now often substituted for benzene because of health concerns.

CHAPTER

22

Benzene and its Homologous

Introduction

In this chapter we are going to discuss the homologous of benzene, they are toluene, xylene, cumene, styrene etc.

TOLUENE

Introduction and History

The name *toluene* was derived from the older name *toluol* that refers to tolu balsam, an aromatic extract from the tropical American tree *Myroxylon balsamum*, from which it was first isolated. It was originally named by Jöns Jakob Berzelius. **Toluene**, also known as **methylbenzene** or **phenylmethane** is a clear, water-insoluble liquid with the typical smell of paint thinners, reminiscent of the sweet smell of the related compound benzene. It is an aromatic hydrocarbon that is widely used as an industrial feedstock and as a solvent.

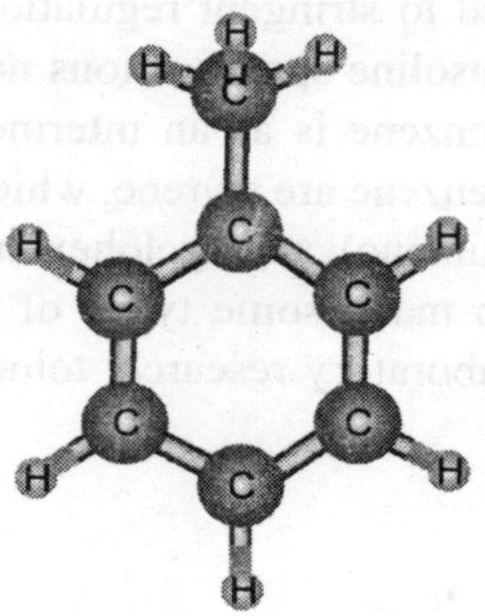

Preparation

Toluene is prepared from petroleum and coal by the same methods as employed for benzene n – heptane derived from petroleum fractions (gasoline and naphtha) is converted to toluene by passing its vapours over platinum catalyst ($Pt - Al_2O_3$)

$$\text{heptane} \xrightarrow[-H_2]{\text{Pt }Al_2O_3} \text{methylcyclohexane} \xrightarrow[-H_2]{\text{Pt }Al_2O_3} \text{toluene}$$

Industrial Preparation of Toluene

The process for synthesis of toluene by reaction of an excess of Benzene with Methanol at medium pressure over an acid zinc silico-phosphate catalyst was originally developed at Leuna. A commercial plant with a design capacity of 5,000 T/month of Toluene was erected at Waldenberg in Silesia where the I.G. have a factory which includes a Methanol plant O.U.D.A. erected the plant, W.I.F.O. financed it and it was operated by the I.G. on a management-fee basis.

The Waldenberg plant started operations in mid-1942. The maximum achieved production was 3,800 T/month were made on account of transport difficulties and consequent shortage of raw materials.

A description of the process and of the Waldenberg plant was provided in the first place at Leuna by Dr. Herold, the head of the Development Department. His information was later confirmed and some additional data produced, also at Leuna, by Dr. Klopfer who was in charge of catalyst testing. A process description was also found in an I.GW.I.F.O. Ageement, a copy of which was removed from Leuna.

General Process Data

Methanol and Benzene used for the process must be as free as possible from impurities, particularly nitrogen compounds. Methanol is specially purified from amines, by means of an organic cation exchanger, down to a maximum nitrogen content equivalent to 2 mgms. NH_3/litre. Nitriles are the chief impurities in the benzene and no satisfactory method has been developed for their removal; formation of zinc chloride addition compounds is not successful. In the end, the I.G. have resorted to careful analytical control of all batches of benzene so that these can be blended to give a feed to the plant which never contains more than 0.5 mgms. NH_3/litre.

Benezene and Methanol are reacted in the molecular ration 4:1. Reaction temperature is 340-380°C and pressure 30-35 ats. Feed rate is 0.25 M^3/M^3 in an active form steam must be added to the reactants. Substantially all in an active form steam must be added to the reactants. Substantially all the Methanol is converted in a single pass.

The catalyst, which has a life of 4-6 weeks, was stated by Dr. Klopfer to have the following composition:

Zno	6%
P_2O_5	60%

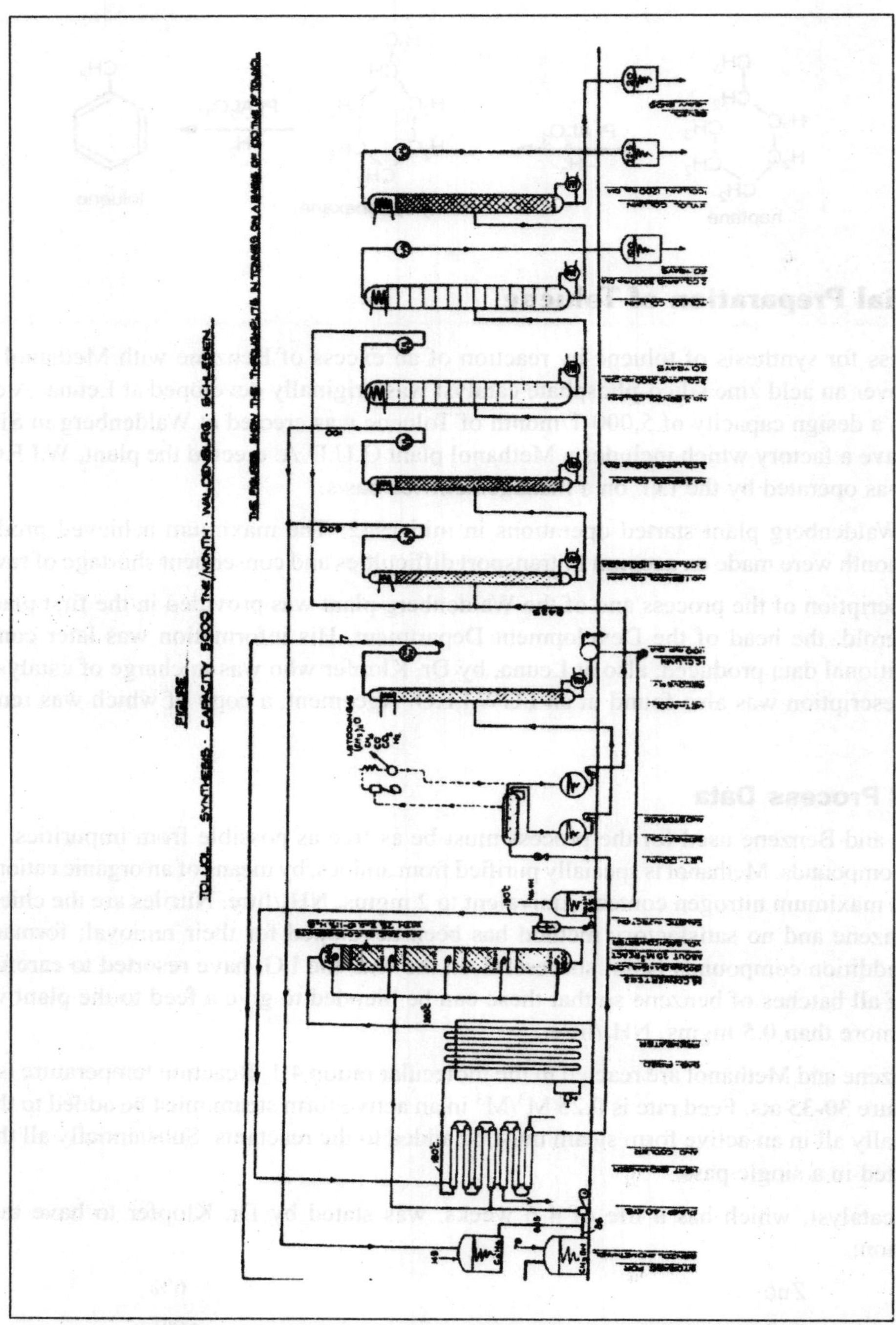

Fig. 22.1. Showing the schematic diagram of a toluene plant.

Kieselguhr	25%
Acid treated Bleaching Earth	3%
Water	6%

It is such that the catalyst was made by mixing orthophosphoric acid, Kieselguhr and zinc oxide, and drying a 2 cm-deep layer of the resultant paste at a temperature below 250°C. The cake obtained is broken up and sieved, pieces of about 20 mm size being used as catalyst. Fine material, amounting to 30% of the charge, is recycled to the pasting stage.Loss of activity of the catalyst is caused by carbon formation during reaction; after 4-6 weeks the catalyst contains about 40% carbon. Attempts to revivify catalyst by burning off this carbon were unsuccessful because of conversion of orthophosphoric into the inactive metaphosphoric acid.

Plant operation

The Waldenberg plant consisted of 14 units, each consisting of 1 converter 1700 mm. External diameter and 16 M long. Each converter contained 15 M^3 of catalyst. Fig 22.1 shows the general layout of a unit. Benzene and Methanol in the weight ratio 9:1 are injected at 35 ats. And pass through a heat exchanger and gas-fired preheater where they are vaporized and heated to 280°C (D. Klopfer) or 350°C (Dr. Herold). They then enter the top of the reaction vessel which is divided into four or five catalyst beds separated by layers of Raschig rings. Cold benzol is injected in controlled quantity at the entry to the converter and to each of the Raschig ring beds in order to maintain a reaction temperature of about 370°C with a maximum of 380°C. the reaction is extremely exothermic, the heat evolution per converter being 100,000 Kg. Cals/hour. There is a layer of calcined lime about 80 cms. Thick at both the top and bottom of each converter. Lining of the converter is acid-resistant refractor brick. A 5% caustic soda solution is injected into the products immediately on leaving the converter in order to prevent corrosion by phosphoric esters. The products then pass via the interchanger and a water-cooled condenser to a separator vessel. Gas, consisting of methane, CO_2, CO dimethylether and benzene vapour, is removed from the top of this vessel and is compressed and cooled to recover benzol (800 grms/M^3). The liquid product separates into water and hydrocarbon layers, the latter having the average composition:

71-80% vol. Benzene

12-17% vol. Toluene

4-6% vol. Xylene

4-6% vol Higher methyl benzenes.

The production of 1 Ton Toluene requires 1.45 T. Benzene and 0.85 T. Methanol.

The water layer is distilled for recovery of a small amount of unreacted Methanol which is returned to feed. The hydrocarbon product is first fractionated (2 packed columns of 1100 mm. diam.) to remove benzol and dimethylether as overheads. The bottoms pass to a second distillation stage (3 packed columns also of 1100 mm. diam.) in which most of the benzol and only a trace of Toluene is distilled over. The next distillation is for complete stripping of Benzene from the product. This is carried out in 3 to 60 plate columns of 2,400 mm. diameter and the overhead contains some 15% Toluene. These overheads are returned to the feed of the first benzene-removal column.

The benzene-free bottoms of the third distillation stage are next fed to two 3,000 mm. diameter 60 plate stills from which a nitration grade Toluene is taken overhead. The fifth stage of distillation consists of the separation of the fourth stage bottoms into Xylene and heavier alkyl benzenes.

Development work on Related Processes

Leuna have experimented with the reaction of Benzene with alcohols containing upwards of 5 carbon atoms, available as by-products of the higher alcohols synthesis. The same catalyst and general conditions as used for Toluene synthesis have been employed.

Long chain alkylates are not formed. Instead, splitting of the alcohol occurs giving simpler alkyl benzenes and some olefins which can be recycled.

Isopropyl benzene was the main product obtained. The above experiments reached the semitechnical stage and a large-scale plant was being considered for Heydebrech.

Physical Properties

Density and phase	0.8669 g/cm^3, liquid
Solubility in water	0.053 g/100 mL (20-25°C)
In ethanol, acetone, hexane, dichloromethane	Fully miscible
Melting point	-93 °C (180 K)/(-135.4°F)
Boiling point	110.6 °C (383.8 K)/ 231.08°F
Critical Temperature	320 °C (593 K)/ 608°F
Viscosity	0.590 cP at 20°C/ 68°F
Dipole moment	0.36 D

Heat constant $(H^o - H_0^0)/T$	14.44 cal/deg mole
Free energy $(F^o - H_0^0)/T$	-61.98 cal/deg mole
Entropy S°	76.42 e.u
Enthalpy $H^o - H^0_0$	4,306 cal/mol
Heat capacity Cp	24.80 cal/deg mole
Heat of formation ΔH°f	29.228 Kg/cal/deg mol
Equilibrium constant of formation $\log_{10}Kf$	21.4236

Chemical Properties

Toluene like benzene may be represented as a resonance hybrid of two canonical forms

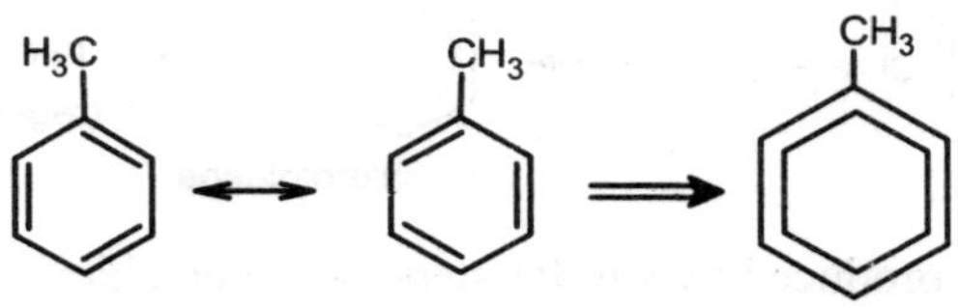

H_3C ... CH_3 ... CH_3

Since CH_3 group is an electron releasing group increases the overall electron density of the ring. Thus, toluene gives all the reactions of benzene ring more readily.

Toluene undergoes three types of reactions (1) Electrophilic substitution in the ring (2) Addition to the ring and substitution in the methyl group

A. Electrophilic Substitution Reaction

Toluene gives all the electrophilic substitution reaction which benzene does. The incoming substituent goes to the ortho and para positions. These reactions are going to be discussed now

Halogenation of Toluene

When toluene is introduced to chlorine or bromine in dark and in the presence of a Lewis acid catalyst ($FeCl_3$ $FeBr_4$) it undergoes halogenation in the ring ortho and para position

$$2\ \text{toluene} + Cl_2 \xrightarrow[\text{dark}]{FeCl_3} \text{o - chlorotoluene} \text{ and } \text{p - chlorotoluene} + 2HCl$$

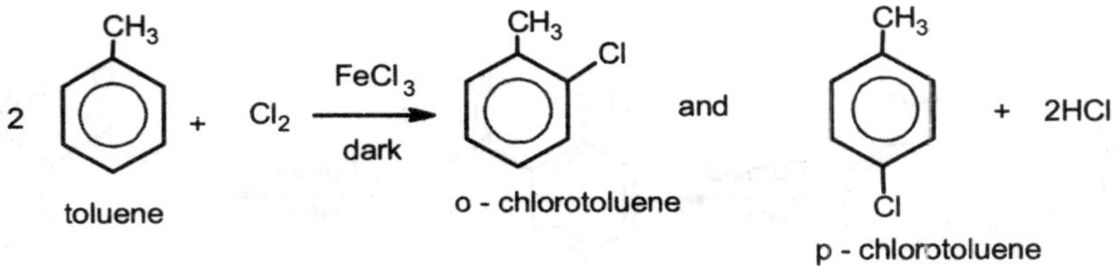

Mechanism

The mechanism of formation of ortho and para chlorination/bromination product in the same as described for benzene. Thus, o-chlorotoluene is produced by the steps

(i) formation of electophile

$$Cl{-}Cl + FeCl_3 \longrightarrow Cl^+ + \overline{FeCl_4}^{+}$$

ELECTROPHILE

(ii) formation of σ complex

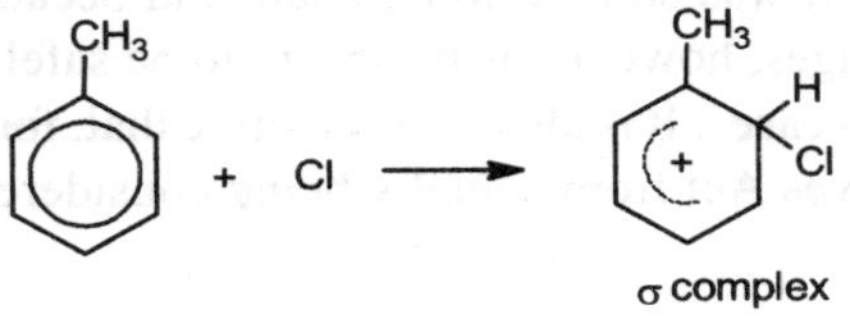

$$\text{toluene} + Cl \longrightarrow \sigma\ \text{complex}$$

σ complex

(iii)

CH3 H Cl + FeCl3 → o - chlorotoluene + FeCl3 (regenerated) + HCl

The para isomer is also produced by similar steps (a) when chlorine or bromine is passed through toluene, it attacks the side chain and the hydrogen atoms of the CH_3 group are successfully replaced

toluene —Cl_2,Δ (-HCl)→ benzyl chloride —Cl_2,Δ (-HCl)→ benzotrichloride

The substitution can be stopped by observing the increase of weight of the product substitution in side is flavoured by sunlight or U.V light and mechanism resembles the halogenation of alkanes.

Nitration of Toluene

Nitration of toluene is easier process than that of benzene because CH_3 is an electron releasing group and makes the π ring system electron – rich. Further nitration of toluene in presence of fuming nitric acid produces 2, 4, 6 trinitrotoluene (TNT)

o - nitrobenzene or (p-nitrotoluene) —Fuming HNO_3→ dinitro toluene —Fuming HNO_3→ 2, 4, 6 trinitrotoluene

TRI-NITROTOLUENE (TNT, Or Trotyl)

History and Introduction

TNT was first made in 1863 by a German chemist Joseph Wilbrand, but its potential was not seen for several years, mainly because it was so hard to detonate and because it was less powerful than other explosives. Among its advantages, however, is its ability to be safely melted using steam or hot water, and so poured molten into shell cases. It is also so insensitive that, for example, in 1910 it was exempted from the British 1875 Explosives Act from actually being considered an explosive for the purposes of manufacture and storage.

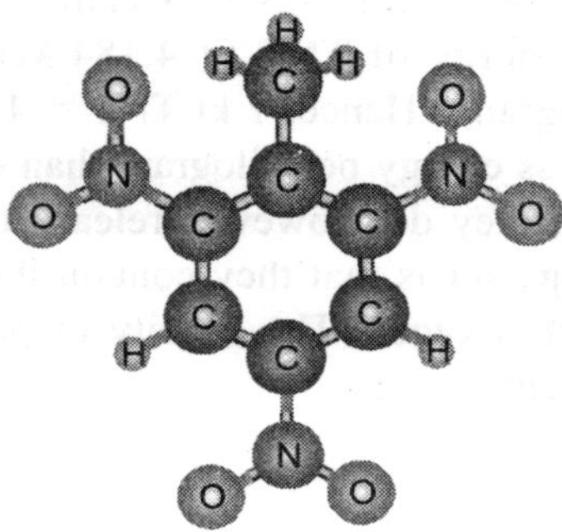

The German armed forces adopted it as an artillery shell filling in 1902, and the British gradually started using it as replacement for lyddite in 1907.

A particular advantage that it gave the German Navy in the First World War was to detonate after the TNT-filled armour-piercing shells had penetrated the armour of British capital ships, whereas the British lyddite-filled shells tended to explode as soon as they struck the German armour, and thus expended their energy outside of the ship.

Because of the insatiable demand for trinitrotoluene during the war, it was frequently mixed with 40%-80% ammonium nitrate, producing an explosive called amatol. Although nearly as powerful as TNT, amatol suffered from the slight disadvantage of being hygroscopic. Another variation called *minol*, consisting of amatol mixed with about 20% aluminum powder, was used by the British in mines and depth charges.

Preparation

The synthesis is done in a stepwise procedure. First toluene is nitrated with a mixture of sulfuric and nitric acid. Even lower-concentrated acid mixtures are capable of doing the first and second introduction of a nitrogroup. The nitrogroups decrease the reactivity of the toluene drastically, because they are electron-withdrawing groups. After separation the mono- and dinitrotoluene is fully nitrated with a mixture of nitric acid and oleum (sulfuric acid with up to 60% dissolved SO_3), this mixture is far more reactive and is capable of introducing the last nitrogroup. The waste acid from this process is used for the first step of the reaction in industrial synthesis. It would be unwise to attempt this at home, in case you aren't sure.

Trinitrotoluene is a pale yellow crystalline aromatic hydrocarbon compound that melts at 354 K (178°F, 81°C). Trinitrotoluene is an explosive chemical and a part of many explosive mixtures, such as when mixed with ammonium nitrate to form amatol. It is prepared by the nitration of toluene $C_6H_5CH_3$; it has a chemical formula of $C_6H_2(NO_2)_3CH_3$, and IUPAC name 2,4,6-trinitrotoluene.

In its refined form, trinitrotoluene is very stable, and, unlike nitroglycerin, it is insensitive to friction, blows, and jarring. This means that it must be set off by a detonator. It does not react with metals or absorb water, and so, unlike dynamite, it can be safely stored for many years. It is however readily acted upon by alkalis to form unstable compounds that are very sensitive to heat and impact.

Amounts of TNT are used as units of energy, especially for expressing nuclear weapon yield, based on a specific combustion energy of TNT of 4.184 MJ/kg (or one calorie—specifically a *thermochemical* calorie—per milligram). Hence, 1 kt TNT = 4.184 TJ, 1 Mt TNT = 4.184 PJ. Note that chemical explosives release less energy per kilogram than everyday household products like fat (38 MJ/kg) and sugar (17 MJ/kg); they do, however, release their combustion energy much more rapidly. One reason for their low power is that they contain their oxidant as well as the fuel — an explosive does not use atmospheric oxygen. The density of pure TNT (without any additives like sawdust or aluminium) is 1.654 g/cm³.

Physcial Properties

MELTING POINT	= - 81°C
Molecular Formula	= $C_7H_5N_3O_6$
Formula Weight	= 227.1311
Composition	= C(37.02%) H(2.22%) N(18.50%) O(42.27%)
Molar Refractivity	= 50.71 ± 0.3 cm^3
Molar Volume	= 141.2 ± 3.0 cm^3
Parachor	= 411.3 ± 4.0 cm^3
Index of Refraction	= 1.637 ± 0.02
Surface Tension	= 71.9 ± 3.0 dyne/cm
Density	= 1.608 ± 0.06 g/cm^3
Polarizability	= 20.10 ± 0.5 $10^{-24}cm^3$
Monoisotopic Mass	= 227.017835 Da
Nominal Mass	= 227 Da
Average Mass	= 227.133136 Da

Chemically it is explosive so no chemical reaction can be carried out with it.

Sulphonation of Toluene

On heating with sulphuric acids toluene produces para and ortho toluene sulphonic acids

CH_3 H_2SO_4 Conc CH_3 O S OH or CH_3 HO S O

2-methylbenzenesulfinic acid
ortho, toluene sulphonic acid

4-methylbenzenesulfinic acid
para, toluene sulphonic acid

Friedel Craft's Reaction

toluene + CH_3Cl (methyl chloride) $\xrightarrow{AlCl_3}$ o-xylene and p-xylene

5. Oxidation

A characteristic property of the toluene is the oxidation. Treatment with dilute nitric acid chromic acid potassium permanganate or potassium ferricyanide oxidizes each side chain however long into a carboxyl group – COOH

toluene + 3[O] $\longrightarrow$ benzoic acid + H_2O

By determining the basicity and nature of the resulting acid the number and the relative position of the side chains in alkyl benzenes can be easily deduced.

Conversion of Toluene to Benzene and Benzene to Toluene

toluene $\xrightarrow{\text{Oxidation}}$ benzoic acid $\xrightarrow[\text{heat}]{CaO}$ benzene + $CaCO_3$

benzene $\xrightarrow[AlCl_3]{CH_3Cl}$ toluene

These reactions are very important students can also try them in lab.

toluene

H_2SO_4

HNO_3 → 1-methyl-2-nitrobenzene → HNO_3 →

CH_3Cl / $AlCl_3$ → o-xylene

Cl(Fe), ordinary temp → 1-chloro-2-methylbenzene

sunlight → (chloromethyl)benzene → (dichloromethyl)benzene → (trichloromethyl)benzene

XYLENE

Introduction

The term **xylenes** refers to a group of 3 benzene derivatives which encompasses *ortho-*, *meta-*, and *para-* isomers of dimethyl benzene. The *o-*, *m-* and *p-* isomers specify to which carbon atoms (of the main benzene ring) the two methyl groups are attached. Counting the carbon atoms from one of the ring carbons bonded to a methyl group, and counting towards the second ring carbon bonded to a methyl group, the *o-* isomer has the IUPAC name of 1,2-dimethylbenzene. The *m-* isomer has the IUPAC name of 1,3-dimethylbenzene. And *p-* isomer has the IUPAC name of 1,4-dimethylbenzene.

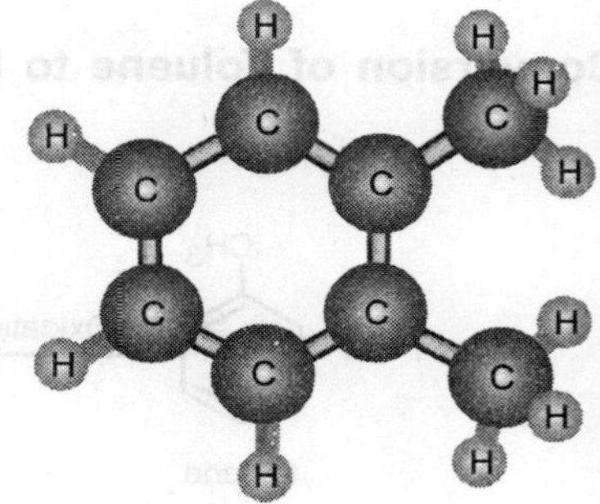

1,2-dimethylbenzene (ortho-xylene) | 1,3-dimethylbenzene (meta-xylene) | 1,4-dimethylbenzene (para-xylene)

Preparation

Xylene is prepared from toluene, by the reaction of methyl chloride and aluminium trichloride

$$C_6H_5CH_3 \xrightarrow[AlCl_3]{CH_3Cl} C_6H_4(CH_3)_2$$

Industrial Preparation

The three isomeric xylenes are now been produced industrially by reforming C_6-C_8 petroleum fraction of light naphtha at 400°C - 500°C at 25 – 35 atm. Pressure over platinum alumina catalyst. The cyclization and aromatization gives a mixture of benzene toluene and xylene. In the BTX plant the aromatic s produced by reforming the aromatics produced by reforming re extracted with diethyl glycol and refractionated. The C_8H_{10} fraction is subjected to distil o-xylene and apparently in separable mixture of the m and p – isomer. When the mixture is cooled to 60°C pure p-xylene separates out as a crystalline solid, leaving behind the mother liquor rich in m-xylene.

Physical Properties

Properties	o-xylene	m-xylene	p-xylene
Freezing Point	–25.182°C	–47.872°C	13.263
Boiling Point	144.411°C	139.103°C	0.04898
Density at 20°C	0.8802 g/ml	0.8642 g/ml	0.8610 g/ml
Refractive Index	1.50545	1.49722	1.49582
Sp heat (water=1)at 30°C	0.411	0.387	0.397
Flash point (closed cup) 0°C	34.4	30.6	30.0
Sp heat dispersion at 20°C	180°C	180.9	182.4
Refractivity intercept at 20°C	1.06535	1.06514	1.06530
Parachor	282.41	284.21	284.31
Auto ignition temp	495°C		466°C
Vapour dipole moment × 10^{18}	0.62		
Viscosity at 20°C	0.809	0.617	0.6783
Thermodynamic Properties			
Critical temperature	359.0°C	346.0°C	345°C
Critical Pressure	36 atmos	35	34

Critical Density	0.28 gm/mol	0.27	0.24
Heat of Vapourization	82.4 cal/gm	82 cal/gm	81.2 cal/gm
Heat of formation at 25°C	–5.841Kg cal/mol	–6.075 Kg cal/mol	-5.838Kg cal/mol
Entropy at 25°C	58.41cal/deg mol	60.27 cal/deg mol	59.12 cal/deg mol
Free Energy at 25°C	26.37 Kg cal/mol	25.73 Kg cal/mol	26.31 Kg cal/mol
Heat of combustion at const pressure	10,250 cal/g	10.247 cal/g	10,277 cal/g
Cryoscopic constant A mol fraction/°C	0.0265	0.0027	0.0028

Chemical Properties

The xylenes form substitute products much in the same way as toluene. On oxidation with alkaline or acidic potassium permanganate or acidic potassium dichromate, the methyl group is uncovered COOH groups.

O-xylene on oxidation with air in the presence of vanadium pentoxide gives phthalic anhydride

$$\text{o-xylene vapour} + 3O_2 \xrightarrow[500\ ^\circ C]{V_2O_5} \text{phthalic acid} + \text{phthalic anhydride}$$

p-xylene is oxidized to terephthalic acid on large scale by one of the most recent methods of passing air into the liquid hydrocarbon under pressure in the presence of a cobalt salt catalyst.

$$\text{p-xylene} + 3O_2 \xrightarrow[200\ ^\circ C,\ 200\ atm]{\text{Cobalt Salt}} \text{terephthalic acid} + 2H_2O$$

Uses

Technical (industrial) xylene is a mixture of the 3 isomers plus ethylbenzene (6-15%) and occasionally toluene, trimethyl benzene and other trace components. It is widely used as a solvent and thinner for paints and varnishes, often in combination with other organic compounds and as a solvent in glues and printing inks. It is used as a process chemical in the rubber and leather industries, in the formulation of pesticides, as an intermediate in the manufacture of certain polymers, in petroleum distillation, and in histology laboratories.

STYRENE

Styrene (also **vinyl benzene**, ethenylbenzene, phenethylene or phenylethene, cinnamene, diarex HF 77, styrolene, styrol, styropol) is an organic compound which is an aromatic hydrocarbon having the chemical formula C_8H_8. At room temperature and pressure, styrene is a liquid. The chemical structure is shown at right. It is colourless, oily, toxic, flammable, and occurs in very small amounts in some plants, but is produced in industrial quantities from petroleum. The production of styrene in the United States was increased dramatically during the 1940's to supply the war needs for synthetic rubber. Because the styrene molecule has a vinyl group with a double bond, it can readily undergo polymerization. It is used as a monomer to make plastics such as polystyrene, ABS, styrene-butadiene (SBS) rubber, styrene-butadiene latex, SIS (styrene-isoprene-styrene), S-EB-S (styrene-ethylene/butylene-styrene), styrene-divinylbenzene (S-DVB), and unsaturated polyesters. It evaporates easily and has a sweet smell. It often contains other chemicals that can result in a sharp, unpleasant smell.

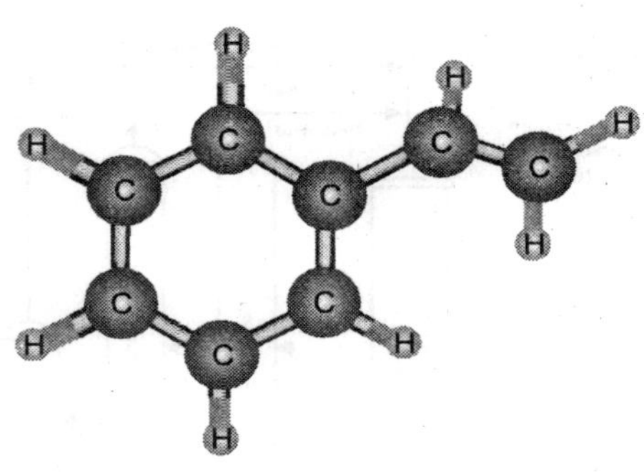

Styrene dissolves in many organic liquids but does not dissolve significantly in water. Millions of tonnes per day are produced to make products such as rubber, plastic, insulation, fiberglass, pipes, automobile parts, food containers, and carpet backing.

Most of these products contain styrene linked together in a long chain (polystyrene) as well as unlinked styrene. Low levels of styrene also occur naturally in a variety of foods such as fruits, vegetables, nuts, beverages, and meats.

Preparation

Styrene is prepared by reaction of benzene with ethylene in liquid state, using Aluminium trichloride $AlCl_3$ as catalyst at 90ºC at moderate pressure. The ethyl benzene thus obtained is dehydrogenated at 600ºC over a catalyst (Fe_2O_3)ZnO or MgO supported on alumina

$$C_6H_6 + H_2C{=}CH_2 \xrightarrow[90\ ^\circ C]{AlCl_3} C_6H_5{-}CH_2{-}CH_3 \xrightarrow[600\ ^\circ C]{Fe_2O_3} C_6H_5{-}CH{=}CH_2$$

benzene ethylbenzene styrene

Industrial Preparation Styrene

In considering details of the technology of styrene manufacture by dehydrogenation it is convenient to treat separately: (1) the ethylbenzene production step (alkylation), (2) the conversion of ethylbenzene to styrene (dehydrogenation), and (3) the separation of the pure monomer (fractionation). These steps are illustrated schematically in Fig. 22.21.

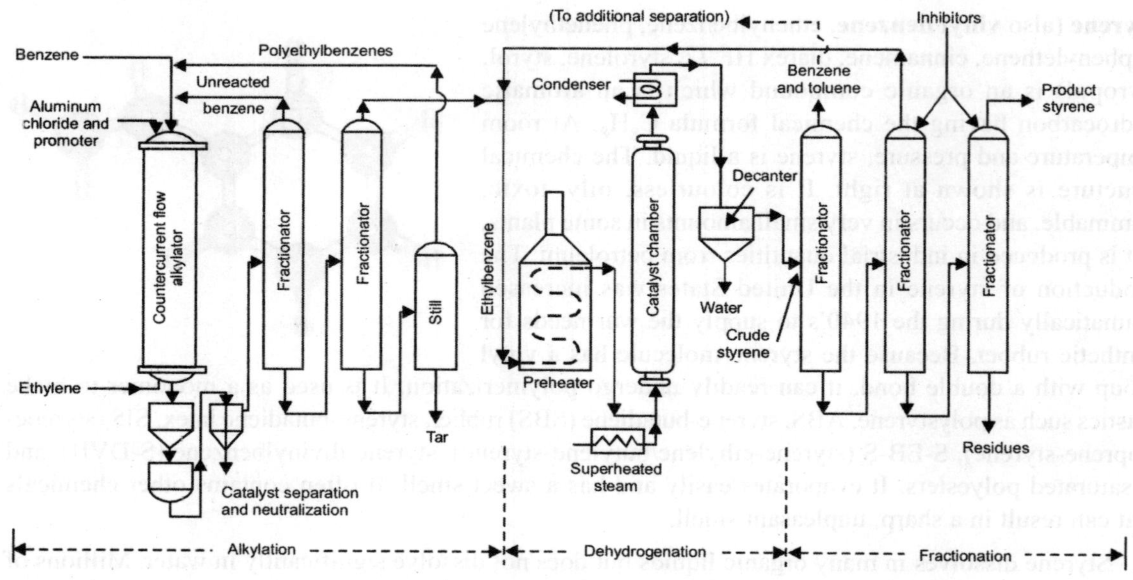

Fig. 22.2.

Physical Properties

Physical form	Colourless oily liquid
Odour	Unpleasant, aromatic
Odour Threshold	0.1 ppm0.433 mg/m³
Relative Molecular mass	104.14
Density at 20°	KG/L906 kg/m³
Boiling point	145°C
Melting point/freezing point	– 31°C
Kinematic Viscosity at 0°C 1.1	cSt1.1 mm²/s
Vapour density (air =1)	3.6
Vapour pressure at 20°C	6.0 mbar0.6 kPa

at 50° C 32 mbar	3.2 kPa
Conductivity : Thermal @ 20°C	0.16 W/m°C
Flash point (ABEL)IP	170 32° C
Autoignition temperature	490°C
Satn .Concentration in air at 20° C	25.6 g/m³
Explosion limits in air	1.1 – 6.1% vol
Miscibility with water	Immiscible
Solubility in water at 20° C	0.29 kg/m³
Solubility of water in substance at 20° C	0.54 kg/m³
Coefficient of Expansion @ 20°	0.979×10^{-3}/°C
Specific Heat @ 20° C	1.73 kJ/kg.° C
Heat of combustion, Hc, @25 °C	- 4265.64 Kj/mol
Heat of formation, Ht,	
Gas, @ 25 °C	35.22 Kcal/mol
Liquid, @ 25 °C	24.72 Kcal/mol
Heat of fusion, Hm	-11011 J/mol
Heat of polymerization, Hp, @ 25 °C	- 70.67 Kj/mol
Refractive index nd20	1.5460
Relative evaporation rate (Diethyl ether = 1 min)	12.4 mins

Chemical Properties

The most important chemical reactions of styrene, is polymerization including co-polymerization. With other monomers

POLYSTYRENE

Polystyrene was accidentally discovered in **1839 by Eduard Simon**, an apothecary in Berlin, Germany. From storax, the resin of *Liquidambar orientalis*, he distilled an oily substance, a monomer which he named styrol. Several days later Simon found that the styrol had thickened, presumably due to oxidation, into a jelly he dubbed styrol oxide ("Stryroloxyd"). By 1845 **English chemist John Blyth and German chemist August Wilhelm von Hofmann** showed that the same transformation of styrol took place in the absence of oxygen. They called their substance metastyrol. Analysis later showed that it was chemically identical to Styroloxyd. In **1866 Marcelin Berthelot** correctly identified the formation of metastyrol from styrol as a polymerization process. About 80 years went by before it was realized that heating of styrol starts a chain reaction which produces macromolecules, following the thesis of German organic chemist **Hermann Staudinger (1881 - 1965).** This eventually led to the substance receiving its present name, polystyrene. The I.G. Farben company began manufacturing polystyrene in **Ludwigshafen, Germany, about 1931**, hoping it would be a suitable replacement for die cast zinc in many applications. Success was achieved when they developed a reactor vessel that extruded polystyrene through a heated tube and cutter, producing polystyrene in pellet form.

Polystyrene is a polymer made from the monomer styrene, a liquid hydrocarbon that is commercially manufactured from petroleum. At room temperature, polystyrene is normally a solid thermoplastic, but can be melted at higher temperature for molding or extrusion, then resolidified. Styrene is an aromatic monomer, and polystyrene is an aromatic polymer.

Pure solid polystyrene is a colourless, hard plastic with limited flexibility. It can be cast into molds with fine detail. Polystyrene can be transparent or can be made to take on various colours. It is economical and is used for producing plastic model assembly kits, plastic cutlery, CD "jewel" cases, and many other objects where a fairly rigid, economical plastic of any of various colors is desired.

The chemical makeup of polystyrene is a long chain hydrocarbon with every other carbon connected to a Phenyl group (an aromatic ring similar to benzene).

Many — Polymerization →

Styrene — Polystyrene

A 3-D model would show that each of the chiral backbone carbons lies at the center of a tetrahedron, with its 4 bonds pointing toward the vertices. Say the -C-C- bonds are rotated so that the backbone chain lies entirely in the plane of the diagram. From this flat schematic, it isn't evident which of the phenyl (benzene) groups are angled toward us from the plane of the diagram, and which ones are angled away. The isomer where all of them are on the same side is called *isotactic* polystyrene, which isn't produced commercially. Ordinary *atactic* polystyrene has these large phenyl groups randomly distributed on both sides of the chain. This random positioning prevents the chains from ever aligning with sufficient regularity to achieve any crystallinity, so the plastic has no melting temperature, T_m. But metallocene-catalyzed polymerization can produce an ordered *syndiotactic* polystyrene with the phenyl groups on alternating sides. This form is highly crystalline with a T_m of 270°C.

Physical Properties

Density	1050 kg/m³
Electrical conductivity (σ)	10-16 S/m
Thermal conductivity	0.08 W/(m·K)
Young's modulus (E)	3000-3600 MPa
Tensilé strength (σt)	46–60 MPa
Elongation @ break	3–4%
Notch test	2–5 kJ/m²
Glass temperature	95 °C
Melting point—	
Heat transfer coefficient (λ)	0.17 W/(m·K)
Linear expansion coefficient (á)	8 10-5 /K
Specific heat (c)	1.3 kJ/(kg·K)

Oxidation of Styrene

Oxidation with polyhydroxybenzoic acid is reported give yields of 70% of styrene oxide. Air oxidation of styrene gives a mixture of products including formaldehyde, benzaldehyde and polyners. Potassium permanganate with dilute alkali oxidizes styrene principally to benzolyformic acid (phenylglyoxalic acid) $C_6H_5COOHCO$

Chlorine, bromine chloride, iodine chloride readily added to the styrene, but addiftion reaction is of quantitative importance.

CH=CH₂ —(CCl_4Br, 0 °C)→ CH=CH—Br / CH_2—Br

Similarly hydrogen halides and hypohalous acids readily add to styrene.

styrene (CH=CH₂) + HCl —(Solvent, pressure)→ (1-chloroethyl)benzene (HC(CH₃)—Cl)

$$C_6H_5CH{=}CH_2 + HOCl \longrightarrow C_6H_4(COOH)(CH_2{-}Cl)$$

styrene → 2-(chloromethyl)benzoic acid

Addition of hydrogen halides normally yields (1 – haloethyl) benzene although in the presence of lauryl peroxide, hydrogen bromide adds to give (2 – bromomethyl)benzene showing the typical peroxide effect.

Various oxygen nitrogen and sulphur compounds reacts with styrene in the presence of aluminium trichloride in toluene solution, phosgene reacts.

CHAPTER

23

Orientation and Isomerism of Aromatic Hydrocarbon

Isomerism of Benzene

It has already been pointed out ion the explanation of the structure of benzene molecule is symmetrical. This is confirmed by studying the isomerism of benzene derivatives.

Isomerism of Monoderivatives

It is possible to substitute 1,2,3,4,5 and 6 monovalent atoms or groups in benzene. It has further been shown that when any one atom of hydrogen is replaced by a univalent atom or group, the same compound is always produced. This proves that all the six hydrogen atoms in benzene molecule are equal and identical and hence the mono-substitution products of benzene exist in one form only when toluene ($C_6H_5CH_3$) is prepared by any method, the compound is always produced and may be represented by any one of the following structured formula.

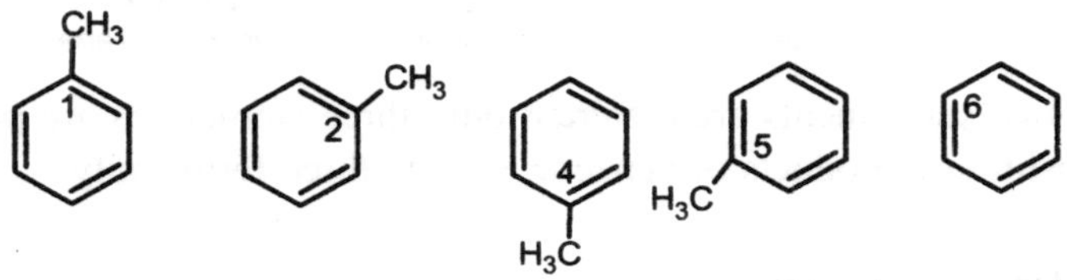

CH_3 1   2 CH_3 4 H_3C H_3C 5   6

That is toluene exists in one form only and shows no isomerism. The same is true of other mono-derivatives of benzene.

Isomerism of Di-derivatives

If two hydrogen atoms of benzene are replaced by two monovalent atoms or groups only three isocyanides are obtained. Thus three dimethyl benzenes (xylenes) $C_6H_4(CH_3)_2$, three dinitrobenzene $C_6H_4(NO_2)_2$ three di hydroxy benzenes $C_6H_5(OH)_2$ etc. are all known.

The existence of such three isomerides can be easily explained by the ring structure numbered shown.

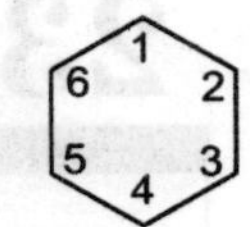

Let any mono-substituted product $C_6H_5 - X$ can be converted into a distributed product $C_6H_6(X_2)$. If the position of 'X' already present is numbered 1, the second atom or group may substitute any one of the hydrogen atoms at 2, 3, 4, 5 or 6 giving compounds the constitution of which may be represented by any one of the following five formations.

(1 : 2)	(1 : 3)	(1 : 4)	(1 : 5)	(1 : 6)
I	II	III	IV	V

In the above structures positions 1:2 ° 1:6. and 1:3 °1:5. That is formulae numbers I and V are identical and also numbers II and IV are similar.

Hence only these isomers (1:2 ° 1:6, 1:3 °1:5 and 1:4) possible. The derivative formed by substitution in position 2 or 6 i.e. adjacent positions are known as ortho compounds, that in positions 3 or 5 or alternate positions. Meta compounds and that in positions 4 of diagonal positions, para compound. They may also be noted by o, m and p or 1:2, 1:3 and 1:4 e.g.

o or 1:2 dinitrobenzene	m or 1:3 dinitrobenzene	p or 1:4 dinitrobenzene

Even when the two substituents are different only three isomerides are possible. Hence we can conclude that di-substitution products of benzene exist in three forms only.

Isomerides of Triderivatives

When more than two hydrogen atoms in benzene are replaced it has been found that the number of isomerides varies according as the substituent atoms or groups are similar or different. If the substituents are similar or same each trisubstitution product exists in three isomeric forms. These are distinguished from one another adjacent (1:2:3) unsaturated (1:2:4) and symmetrical (1:3:5) derivatives of benzene. Thus trichlorobenzene may be represented as.

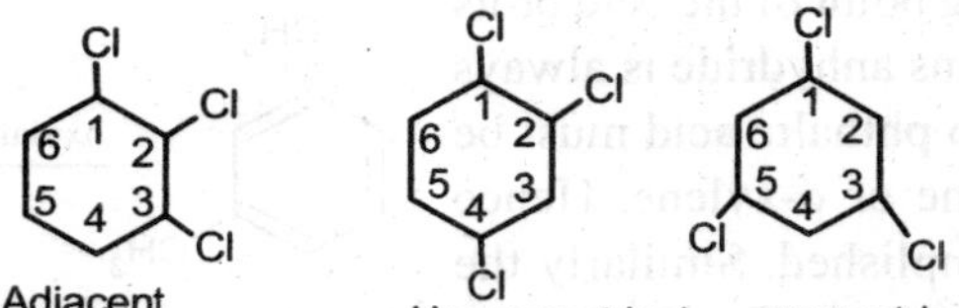

Adjacent or 1 : 2 : 3 trichlorobenzene — Unsymmetrical 1 : 2 : 4 trichlorobenzene — symmetrical 1 : 3 : 5 trichloro benzene

On the other hand when the three substituents are not alike the number of isomerides is greater than three.

Isomerism of the Tetra Derivatives

The tetra-substitution products of benzene which all the substituents are identical exists in the three isomeric forms

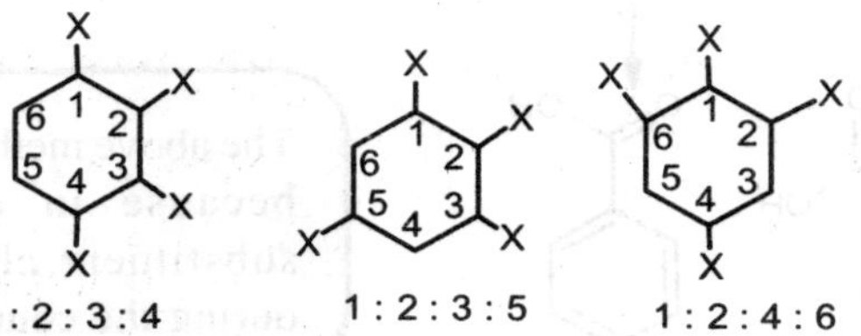

1 : 2 : 3 : 4 — 1 : 2 : 3 : 5 — 1 : 2 : 4 : 6

If the four substituents are different the number of isomers is even greater than in the case of three unlike substituents.

Isomerism of Penta and Hexa Derivatives

When, however, five of all the six atoms of hydrogen are displaced by identical substituents one compound is formed. If the substituents are different in number of possible isomers in each case become much higher.

Orientation of Benzene Derivatives

Orientation is the process by which the relative positions of the substitution agents ion the benzene nucleus can be located. There are two different methods to accomplish this.

Relative Method

If the position of the atoms or groups in certain fundamental compounds is known, the structure of substances directly related to them can be easily ascertained. Consider, for example, three xylenes which give rise to three phthalic acids on complete oxidation. If the position of the carbonyl groups in the three phthalic acids could be ascertained, that of the methyl groups in the three xylenes would be known. The positions of the carboxyl groups in different phthalic acids (i.e. o, m, and p -) can be confirmed by determining the melting point of the acid or its methyl ester. Also which readily forms anhydride is always ortho acid. Thus it follows that ortho phthalic acid must be derived from ortho dimethyl benzene or o-xylene. Hence the orientation of o- xylene is accomplished. Similarly the constitution of meta and para xylenes is confirmed.

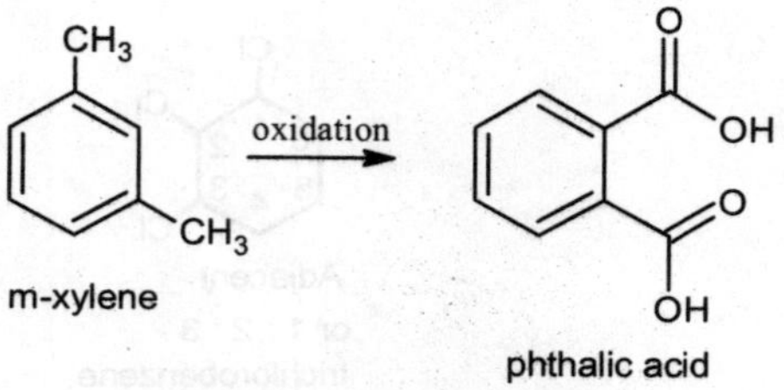

The conversions of xylenes into phthalic acid are shown below:

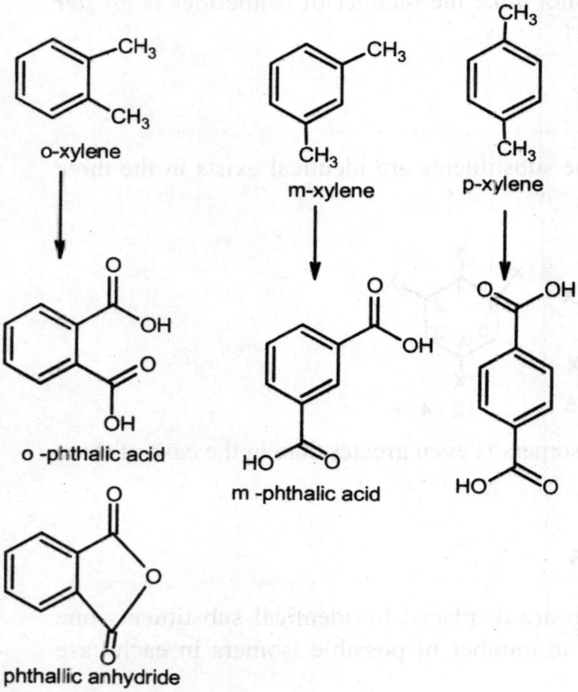

The above method is not satisfactory because in certain cases the substituent changes its position during the course of reaction

Korner's Absolute Method

It is based on the principle followed: A di-derivative will yield a different number of tri-derivatives according to whether the original compound is ortho, meta and para. It has been seen that ortho di-derivatives give mixture of two meta-di-derivatives give a mixture of three and para-di-derivatives give only one tri-substitution product. This can be more clearly be understood from the following scheme

(a) 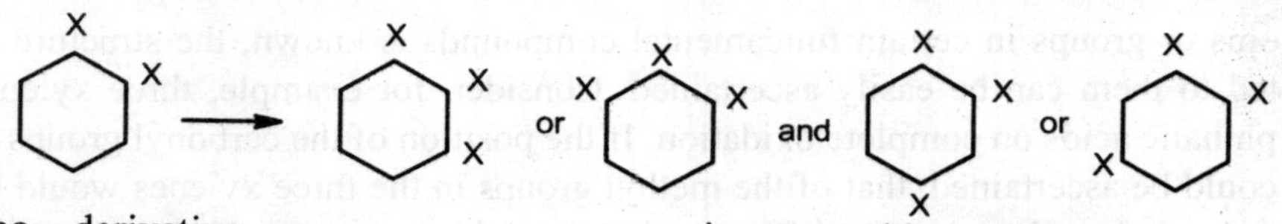

ortho – derivative ——— gives two triderivatives

(b) 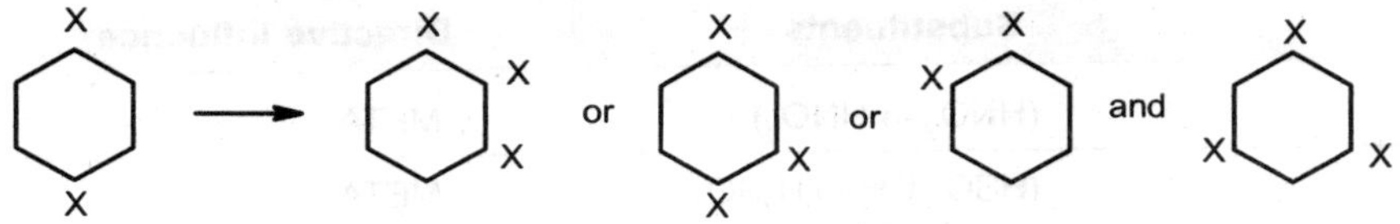 para derivative ——— gives tri-derivative

(c) 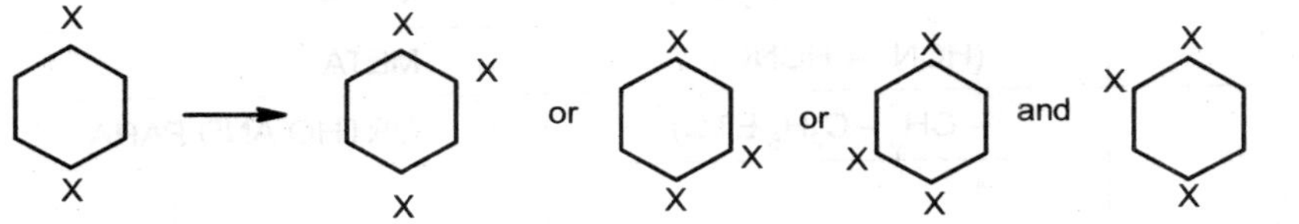 para derivative All the four para-derivatives are identical. It gives only one tri derivative

It is immaterial whether the new group is the same or different. The Korner's Absolute theory becomes difficult some of the isomers cannot be easily indicated.

Directive Influence of Substituents

The nature of a substituent already present in the benzene ring determines the positions taken up by the second entrant. There are two distinct types of substituents: -

Type I Substituents which direct the entrant to the ortho and para positions

Type II Substituents which direct the new entrants to the meta positions.

The groups which direct new entrants to the ortho and para positions are: -

$$-OH - NH_3\,Cl, Br, I - CH_3\text{ -}C = C\text{ - , - }N = N\text{ - etc}$$

It must be suppose that no meta isomer will be formed or the ratio of the ortho and para isomers formed and are influences by temperature, nature of solvent and reagents used. The groups that possess meta – directive influence are -NO_2, -COOH, -SO_2.OH, -CHO and para – substitution proceeds rapidly where as meta-substitution takes place slowly.

Crum Brown and Gibson's Rule

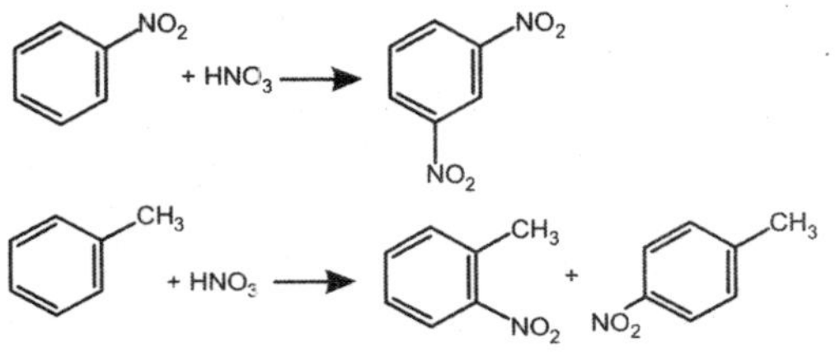

The directive influence of a substituent can also be prepared with the help of Crum Brown and Gibson rule. The rule stales that if a group already present (say I) forms a compound with hydrogen (HI) which can readily be converted by direct oxidation into the corresponding hydroxy compound (HOX) the second substituent with occupy the meta position otherwise will enter the ortho and para positon:

NO_2 + HNO_3 ⟶ NO_2, NO_2

CH_3 + HNO_3 ⟶ CH_3, NO_2 + CH_3, NO_2

Rule has got its limitations

The directive influence of some of the important substituents are given below

	Substituents	Directive Influence
1. $-NO_2$	$(HNO_2 \rightarrow HNO_3)$	META
2. $-SO_2H$	$(HSO_2.OH \rightarrow H_2SO_4)$	META
3. –CHO	(HCO → HCOOH)	META
4. –COOH	$(HCOOH \rightarrow H_2CO_3)$	META
5. –CN	(HCN → HCNO)	META
6. –Cl Br or I	$- CH_3 - C_2H_5$ ETC)	ORTHO AND PARA
7. Alkyl group i.e.	"	"
8. $-NH_2$	"	"
9. –OH	"	"
10. $-CH_3Cl$	"	"

CHAPTER

24

Aromatic Halogen Derivative

Introduction

We have already studied in preceding chapters that benzene and its homologous react with halogens to give either "additive or substituted compounds" The additive derivatives (such as $C_6H_5Cl_6$) are of little importance.

The halogen substitution products of the aromatic hydrocarbons are more important and fall in two classes depending on whether the halogen is in the ring or in the side chain

Class I Nuclear Halogen Compounds

In these compounds the halogen atom is directly attached to the nucleus

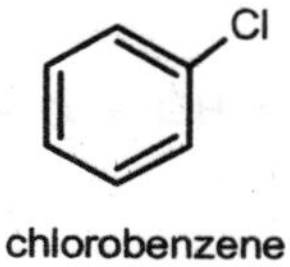

chlorobenzene 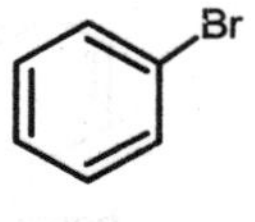 bromobenzene 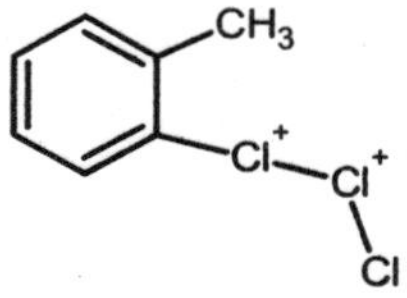 o - chlorotoluere

The side chain, halogen atom is quite reactive and can be easily exchanged for -OH, NH_2 or -CN groups by reactions. They closely resemble alkyl halides in chemical properties

Nuclear Halogen Compounds

CHLOROBENZENE

Introduction

Chlorobenzene is an aromatic organic compound with the chemical formula C_6H_5Cl. It is a colorless, flammable liquid first made in 1851 by reacting phenol and phosphorus pentachloride.

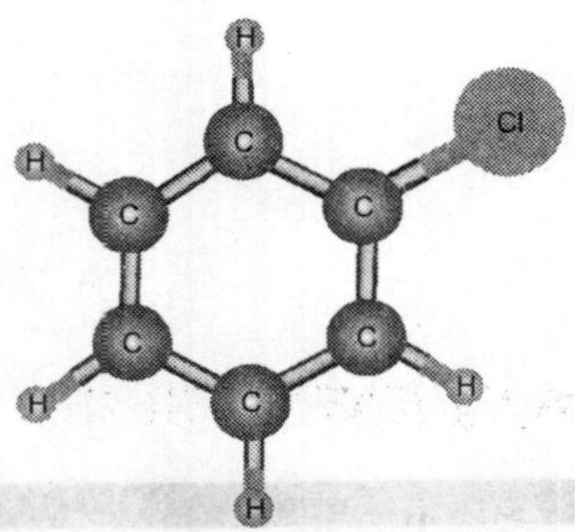

Preparation

Chlorobenzene is prepared by the following methods.

1. From Benzene: By direct chlorination, in the presence of 'halogen carrier' which acts as catalyst.

$$\text{benzene} + Cl_2 \xrightarrow[\text{carrier}]{\text{halogen}} \text{chlorobenzene} + HCl$$

The preparation is important for the manufacture of benzene. Benzene can also be converted industrially into Chlorobenzene, by passing the mixed vapours of benzene, air and hydrogen chloride, over heated cupric chloride.

$$\text{benzene} + HCl + O \longrightarrow \text{chlorobenzene} + H_2O$$

2. From Benzene Diazomium Chloride : By warming with cuprous chloride and HCl. The reaction is known as Sandmayer's Reaction.

Cuprous chloride is used as a catalyst

$$\text{benzene diazomium chloride } (C_6H_5{-}N{=}N{-}Cl) \xrightarrow[HCl]{Cu_2Cl_2} \text{chlorobenzene} + N_2 + HCl$$

3. From Phenol: By the action of phosphorous pentachloride.

phenol + PCl_5 → chlorobenzene + HCl + $POCl_3$

4. From Benzene Sulphonyl Chloride : By the action of phosphorous pentachloride on benzene Sulphonyl chloride.

benzenesulfonyl chloride + PCl_5 → chlorobenzene + $SOCl_2$ (thionyl chloride) + $POCl_3$

Physical Properties

Boiling Point	131 – 132 °C
Melting Point	– 45 °C (solidifies at -55 oC)
Flash Point	85 °F (closed cup)
Vapor Density	3.88 (air = 1)
Vapor Pressure	11.9 mm Hg at 25 °C
Density/Specific Gravity	1.107 at 20/4 °C (water = 1)
Log Octanol/Water Partition Coefficient	2.84
Conversion Factor	1 ppm = 4.6 mg/m^3

Chemical Properties

It is stable and does not react with boiling alkalies, moist silver oxide alcoholic ammonia, KCN etc. it however, shows the following important reactions

1. With Sodium Hydroxide: When heated with sodium methoxide at high temperature, it gives phenyl methyl ether (anisole).

chlorobenzene + $H_3C-O-Na$ (methanolato)sodium → $C_6H_5-O-CH_3$ + NaCl

2. *Witz Fittig's Reaction*: Homologous of benzene are obtained when a mixture of aryl halide and alkyl halide is treated with sodium in the presence of ether.

chlorobenzene + 2Na + CH_3Br —ether→ toluene + NaCl + NaBr

3. *With Magnesium:* Dissolved in ether it yields Grignard's reagent.

chlorobenzene + Mg → phenyl magnesium Chloride (Mg—Cl)

The compounds is important in synthetic chemistry and may be used for the preparation alcohols ketones, acids etc.

4. *With Ammonia:* in the presence of a little copper sulphate at 200ºC under presence it gives aniline.

chlorobenzene + NH_3 —$CuSO_4$, 200 °C (under pressure)→ aniline + HCl

5. *Like Benzene*: It can also be nitrated sulphonated. Chlorinated giving mainly ortho, para and meta substituted products.

chlorobenzene —sulphuric acid, sulphonation→ 2-chlorobenzenesulfinic acid and 4-chlorobenzenesulfinic acid

chlorobenzene —nitric acid, nitration→ 1-chloro-2-nitrobenzene and 1-chloro-4-nitrobenzene

Chlorobenzene $\xrightarrow[\text{chlorination}]{Cl_2}$ 1,2-dichlorobenzene and 1,4-dichlorobenzene

Uses

Chlorobenzene has been used in the manufacture of certain pesticides, most notably DDT by reaction with chloral (trichloroacetaldehyde). It also once found use in the production of phenol. However, use of these manufacturing processes has declined significantly in the past few decades. Today the major use of chlorobenzene is as an intermediate in the production of nitrochlorobenzenes and diphenyl oxide, which are important in the production of commodities such as herbicides, dyestuffs, and rubber.Chlorobenzene is also used as a high-boiling aprotic solvent in organic chemistry, as a solvent for paints, and for degreasing automobile parts.

DICHLORODIPHENYLTRICHLOROETHANE, DDT

History

DDT was first synthesized in 1874 by Othmar Ziedler, but its insecticidal properties were not discovered until 1939, by the Swiss scientist Paul Hermann Müller, who was awarded the 1948 Nobel Prize in Physiology and Medicine for his efforts. DDT is the best-known of a number of chlorine-containing pesticides used in the 1940s and 1950s. It was used extensively during World War II by Allied troops and certain civilian populations to control insect typhus and malaria vectors (nearly eliminating typhus as a result). Civilian suppression used a spray on interior walls, which kills mosquitoes that rest on the wall after feeding to digest their meal; resistant strains are repelled from the area. Entire cities in Italy were dusted to control the typhus carried by lice. DDT also sharply reduced the incidence of biting midges in Great Britain, and was used extensively as an agricultural insecticide after 1945.

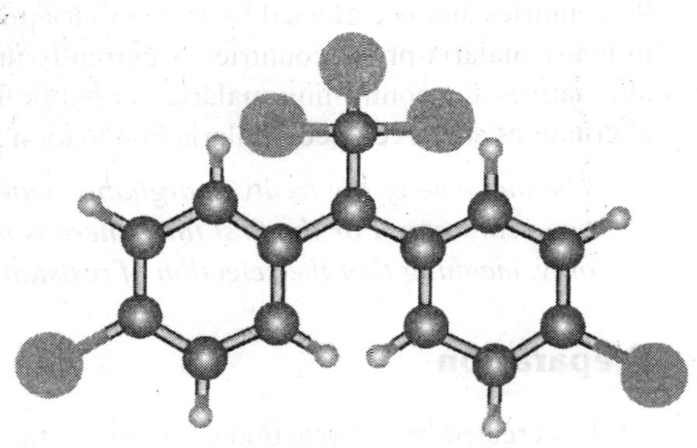

DDT was responsible for eradicating malaria from Europe and North America. Though today malaria is considered a tropical disease, it was more widespread prior to an extensive malaria eradication program carried out in the 1950s. Though this program was initially highly successful worldwide (reducing mortality rates from 192 per 100,000 to a low of 7 per 100,000), resistance emerged in many insect populations over time. DDT was less effective in tropical regions due to the continuous life cycle of mosquitoes and poor infrastructure. It was not pursued aggressively in sub-Saharan Africa due to these perceived difficulties, with the result that mortality rates in the area were never reduced to the same dramatic extent, and now constitute the bulk of malarial deaths worldwide, especially following the resurgence of the disease as a result of microbe resistance to drug treatments and the spread of the deadly malarial variant caused by Plasmodium falciparum.

In the 1970s and 1980s, agricultural use of DDT was banned in most developed countries, and DDT was replaced in most antimalarial uses by less persistent, but more expensive, alternative insecticides. DDT was first banned from use in Norway and Sweden in 1970, but was not banned in the United Kingdom until 1984.

As of 2006, DDT continues to be used in other (primarily tropical) countries where mosquito-borne malaria and typhus are serious health problems. Use of DDT in public health to control mosquitoes is primarily done inside buildings and through inclusion in household products and selective spraying; this greatly reduces environmental impact compared to the earlier widespread use of DDT in agriculture. It also reduces the risk of resistance to DDT.[2] This use only requires a small fraction of that previously used in agriculture; for the whole country of Guyana, covering an area of 215,000 km², the required amount is roughly equal to the amount of DDT that might previously have be used to spray 4 km² of cotton during a single growing season.

The Stockholm Convention, ratified in 2001 and effective as of 17 May 2004, calls for the elimination of DDT and other persistent organic pollutants, barring health crises. The Convention was signed by 98 countries and is endorsed by most environmental groups. However, a total elimination of DDT use in many malaria-prone countries is currently unfeasible because there are few affordable or effective alternatives for controlling malaria, so public health use of DDT is exempt from the ban until such alternatives are developed. Malaria Foundation International states:

The outcome of the treaty is arguably better than the status quo going into the negotiations over two years ago. For the first time, there is now an insecticide which is restricted to vector control only, meaning that the selection of resistant mosquitoes will be slower than before.

Preparation

DDT is created by the reaction of trichloroethanol with chlorobenzene (C_6H_5Cl).

1,2,2-trichloroethanol

dichloroethyl trichlorobenzene

DDT is a white crystalline solid with no apparent odour or taste. It is a man-made organochlorine consisting of a mixture of DichloroDiphenylTrichloroethane isomers. It is made by a reaction between chlorine gas and the double benzene ring structure under optimal temperature and pressure conditions. When the two compounds are reacted, DDT is formed easily because of the high reactivity of chlorine gas. DDE and DDD are also made in the same reaction and often contaminate DDT, but are also both very toxic to living organisms.

Physical Properties

Melting point	= 108.5 °C
Boiling point	= 260 °C
Molecular Formula	= $C1_4H_9Cl_5$
Formula Weight	= 354.48626
Composition	= C(47.43%) H(2.56%) Cl(50.01%)
Molar Refractivity	= 84.50 ± 0.3 cm^3
Molar Volume	= 244.1 ± 3.0 cm^3
Parachor	= 638.9 ± 4.0 cm^3
Index of Refraction	= 1.608 ± 0.02
Surface Tension	= 46.8 ± 3.0 dyne/cm
Density	= 1.451 ± 0.06 g/cm^3
Polarizability	= 33.50 ± 0.5 $10^{-24}cm^3$
Monoisotopic Mass	= 351.914689 Da
Nominal Mass	= 352 Da
Average Mass	= 354.48999 Da

Chemical Properties of DDT

Molecular Weight	355 g/mol
Concentration in Gas Phase	1.9 x 10-6 mg/L
Solubility in Water Very Low	(0.001-0.04 mg/L)
Half-Life in Soil Approx.	2.8 yrs
Potential for Entry into Fresh Water	Strong
Aquatic Toxicity	High
Aquatic Persistence	Prolonged
Bioaccumulation Potential	Strong

Purpose, Uses of DDT

DDT was developed during the second World War to control a wide variety of insects. It was hailed as the "miracle insecticide" and quickly became one of the most widely used pesticides in the world to control insects on agricultural crops. It is also used to control insects that carry malaria, typhus, and other harmful diseases in third world countries. The use of DDT was banned in the U.S.A. in 1972, although it is still being used in many underdeveloped countries.

Major Sources of DDT and its Route of Entry into Fresh Waters

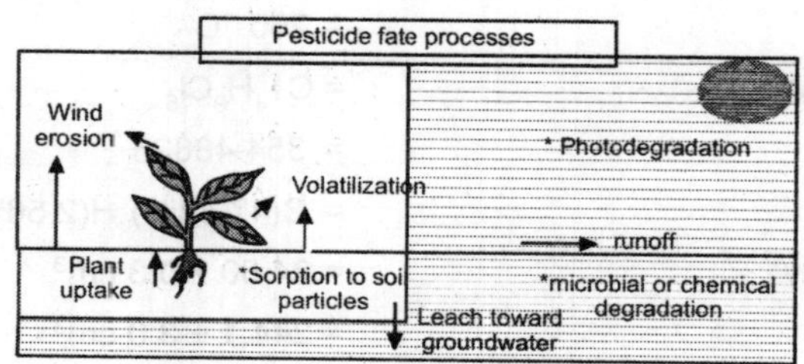

Fig. 24.1. The Fate of DDT and Other Pesticides in the Environment.

The major source of DDT in water is agricultural run-off from fields that were once heavily sprayed with DDT for pest control. Some DDT was also sprayed directly into lakes and streams through aerial spraying of crops. DDT binds very strongly to soil particles and is very slow in reaching ground water, thus soil run-off is a key contributor to its distribution to the aquatic environment. Industrial effluents and waste material from pesticide factories may also result in DDT reaching aquatic environments. Some DDT evaporates from the soil and surface water into the air and some is broken down by sunlight or by microorganisms in the soil or surface water. When DDT is broken down in soil, it usually forms DDE or DDD. Levels of DDT in North America are due to contaminated sediments from lake bottoms and tributaries due to runoff from sites of historical use, leaking landfill sites, illegal use of old stock, and long range transportation through the atmosphere from countries still using DDT. DDT builds up in plant tissues and in the fatty tissues of fish, birds, and animals. Most DDT enters the body through ingestion of contaminated foods and sediments, but minute amounts may pass through the skin or lungs. DDT levels increase in animals that are high on the food chain due to bioaccumulation.

BROMOBENZENE

Introduction

Bromobenzene are a group of halobenzenes formed in a substitution reaction between bromine and benzene with a hydrogen bromide biproduct. The name strictly refers to monobromobenzene, a benzene with a single bromine; however it can be used to refer to a benzene containing any number of bromine molecules.

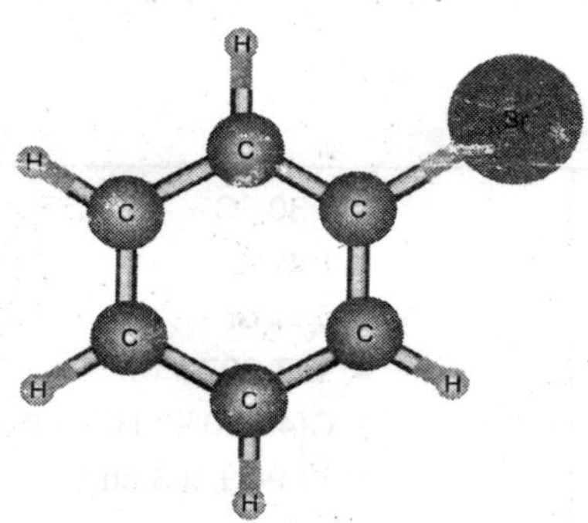

Preparation

1. From Benzene: By heating with benzene under reflux condenser in presence of a halogen carrier (iron fillings).

$$\text{benzene} + Br_2 \xrightarrow[\text{iron fillings}]{\text{reflux}} \text{bromobenzene} + HBr$$

Bromobenzene can also be prepared by the action of hypobrous acid on benzene.

$$\text{benzene} + HBrO \longrightarrow \text{bromobenzene} + H_2O$$

2. Sandmayer's Reaction: By warming benzene diazomium chloride with cuprous bromide and hydrogen peroxide.

$$C_6H_5N{=}NCl \xrightarrow[Cu_2Br_2]{HBr} \text{bromobenzene} + N_2 + HCl$$

Physical Properties

Melting point	= – 30 °C
Boiling point	= 156 °C
Molecular Formula	= C_6H_5Br
Formula Weight	= 157.0079
Composition	= C(45.90%) H(3.21%) Br(50.89%)
Molar Refractivity	= 33.94 ± 0.3 cm^3
Molar Volume	= 105.6 ± 3.0 cm^3
Parachor	= 257.7 ± 4.0 cm^3
Index of Refraction	= 1.555 ± 0.02
Surface Tension	= 35.4 ± 3.0 dyne/cm
Density	= 1.486 ± 0.06 g/cm^3
Polarizability	= 13.45 ± 0.5 10-24cm^3
Monoisotopic Mass	= 155.957455 Da
Nominal Mass	= 156 Da
Average Mass	= 157.009625 Da

Chemical Properties

It is a colourless liquid, closely resembles chlorobenzene in chemical properties.

Uses

The commercial bromobenzenes are used as heavy liquid solvents, motor oil additives. and as an intermediates to manufacture organic chemicals including pharmaceuticals, pesticides and flame retardants for polymeric materials.

IODOBENZENE

Iodobenzene resembles Chlorobenzene very much

Preparation

It is conveniently prepared by the action of benzene diazomium chloride and potassium chloride.

+ KI ⟶

iodobenzene

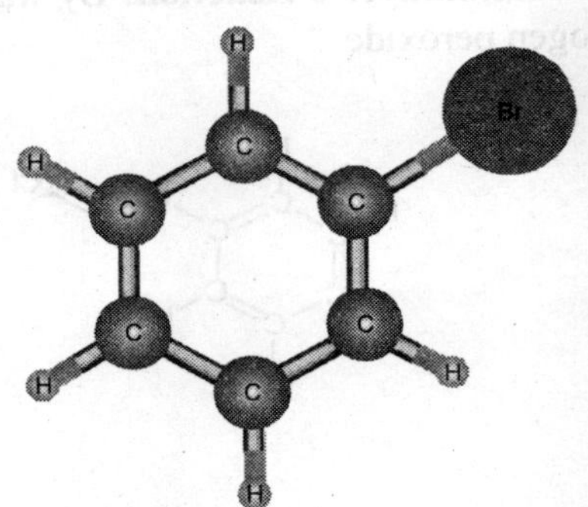

It is a colourless liquid more reactive than benzene chloride and closely resembles it in other properties.

Physical Properties

Density	= 1.831 g/cm³
Melting point	= − 29 °C
Boiling point	= 188 °C
Molecular Formula	= C_6H_5I
Formula Weight	= 204.00837
Composition	= C(35.32%) H(2.47%) I(62.21%)
Molar Refractivity	= 39.15 ± 0.3 cm³
Molar Volume	= 111.5 ± 3.0 cm³
Parachor	= 282.8 ± 4.0 cm³
Index of Refraction	= 1.619 ± 0.02
Surface Tension	= 41.4 ± 3.0 dyne/cm
Dielectric Constant	= Not available
Polarizability	= 15.52 ± 0.5 10-24 cm³
Monoisotopic Mass	= 203.943585 Da
Nominal Mass	= 204 Da
Average Mass	= 204.010561 Da

Chemcial Properties

Same as the bromo and chloro benzene

Side Chain Halogen Derivatives

BENZYL CHLORIDE *OR* CHLOROMETHYL BENZENE

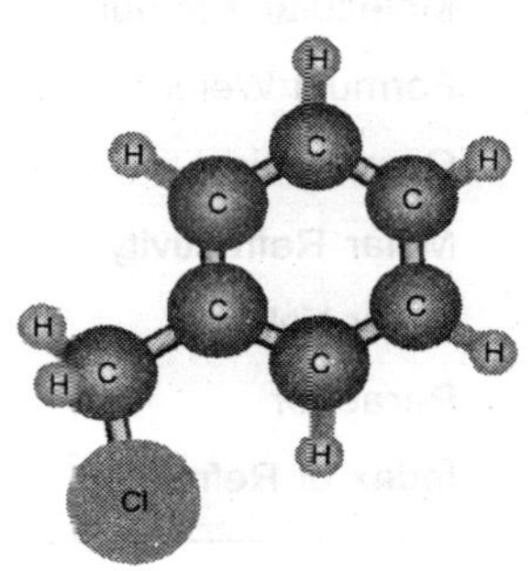

Introduction

Benzyl Chloride is a typical member among the side chain halogen compounds.

Preparation

Benzyl chloride is prepared by following methods.

The prouct is prepared by fractional distillation.

1. From Toluene

By passing dry chlorine into boiling toluene, in sunlight until it has obtained the required specific gravity.

toluene + Cl_2 → (chloromethyl)benzene

2. From Benzyl Alcohol

By the action of phosphorous pentachloride or hydrogen chloride on benzyl alcohol.

phenylmethanol (benzyl alcohol) + PCl_5 → (chloromethyl)benzene (benzyl chloride) + $POCl_3$ + HCl

Physical Properties

Boiling point:	= 179°C
Melting point:	= -43°C
Molecular Formula	= C_7H_7Cl
Formula Weight	= 126.58348
Composition	= C(66.42%) H(5.57%) Cl(28.01%)
Molar Refractivity	= 36.01 ± 0.3 cm3
Molar Volume	= 117.1 ± 3.0 cm3
Parachor	= 282.5 ± 4.0 cm3
Index of Refraction	= 1.527 ± 0.02

Surface Tension	= 33.8 ± 3.0 dyne/cm
Density	= 1.080 ± 0.06 g/cm^3
Dielectric Constant	= Not available
Polarizability	= 14.27 ± 0.5 10^{-24} cm^3
Monoisotopic Mass	= 126.023628 Da
Nominal Mass	= 126 Da
Average Mass	= 126.585828 Da
Solubility in water:	= none (<0.1 g/100 ml)
Vapour pressure, at 20°C:	= 120Pa
Relative vapour density (air = 1):	= 4.4
Relative density of the vapour/air-mixture at 20°C (air = 1): 1.00	
Flash point:	= 67°C c.c.
Auto-ignition temperature:	= 585°C
Explosive limits, vol% in air:	=1.1-14.0
Octanol/water partition coefficient as log Pow: 2.3	

Nucleophilic Substutution Reaction of Benzyl Chloride

An examination of the kekule structure of benzyl chloride shows a remarkable electronic similarity to that of alkyl chloride.

$C_6H_5CH_2Cl$ — benzyl chloride

$CH_2=CH-CH_2Cl$ — allyl chloride

Thus benzyl chloride gives substitution reactions at the halogen as readily as allyl chloride, and even more readily than ethyl chloride. This reactivity of the chlorine in benzyl is due to the fact that the residual carbonium ion $C_6H_5CH_2^+$ is resonance stabilized.

$$C_6H_5CH_2-Cl \longrightarrow C_6H_5CH_2^+ + Cl^-$$

benzyl chloride → carbonium ion

The carbonium ion is stable as it exists in the following five carbonical forms

$\overset{+}{C}H_2$ (I) ⟷ CH_2 (II) ⟷ CH_2 (III) ⟷ CH_2 (IV) ⟷ $\overset{+}{C}H_2$ (V)

As shown in the fig above benzyl carbonium ion can be stabilized by resonance since the benzyl carbon atom has a vacant p –orbital. Which can overlap with six p-orbitals of the ring carbons. This permits the six π electrons to become dissociated with all the seven carbons in an extended π-orbital.

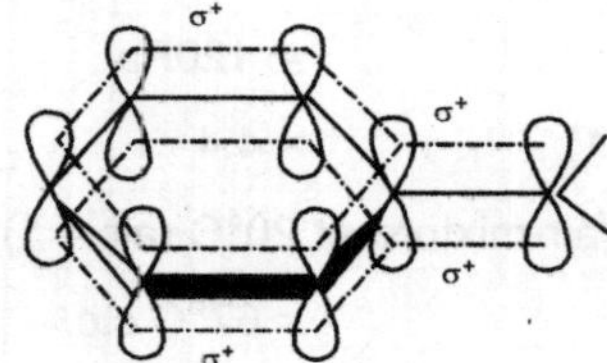

Fig. 24.2. Showing the stabilized benzyl carbonium ion due to delocalization of electrons and dispersal of positive charge over all seven carbons

Chemical Properties

1. Hydrolysis: On hydrolysis with boiling water or dilute alkalies (K_2CO_3) benzyl alcohol is produced.

$$C_6H_5CH_2Cl + H_2O \xrightarrow{K_2CO_3} C_6H_5CH_2OH + HCl$$

(chloromethyl)benzene
benzyl chloride

phenylmethanol
benzyl alcohol

2. With KCN: Benzyl cyanide is produced

$$C_6H_5CH_2Cl + KCN \longrightarrow C_6H_5CH_2CN$$

(chloromethyl)benzene
benzyl chloride

phenylacetonitrile
benzyl cyanide

POISON
6

3. With alcoholic ammonia: Amine is formed.

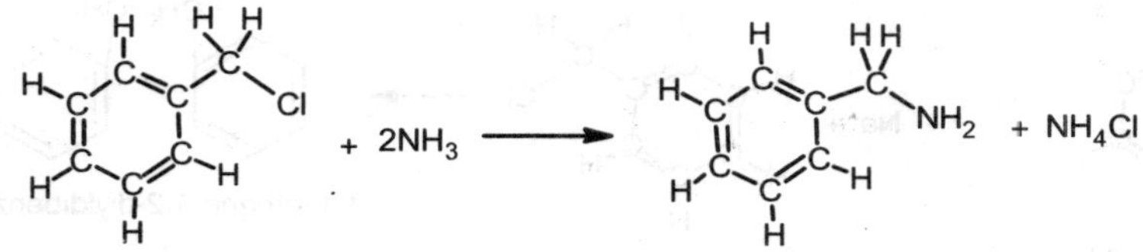

+ $2NH_3$ → NH_2 + NH_4Cl

(chloromethyl)benzene
benzyl chloride

1-phenylmethanamine
benzyl amine

4. (a) Like aromatic compounds in general, it can also under go Nitration, Sulphonation and halogenation.

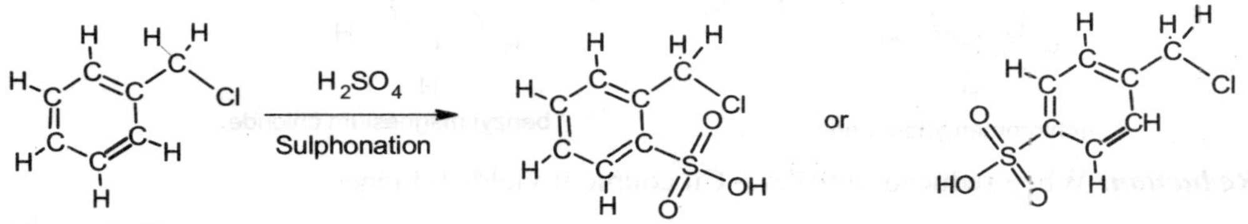

H_2SO_4 Sulphonation → or

(chloromethyl)benzene
benzyl chloride

2-(chloromethyl)benzenesulfonic acid

4-(chloromethyl)benzenesulfonic acid

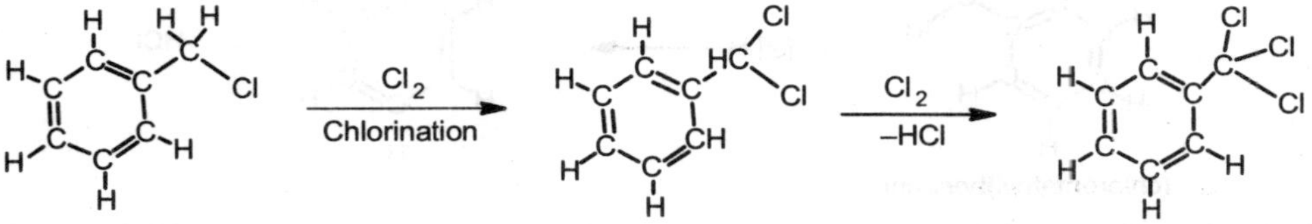

Cl_2 Chlorination → Cl_2 –HCl →

(chloromethyl)benzene
benzyl chloride

(dichloromethyl)benzene

(trichloromethyl)benzene

6. Oxidation: The side chain is oxidized to either –CHO or COOH group.

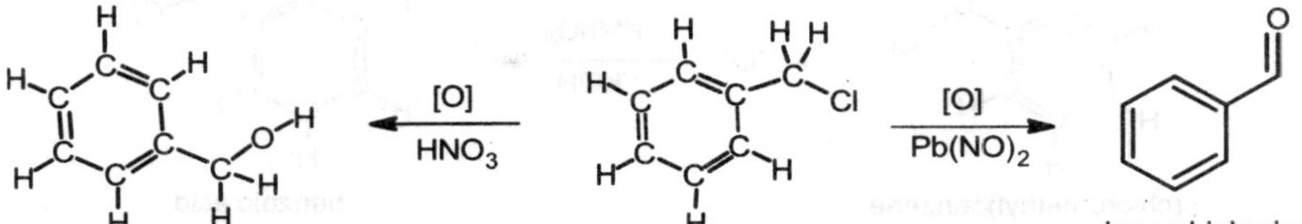

← [O] HNO_3 ; [O] $Pb(NO)_2$ →

phenylmethanol
bnenzyl alcohol

(chloromethyl)benzene
benzyl chloride

benzaldehyde

7. Friedel-Craft's Reaction: Higher homologous are formed.

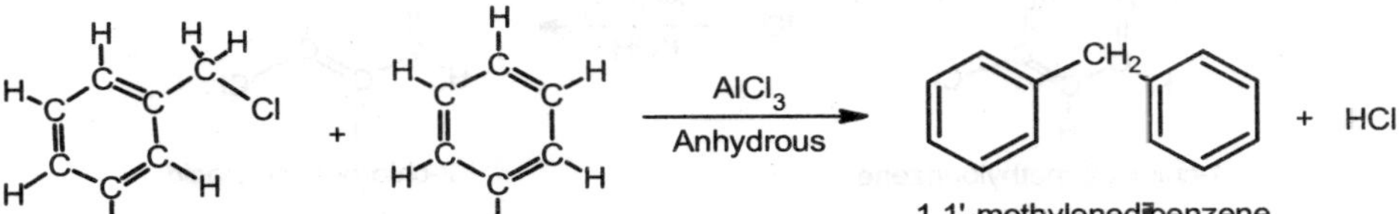

+ → $AlCl_3$ Anhydrous → + HCl

1,1'-methylenedibenzene

8. Witz Fittig's Reaction: In this preparation dibenzyl is formed

(chloromethyl)benzene + Na + (chloromethyl)benzene → 1,1'-ethane-1,2-diyldibenzene + 2NaCl

9. Formation of Grignard's Reagent:

(chloromethyl)benzene + Mg —Ether→ benzyl magnesium chloride

10. Reduction: When reduced with Zn – Cu couple it yields toluene.

(chloromethyl)benzene + [2] H → toluene + HCl

11. Oxidation: When oxidized with dilute nitric acid or alkaline potassium permanganate the CH_2Cl group is oxidized to COOH group. Benzoic acid is produced.

(chloromethyl)benzene + [O] —$KMnO_4$ / KOH→ benzoic acid

This reduction can be used for distinguishing benzyl chloride from the isomeric chlorotoluenes.

1-chloro-2-methylbenzene + [O] —$KMnO_4$ / KOH→ 2-chlorobenzoic acid

The product obtained from o- chlorotoluene contains the chlorine atom. While that from benzyl chloride does not.

Uses

Benzyl Chloride is widely used as an intermediate for manufacturing organic compounds including benzyl alcohol, benzyl cyanide and other benzyl compounds used in the end applications of perfumery, dyes, pharmaceuticals, synthetic resins, photographic chemicals, warfare chemicals, penicillins, quaternary ammonium compounds, plasticizer and esters. It is used in fuel as a gum inhibitor.

BENZAL CHLORIDE

Benzal chloride may be prepared

(1) By passing Chlorine into boiling toluene, until the required weight increase takes place.

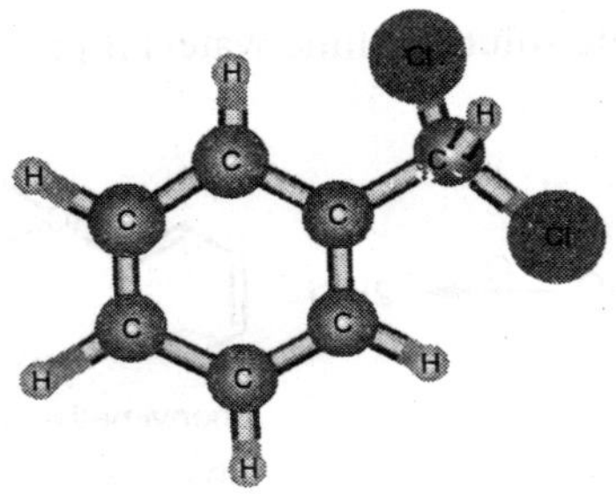

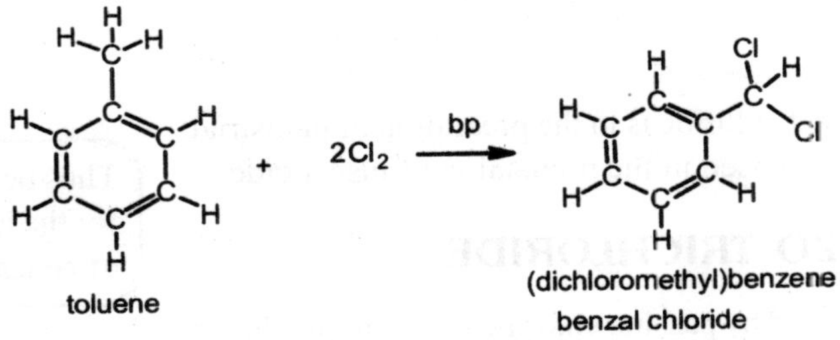

toluene + $2Cl_2$ $\xrightarrow{bp}$ (dichloromethyl)benzene

toluene

(dichloromethyl)benzene

benzal chloride

(2) By the action of phosphorous pentachloride on benzaldehyde

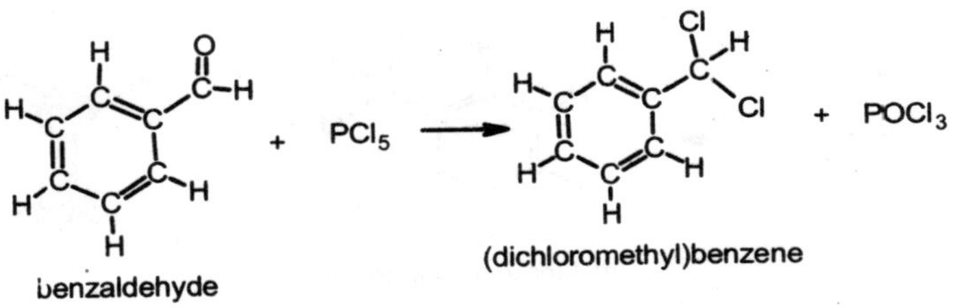

benzaldehyde + PCl_5 $\rightarrow$ (dichloromethyl)benzene + $POCl_3$

benzaldehyde

(dichloromethyl)benzene

Physical Properties

Boiling point:	205°C
Melting point:	– 17°C
Relative density (water = 1):	1.26
Solubility in water:	none
Vapour pressure, kPa at 35.4°C:	0.13
Flash point:	93°C c.c.
Auto-ignition temperature:	525°C
Explosive limits, vol% in air:	1.1-11
Octanol/water partition coefficient as log Pow: 3.22	

Chemical Properties

When heated with calcium hydroxide solution (lime water) it produces benzaldehyde by hydrolysis.

$$C_6H_5CHCl_2 + 2H_2O \xrightarrow{O} 2HCl + C_6H_5CH(OH)_2 \longrightarrow C_6H_5CHO$$

(dichloromethyl)benzene — phenylmethanediol unstable — benzaldehyde

Uses

The main use of the benzal chloride is in the prearation of industrial benzaldehyde and it is also used in the prepataion of insecticide.

Thus benzyl chloride is used for the industrial preparation of benzaldehyde.

BENZO TRICHLORIDE

Benzo trichloride is prepared by passing chlorine into boilign toluene till no further increase in weight takes place.

$$C_6H_5CH_3 + 3Cl_2 \xrightarrow{bp} C_6H_5CCl_3$$

toluene — (trichloromethyl)benzene benzo trichloride

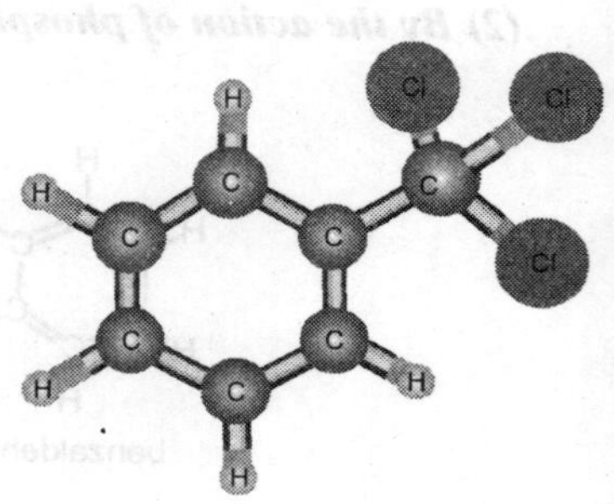

Physical Properties

Boiling point	= 221°C
Melting point	= –5°C
Molecular Formula	= $C_7H_5Cl_3$
Formula Weight	= 195.4736
Composition	= C(43.01%) H(2.58%) Cl(54.41%)
Molar Refractivity	= 45.28 ± 0.3 cm^3
Molar Volume	= 140.7 ± 3.0 cm^3
Parachor	= 351.2 ± 4.0 cm^3
Index of Refraction	= 1.556 ± 0.02
Surface Tension	= 38.7 ± 3.0 dyne/cm
Density	= 1.388 ± 0.06 g/cm^3
Polarizability	= 17.95 ± 0.5 10^{-24} cm^3
Monoisotopic Mass	= 193.945683 Da
Nominal Mass	= 194 Da

Chemical Properties

Since the halogen atoms are attached to an electron donating phenyl group, these are more reactive than in chloroform. Thus when boiled with calcium hydroxide solution, it is hydrolysed to yield benzoic acid.

This reaction if sued for the industrial preparation of benzoic acid.

$$C_6H_5CCl_3 + 3H_2O \xrightarrow{\Delta} 3HCl + C_6H_5C(OH)_3 \longrightarrow C_6H_5COOH + H_2O$$

(trichloromethyl)benzene   phenylmethanetriol   benzoic acid

Table Showing the Differences and Similarities of Nuclear and Side Chain Halogen Derivative

Nuclear Halogen Derivatives	Side Chain Derivatives
SIMILARITY	
1. They show Witz Fittig's Reactions with magnesium	1. They also show Witz Fittig's Reaction, also form Grignard's Reagents with magnesium
2. They can be nitrated sulphonated and chlorinated giving o – and p – substitution products	2. Same as nuclear derivatives
DISSIMILARITY	
1. They have pleasant odour	1. They have unpleasant odour
2. The halogen atom is non-reactive and **cannot be replaced by –OH, NH_2 or –CN groups** under ordinary conditions.	2. The halogen atoms are quite reactive **and can be easily be replaced by – OH, NH_2 or – CN groups by usual methods.**
3. They do not give Friedel Craft's Reaction	3. They give Friedel Craft's Reaction.
4. The halogen atom remains intact during oxidation.	4. The whole side chain includes the halogen atom is converted into carbolic group on oxidation

Chlorotuluene and benzyl chlorides are isomeric compounds

CHAPTER

25

Aromatic Sulphonated Compounds

Introduction

One of the most important characteristics of aromatic hydrocarbon and their derivatives is the ease with which they form sulphuric acid or chlorosulphuric acid (Cl.SO_2OH). The process is known as *Sulphonation*. During Sulphonation hydrogen atom of the nucleus are replaced by the sulphonic groups. (–SO_2OH)

(*a*)

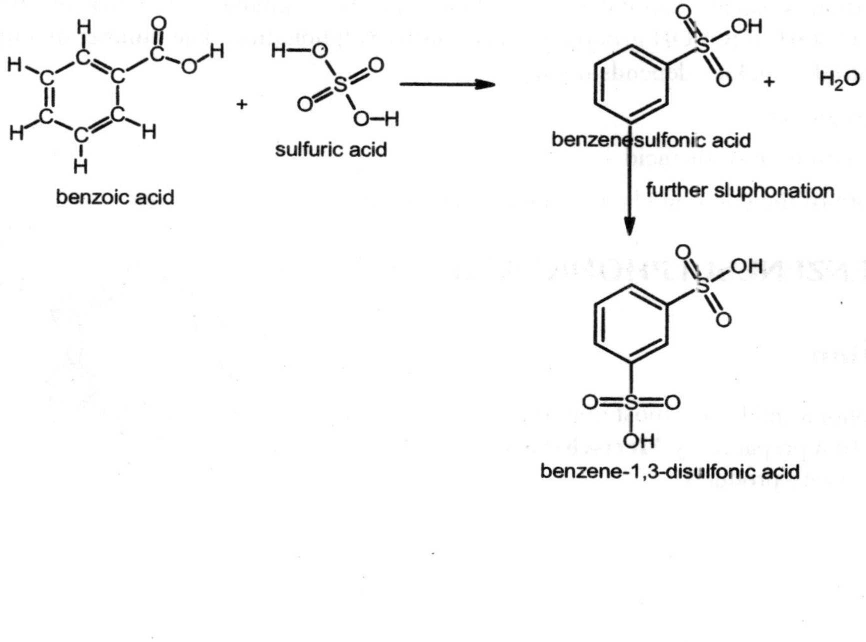

benzoic acid + sulfuric acid → benzenesulfonic acid + H_2O

further sluphonation

benzene-1,3-disulfonic acid

(b)

aniline + sulfuric acid —180 °C→ 4-aminobenzenesulfonic acid

(c)

toluene + sulfuric acid —heat→ 2-methylbenzenesulfonic acid and 4-methylbenzenesulfonic acid

The distribution takes place according to Crum Brown's Rule

(d)

benzene + chlorosulphonic acid → + HCl

Sulphonation is usually facilitated if the nucleus already contains $-CH_3$, NH_2 or OH groups. The presence of –COOH or SO_3OH group however retards Sulphonation. The number of sulphonic group introduced into the nucleus depends upon:

1. Temperature
2. Concentration of the acid
3. Nature of the compounds undergoing Sulphonation

BENZENESULPHONIC ACID

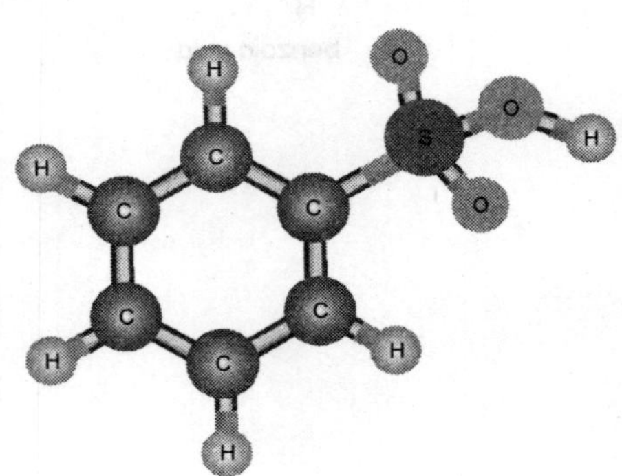

Introduction

Benzene sulphonic acid is the most important member of this class. It was first prepared by **Mertscherlich** in 1834.It was found in the waste products.

Preparation

Benzenesulphonic acid is prepared by heating benzene and sulphuric acid at 80°C

$$\text{benzene} + H_2SO_4 \xrightarrow{80^\circ C} C_6H_5SO_3H + H_2O$$

Lab Preparation

A mixture of benzene and sulphuric acid (1:2) is heated in a round bottom flask under reflux condenser. The resulting mixture is poured into solution of NaCl. When sodium benzenesulphonate separates. This is used as such for most purpose. If pure benzenesulphonate is required, the Sulphonation mixture is neutralized with lead carbonate. The precipitate of sulphate formed by excess of sulphuric acid. The filtrate containing benzenesulphonate is decomposed with calculated amount of dilute sulphuric acid, filtered and filtrate concentrated to crystallization.

(*a*) $H_2SO_4 + PbCO_3 \rightarrow PbSO_4 + CO_2 + H_2O \uparrow$

(*b*) $C_6H_5SO_3H + PbCO_3 \rightarrow (C_6H_5SO_3)_2\,Pb + H_2O + CO_2$

(*c*) $(C_6H_5SO_3)_2\,Pb + H_2SO_4 \rightarrow PbSO_4 + 2C_6H_5SO_3H$

Benzenesulphonic acid crystallizes from white leaflets containing $1\frac{1}{2}H_2O$. the anhydrous compounds melts at 51°C. it is very hydroscopic and formed syrupy liquid. When exposed to mist air. It is highly soluble in water and the solution is strongly acidic (about as strong as sulphuric acid). The detail physical properties are shown below.

Industrial Preparation

On large scale the Sulphonation of benzene is carried out in large enameled or lead lined pans heated by steam coils and fitted with mechanical stirrers (see fig. 25.1). The details of the process is same as in the lab method.

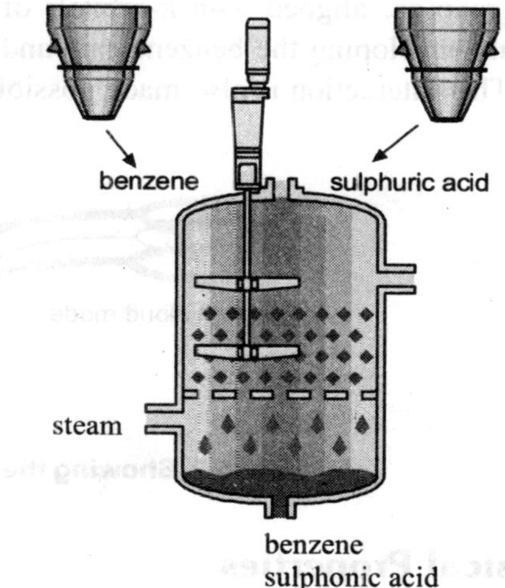

Fig. 25.1. Showing the Industrial Process of Benzenesulphonic acid.

Physical Properties

Melting Point	= 10ºC
Boiling Point	= <204.5°C
Molecular Formula	= $C_6H_6O_3S$
Formula Weight	= 158.17504
Composition	= C(45.56%) H(3.82%) O(30.34%) S(20.27%)
Molar Refractivity	= 37.10 ± 0.4 cm^3
Molar Volume	= 112.2 ± 3.0 cm^3
Parachor	= 300.2 ± 6.0 cm^3
Index of Refraction	= 1.575 ± 0.02
Surface Tension	= 51.2 ± 3.0 dyne/cm
Density	= 1.409 ± 0.06 g/cm^3
Polarizability	= 14.70 ± 0.5 $10^{-24}cm^3$
Monoisotopic Mass	= 158.003764 Da
Nominal Mass	= 158 Da

Structure

The structure of Benzenesulphonic acid explained below. The π-orbitals in one of the p-bonds of the -SO_3H group are aligned with π-orbitals of the benzene ring. Since these interact to form extended π-orbitals enveloping the benzene ring and one of the double bonds (S == O) in the sulphonic acid group. This interaction is also made possible by the shortness of C – S bonds.

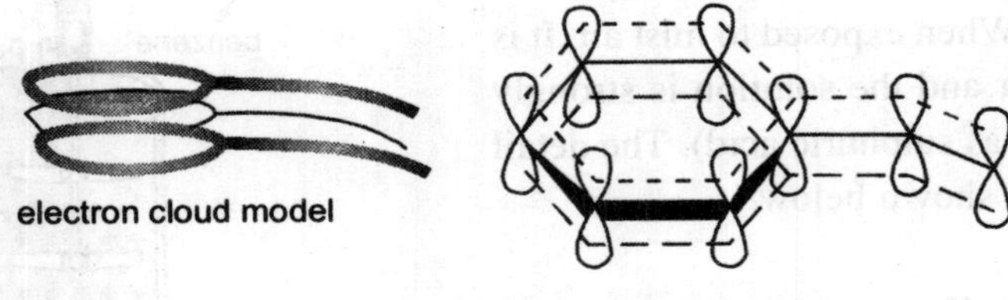

Fig. 25.2. Showing the structure of benzenesulphonic acid.

Chemical Properties

In general its behaviour is similar to the acids chlorides of arylcarboxilic acids, although it is not as reactive.

1. Hydrolysis: It is decomposes slowly by the action of cold water. The hydrolysis is faster in alkali, when the salt is formed.

$$C_6H_5 - SO_2 - Cl + H_2O \rightarrow C_6H_5 - SO_2 - OH + HCl$$

Benzenesulphonyl Chloride — Benzenesulphonic acid

2. Formation of Ester: It reacts with alcohols and phenols in the presence of an alkali to form esters i.e. alkyl benzene sulphonates

$$C_6H_5 - SO_2 - Cl + H - O - C_6H_5 \xrightarrow{NaOH} C_6H_5 - SO_2 - O - C_6H_5 + HCl$$

Phenyl Benzenesulphonate

3. Formation of Amide: It reacts with primary amines and secondary amines to form n – alkyl and N- dialsulphonamides

$$C_6H_5SO_2Cl + H-NH-R \longrightarrow C_6H_5SO_2-NH-R + H^+ + Cl^-$$

n - alkyl sulphonamide
(soluble in alkali)

$$C_6H_5SO_2Cl + H-NR-R \longrightarrow C_6H_5SO_2-NR-R + H^+ + Cl^-$$

n - dialkyl sulphonamide
(soluble in alkali)

This reaction forms the basic of Hinber's method for the separation and identification of 1°, 2° and 3° amines.

(4) Reduction: While Benzenesulphonic acid is resistant to reduction, the Sulphonyl chloride is reduced to thiophenol in two stages

(*i*) With a mild reducing agent such as zinc dust and water, Benzenesulphonic acid is formed.

$$C_6H_5SO_2Cl + 2H \xrightarrow{Zn/H_2O} C_6H_5SO_2OH + HCl$$

benzenesulfonyl chloride — benzenesulfonic acid

(*ii*) With a stronger reducing agent such as zinc and hydrochloric acid, or lithiumaluminium hydride, thiophenol is formed

$$C_6H_5SO_2Cl + 6H \xrightarrow{LiAlH_4} C_6H_5SH + HCl$$

benzenesulfonyl chloride — benzenethiol (thiophenol)

(5) Friedel Craft's Reaction: Benzenesulphonyl chloride undergoes Friedel Craft's Reaction with aromatic compounds in the presence of anhydrous aluminium chloride to give diphenyl sulphone.

$$C_6H_6 + C_6H_5SO_2Cl \longrightarrow C_6H_5SO_2C_6H_5 + HCl$$

benzene + benzenesulfonyl chloride → 1,1'-sulfonyldibenzene (diphenyl sulphone)

Uses

The substance is toxic to aquatic organisms. Mainly used as resistant and relative purity protective agent of dyes intermediate and vulcanization dye; and widely used in shipping industry, perfume industry; electroplate industry, analytical chemistry, etc.

BENZENE DISULPHONIC ACID

Introduction

When benzene is heated with excess sulphuric acid at 200ºC, it yields Benzene m - disulphonic acid, along with a small amount of benzene – p – disulphonic acid

$$2\,C_6H_6 + H_2SO_4 \text{ (fuming)} \xrightarrow{200\,°C} \text{benzene-1,3-disulfonic acid or benzene-1,4-disulfonic acid}$$

benzene-1,3-disulfonic acid — benzene-1,4-disulfonic acid

The yield of the p – isomer is increased to 30% in the presence of mercuric sulphate as catalyst. Benzene – o disulphonic acid may be prepared by the following scheme.

3-aminobenzenesulfonic acid $\xrightarrow[\text{sulphonation}]{H_2SO_4}$ 3-aminobenzene-1,2-disulfonic acid $\xrightarrow{NaNO_2/HCl}$ 1-chloronio-2-(2,3-disulfophenyl)diazenium $\xrightarrow{2H}$

When fused with potassium hydroxide benzene – m – disulphonic acid (also p – isomer) yields resorcinol. The o – isomer forms calechol

benzene-1,3-disulfonic acid $\xrightarrow[\text{fuse}]{NaOH}$ disodium benzene-1,3-diolate $\longrightarrow$ resorcinol

benzene-1,4-disulfonic acid $\xrightarrow[\text{fuse}]{NaOH}$ disodium benzene-1,4-diolate $\longrightarrow$ hydroquinone cathechol

TOLUENESULPHONIC ACIDS

Preparation

Toluene on Sulphonation with concentrated sulphuric acid forms o and p toluene sulphonic acid.

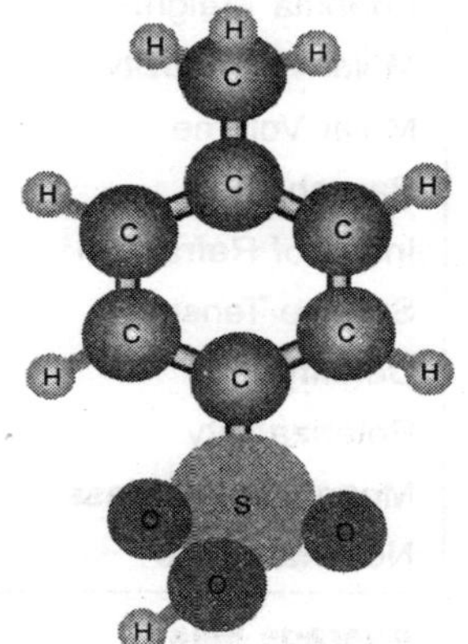

toluene → (Conc H_2SO_4, Δ) → 2-methylbenzenesulfonic acid + 4-methylbenzenesulfonic acid

If the reaction is carried at 110°C - 120°C the p – isomer predominates in the product, while at room temperature in the product. While at room temperature the product is mainly the o – isomer.

The m- toluenesulphonic acid may be obtained from p-toluene – m – sulphonic acid by replacing NH_2 group by H atom.

($NaNO_2$/HCl, 0 °C) → 1-chloronio-2-(4-methylphenyl)diazenium → (2H) → 3-methylbenzenesulfonic acid

Both o and p – toluenesulphonic acid, are white crystalline solid, the detail physical properties are gene below.

	P – isomer	o - isomer
Melting point	*106°C*	*67.5°C*
Formula Weight	172.20162	172.20162
Molar Refractivity	41.72 ± 0.4 cm^3	41.72 ± 0.4 cm^3
Molar Volume	128.4 ± 3.0 cm^3	128.4 ± 3.0 cm^3
Parachor	338.5 ± 6.0 cm^3	338.5 ± 6.0 cm^3
Index of Refraction	1.562 ± 0.02	1.562 ± 0.02
Surface Tension	48.2 ± 3.0 dyne/cm	48.2 ± 3.0 dyne/cm
Density	1.340 ± 0.06 g/cm^3	1.340 ± 0.06 g/cm^3
Polarizability	16.54 ± 0.5 10^{-24} cm^3	16.54 ± 0.5 10^{-24} cm^3
Monoisotopic Mass	172.019414 Da	172.019414 Da
Nominal Mass	172 Da	172 Da
Average Mass	172.203368 Da	172.203368 Da

Uses

Toluenesulphonic acid is used in the preparation of saccharin and chloramine T

CHLORAMINE T

It is the sodium salt of N – Chloro – p – toluenesulphonamide. It is made by the action of sodium hypochlorite on p - toluenesulphonamide.

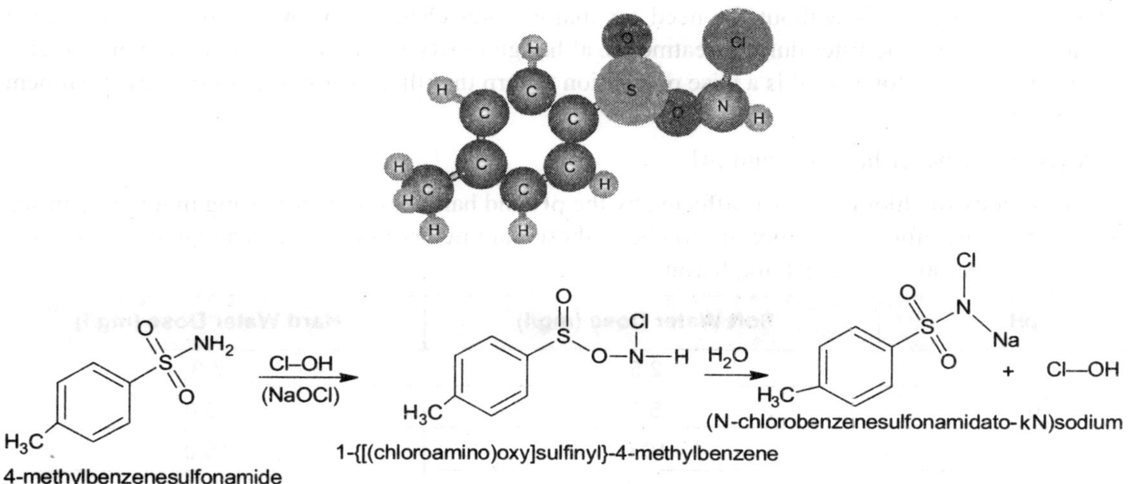

Physcial Properties

Molecular Formula	= $C_7H_8ClNO_2S$
Formula Weight	= 205.66192
Composition	= C(40.88%) H(3.92%) Cl(17.24%) N(6.81%) O(15.56%) S(15.59%)
Molar Refractivity	= 50.47 ± 0.4 cm^3
Molar Volume	= 142.3 ± 5.0 cm^3
Parachor	= 402.4 ± 6.0 cm^3
Index of Refraction	= 1.627 ± 0.03
Surface Tension	= 63.9 ± 5.0 dyne/cm
Density	= 1.44 ± 0.1 g/cm^3
Polarizability	= 20.00 ± 0.5 $10^{-24}cm^3$
Monoisotopic Mass	= 204.996426 Da
Nominal Mass	= 205 Da
Average Mass	= 205.663524 Da

Uses

Chloramine-T and fish disease.

As chloramine-T dissolves it slowly breaks down to produce hypochlorous acid (HOCl), which in turn releases chlorine and oxygen. There is some uncertainty as to the active species of the breakdown products. Certainly, any chlorine present is liable to have an effect against flukes and parasites such as Costia, Chilodonella, white spot and Trichodina, as well as bacteria. However, it is now believed that the chloramine-T ion is the major active species. It breaks down fairly rapidly and treatments can be repeated on a daily basis without the need for major water changes. As with most treatments, it is advisable to by-pass the filter during treatments, although nitrifying bacteria seem to be fairly tolerant of of this treatment. However it is a wise precaution to turn the filter off for 3-4 hours while treatments are in progress.

Affected by water hardness and pH

The toxicity of chloramine-T is affected by the pH and hardness of water, being more toxic in soft, acidic water. Therefore it is important to check these parameters before use. The table below shows typical dose rates at various pH and hardness

pH	Soft Water Dose (mg/l)	Hard Water Dose (mg/l)
6.0	2.5	7.0
6.5	5.0	10.0
7.0	10.0	15.0
7.5	18.0	18.0
8.0	20.0	20.0

Table Dosage rates of chloramine-T at different water hardness and pH

How effective is it?

My own experience of pond treatments (usually doses <5mg /litre in hard, alkaline water) are varied and often disappointing. Follow-up examinations often show targeted parasites such as Trichodina and flukes are still present in substantial numbers after treatment. The usually recommended dose rate for ponds is given as 2mg/litre by most aquatic sources. I have serious doubts as to the efficacy of this dosing level in the typical hard alkaline water found in southern England.Controlled treatments in clean, bare treatment tanks using doses of 10-16mg/ litre have shown more promise, having an effect against Gyrodactylus (skin flukes) and Dactylogyrus (gill flukes). This regime works well against Costia, but not quite as well against Trichodina. These trials were only started last year and more data is needed complete the picture. This page will be updated as more tests are carried out. My initial conclusions are that it is probably better suited for tank treatments rather than pond treatment as conditions and dosages can be better controlled.

In the case of the pond treatments, more investigation needs to be carried out to determine an effective dose that allows for the effects of hard water and organic pollution. It is likely that current pond dosage rates of < 5mg/litre, particularly in hard, alkaline pond water, need to be increased to > 10 mg/litre and a follow up examination carried to determine the effectiveness of the treatment, with a view to re-dosing as necessary.

More on chloramine-T

Do not use it as a high dose dip. Its chemical action will cause serious damage to fish

Do not allow it to come into contact with metal surfaces as toxic compounds can be formed

Wear a particle mask and goggles when handling to prevent injury to skin or eyes.

Toxicity is greatly increased in soft acidic water – only use the low dose.

Aerate the water vigorously during treatment

Turn off any UV lamps when using any pond or tank treatments

Ponds should be cleaned and vacuumed prior to treatment in order to reduce free organics that would affect treatment efficacy

The breakdown of chloramine-T is speeded up in sunlight, so treatments are best carried out evenings or on cloudy days.

Useful conversions are:

ppm = mg/litre	i.e. 5 ppm = 5 mg / litre
mg / litre x 3.785 = mg / gall (US)	i.e . 5 mg / litre = 18.9 mg / gall (US)
mg/ litre x 4.546 = mg / gall (UK)	i.e. 5 mg / litre = 22.7 mg / gall (UK)

To convert imperial gallons to US gallons multiply by 1.2

Other useful figures:
1 ounce = 28.35 grams

1% solution =
10 ml per litre
10 gram per litre
38 gram per gall (US)
45 gram per gall (UK)

SACCHARIN
O-Sulphobenzoic Acid

Introduction and History

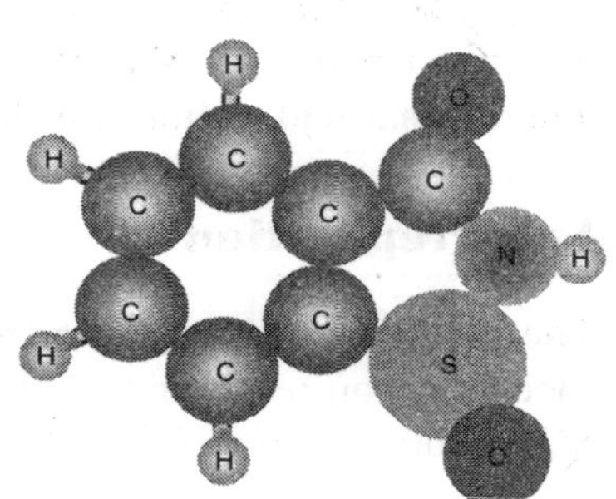

The word *saccharin* has no final "e". The word *saccharine*, with a final "e", is much older and is an adjective meaning "sugary". Both words are derived from the Greek word *σακχαρον* (sakcharon, German ch sound), which ultimately derives from Sanskrit for sugar, *sharka*, which literally means gravel. The word's connection with sugar means the term is used metaphorically, often in a derogative sense, to describe something "unpleasantly over-polite" or "overly sweet".

The history of saccharin and the quest for sweeteners goes back thousands of years. As early as 2500 B.C. Egyptians were actively producing mass quantities of sweetener solely for culinary purposes. Early civilizations used honey, fruits and date syrup to sweeten foods. Natural cane sugar first came about in India circa 4000 B.C. In the Indian culture sugar was held as one of the seven stable necessities of mankind. In Hindu myth fresh water, salt, butter, milk, yogurt, wine, and sugar represented seven mystical seas of produce that sustain life. Upon its introduction into the Roman empire sugar was at first viewed as more of a medicinal substance and not as qualified to sweeten foods as honey. Sugar slowly gained popularity and spread across Europe, Asia, Africa, and eventually to the New World. One of the main reasons for the massive Caribbean slave trade was to sustain sugar plantations.

Saccharin's sweetness was accidentally discovered by **Ira Remsen, a professor at Johns Hopkins University**, and **Constantin Fahlberg**, a research fellow working in Remsen's lab. In 1879, while working with coal tar derivatives (toluene), Remsen discovered saccharin's sweetness at dinner after not thoroughly washing his hands, as did Fahlberg during lunch. Remsen and Fahlberg jointly published their discovery in 1880 (Fahlberg, C.; Remsen, I. *Über die Oxydation des Orthotoluolsulfamids*. Chem. Ber. 1879, 12, 469-473).

Saccharin was an important discovery, especially for diabetics. Saccharin goes directly through the human digestive system without being digested. It does not affect blood insulin levels, and has effectively no food energy.

Although saccharin was commercialized not long after its discovery, it was not until sugar rationing during World War I that its use became widespread. Its popularity further increased during the 1960s and 1970s among dieters, since saccharin is a calorie-free sweetener. In the United States saccharin is often found in restaurants in pink packets; the most popular brand is "Sweet'N Low". A small number of soft drinks are sweetened with saccharin, the most popular being the Coca-Cola Company's cola drink Tab, introduced in 1963 as a diet soft drink.

Preparation

Saccharin can be produced in various ways. Remsen & Fahlberg's original route starts with toluene, but yields from this starting point are small. In 1950, an improved synthesis was developed at the Maumee Chemical Company of Toledo, Ohio. In this synthesis, anthranilic acid successively reacts with nitrous acid, sulfur dioxide and chlorine, and then ammonia to yield saccharin.

Lab Preparation

The production of saccharine is not only easy, but inexpensive and can be done in bulk. To obtain saccharin you must first have toluene ($C_6H_5CH_3$.) Cl_2SO_2OH is added making two products- one SO_2Cl formation is attached to the ortho site of the phenyl group, another SO_2Cl formation is attached to the para site of the phenyl group. Ammonium chloride is then added to the mixture and replaces the chlorine in the SO_2Cl group on the ortho site. The substance is then purified, separated, and placed in a high concentration of potassium permanganate and heated to 150° C. The methyl group will then change to a carboxyl group (COOH.) This change will cause the molecule to "self-cyclize" and release saccharin and water. Lastly, a base is added to swap the hydrogen atom bonded to the nitrogen atom with a metal ion such as calcium or sodium.

toluene + chlorosulfurous acid → benzene-1,4-disulfonyl dichloride + 2-methylbenzenesulfonyl chloride —filtration→ 2-methylbenzenesulfonyl chloride

NH_4Cl

—aq $KMnO_4$→ 2-(aminosulfonyl)benzoic acid —Δ, $-H_2$→ saccharine → sodium saccharine

Physical Properties

Molecular Formula	$= C_7H_5NO_3S$
Formula Weight	= 183.1845
Composition	= C(45.90%) H(2.75%) N(7.65%) O(26.20%) S(17.50%)
Molar Refractivity	$= 41.91 \pm 0.4\ cm^3$
Molar Volume	$= 117.6 \pm 3.0\ cm^3$
Parachor	$= 325.4 \pm 6.0\ cm^3$
Index of Refraction	$= 1.631 \pm 0.02$
Surface Tension	$= 58.6 \pm 3.0$ dyne/cm
Density	$= 1.557 \pm 0.06\ g/cm^3$
Polarizability	$= 16.61 \pm 0.5\ 10^{-24} cm^3$
Monoisotopic Mass	= 182.999013 Da
Nominal Mass	= 183 Da
Average Mass	= 183.186163 Da

The chemical formula for saccharin is $C_7H_5NO_3S$. The molar mass for saccharin is 183.19 grams per mole. Saccharin consists mostly of carbon; carbon makes up 45.89% of the composition of saccharin by mass. The mass of saccharin is also 2.75% hydrogen, 7.65% nitrogen, 26.20% oxygen, and 17.50% sulfur. By number saccharin is 41% carbon, 29% hydrogen, 6% nitrogen, 18% oxygen, and 6% sulfur. Saccharin melts at 228.8° to 229.7° C. Structurally, saccharin is made up of two connected rings; the first ring is a phenyl ring and the second is a 5 membered ring with a nitrogen, a carboxyl group, and a sulfone group beside the nitrogen. Due to the presence of a nitrogen, the molecule is heterocyclic.

If temperatures rise above 300° C saccharin will boil. Saccharin's relative low level of reactance makes this substance ideal for cooking because it remains stable at high temperatures. Saccharin can also be dissolved easily into water, another good quality for food preparation. It is non-reactive in water at temperatures of as much as 150° C and at common pH levels measuring anywhere from 2 to 7. It's natural density is 0.828 g/mL. The hydrogen bond is what reacts to cause the sensation of a sweet taste in the mouth. Upon digestion saccharin does not break down; it remains unchanged throughout digestion.

Uses

It is mainly used as a sweetener for the diabetic patients

SULPHANILIC ACID

Preparation

It is prepared by sulphonation of aniline, with concentrated sulphuric at 180°C in high yield. It is believed that *N*- phenyl sulphonic acid is produced as an intermediate in the Sulphonation.

It may be noted that the intermediate sulpahmic acid undergoes rearrangement involving Sulphonation on the ring and desulphonation on nitrogen.

aniline —con H_2SO_4, Δ→ anilium hydrogen sulphate —$-H_2O$→ phenylsulfamic acid —rearrange→

Physical Properties

Molecular Formula	$= C_6H_7NO_3S$
Formula Weight	= 173.18968
Composition	= C(41.61%) H(4.07%) N(8.09%) O(27.71%) S(18.51%)
Molar Refractivity	$= 40.71 \pm 0.4\ cm^3$
Molar Volume	$= 114.4 \pm 3.0\ cm^3$
Parachor	$= 328.2 \pm 6.0\ cm^3$
Index of Refraction	$= 1.629 \pm 0.02$
Surface Tension	$= 67.5 \pm 3.0$ dyne/cm
Density	$= 1.512 \pm 0.06\ g/cm^3$
Polarizability	$= 16.14 \pm 0.5\ 10^{-24} cm^3$
Monoisotopic Mass	= 173.014663 Da
Nominal Mass	= 173 Da
Average Mass	= 173.191078 Da

Chemical Properties

Sulphanilic acid decomposes at 280°C. It is almost insoluble in cold water and many other solvents.

These unusual properties arise from salt like character of the molecule. The Sulphanilic acid molecule has an acid group (SO_3H) and a basic group (NH_2) existing side by side. These groups interact to form internal salt or dipolar ion called ***Zwitterion.***

Sulphanilic acid dissolves in aqueous sodium hydroxide to form water soluble salt because a base can extract a proton from the $-H_3N^+$ group of the Zwitterions

The SO_3H is loosened by the presence of the amino group in the para position and can be replaced by –H or $-NO_2$ group

4-aminobenzenesulfonic acid + 2NaOH $\xrightarrow{\text{fuse}}$ aniline + $NaSO_4$ + H_2O

4-aminobenzenesulfonic acid + HNO_2 $\longrightarrow$ 4-aminophenyl nitrate + sulfuric acid

Both Sulphanilic acid and p – nitroaniline can be dissolved and with phenolic to form dyes. Sulphanilic acid is valuable dye intermediate. It is substituted amide forms numerous sulphanilimide.

Uses

The most famous applicaton of sulfamic acid is in the synthesis of compounds that taste sweet. Reaction with cyclohexylamine followed by addition of NaOH gives $C_6H_{11}NHSO_3Na$, sodium cyclamate.Dulfamates (O-substituted-, N-substituted-, or di-/tri-substituted derivatives of sulfamic acid) have been used in the design of many types of therapeutic agents such as antibiotics, nucleoside/ nucleotide human immunodeficiency virus (HIV) reverse tanscriptase inhibitors, HIV protease inhibitors (PIs), anti-cancer drugs (steroid sulfatase and carbonic anhydrase inhibitors), anti-epileptic drugs, and weight loss drugs.Sulfamic acid is used as an acidic cleaning agents, typically for metals and ceramics. It is a replacement for hydrochloric acid for the removal of rust.

- Catalyst for esterification process
- Dye and pigment manufacturing
- Herbicide
- Coagulator for urea-formaldehyde resins
- Ingredient in fire extinguishing media
- Pulp and paper industry as a chloride stabilizer
- Synthesis of nitrous oxide by reaction with nitric acid

SULPAHNILIMIDE
P-AMINOBENZENESULPHONAMIDE

Introduction and History

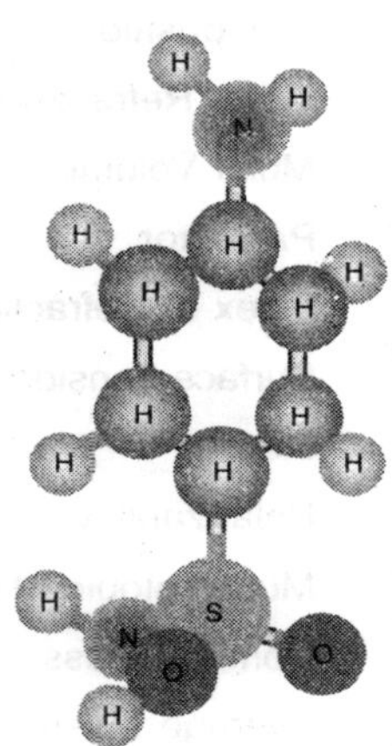

Gerhard Johannes Paul Domagk (1895-1964), a German biochemist, whose research with antibacterial chemicals resulted in the discovery of a new class of drugs that provided the first effective treatments for pneumonia, meningitis, and other bacterial diseases. Domagk's research involved analyzing thousands of chemicals for their antibacterial properties. In 1932 he tested a red dye, Prontosil. The dye itself had no antibacterial properties, but when Domagk slightly changed its chemical makeup, Prontosil showed a remarkable ability to arrest infections in mice caused by streptococcal bacteria. Domagk tested the drug on his daughter, who was near death from a streptococcal infection and had failed to respond to other treatments. She subsequently made a complete recovery

It is amide of sulphanilic acid

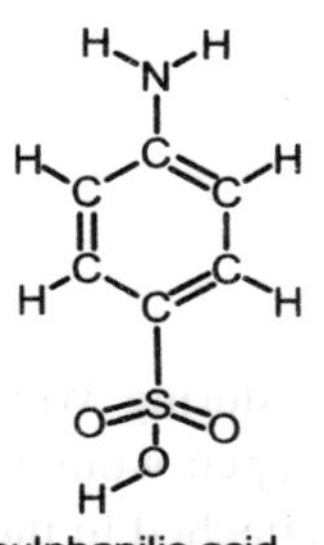

sulphanilic acid

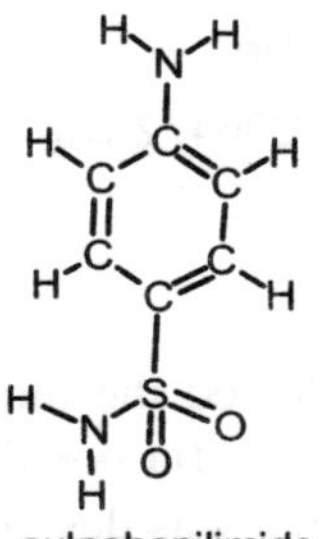

sulpahanilimide

Preparation

Sulphanilimide is obtained from aniline by (acetylation with acetic anhydride (to protect NH_2 from chlorosulphuric acid to form chlorosulphuric acid). When acetanilide is formed (2) Reaction with excess chlorosulphuric acid to form N – acetylsulphophenenilyl chloride. (3) Reaction with ammonia to get N – acetylsulpahnilide; (4) regeneration of NH_3 group by hydrolysis of acetyl group by aqueous acid.

aniline $\xrightarrow{\text{acetic anhydride}}$ N-phenylacetamide $\xrightarrow{ClSO_2OH}$ 4-(acetylamino)benzenesulfonyl chloride $\xrightarrow{NH_3}$ $\xrightarrow[H^+]{H_2O}$ 4-aminobenzenesulfonamide

Physical Properties

Melting Point	= 156°C
Molecular Formula	= $C_6H_8N_2O_2S$
Formula Weight	= 172.20492
Composition	= C(41.85%) H(4.68%) N(16.27%) O(18.58%) S(18.62%)
Molar Refractivity	= 42.80 ± 0.4 cm^3
Molar Volume	= 120.6 ± 3.0 cm^3
Parachor	= 340.9 ± 6.0 cm^3
Index of Refraction	= 1.627 ± 0.02
Surface Tension	= 63.7 ± 3.0 dyne/cm
Density	= 1.427 ± 0.06 g/cm^3
Polarizability	= 16.97 ± 0.5 $10^{-24}cm^3$
Monoisotopic Mass	= 172.030648 Da
Nominal Mass	= 172 Da
Average Mass	= 172.206472 Da

Chemical Properties

The discovery of Sulfanilamide greatly affected the mortality rate during World War II. American soldiers were taught to immediately sprinkle sulfa powder on any open wound to prevent infection. Every soldier was issued a first aid pouch that was designed to be attached to the soldier's waist belt. The first aid pouch contained a package of sulfa powder and a bandage to dress the wound. One of the main components carried by a combat medic during World War II was sulfa powder and sulfa tablets.

Bacteriostatic Action

p – Aminobenzoic acid is produced in our body by the hydrolysis of proteins. It is an intermediate necessity for the growth of the certain bacterial enzymes which cause disease. The enzyme forms a

4-aminobenzenesulfonamide + 4-aminobenzoic acid ⇌ 4-aminobenzoic acid + 4-aminobenzenesulfonamide +

complex with p –amino benzoic acid in one of the metabolic reactions for its growth. When sulphanilimide which is structurally similar is added in excess; the enzyme (i) forms compounds with its preferences.

Thus the complex formation of the enzyme with p-aminobenzoic acid becomes static and the bacterial growth stops because of this Bacteriostatic action sulphanilamide and its derivatives the sulpha drugs

SULPHA DRUSGS

History

The first sulfonamide was trade named prontosil, which is a prodrug. The first experiments with prontosil began in 1932 by the German chemist Gerhard Domagk, and the results were published in 1935 (after his employer, IG Farben, had obtained a patent on the compound). Prontosil was a red azo dye, and in mice had a protective action against streptococci. It had no effect in the test tube, and only exerted an antibacterial effect in the live animal itself. The drug is connected with an azo bond, which is a N=N bond. This bond is easily broken in the GI tract to release the active drug sulfanilamide. And, with war on the horizon, there was interest among the Allies to break the German patent. Soon it was discovered (1936) that prontosil's active agent was a smaller, more effective compound known as sulfanilamide:

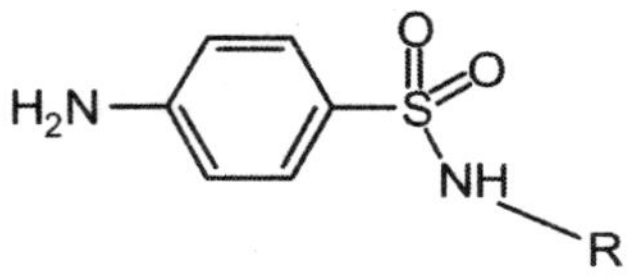

$H_2N-C_6H_4-SO_2-NH-R$

General Formula of Sulpha Drugs

Sulfanilamide had a central role in preventing infection during World War II. American soldiers were issued a first aid kit containing sulfa powder and were told to sprinkle sulfa on any open wound. During the years 1942 to 1943, Nazi doctors conducted sulfanilamide experiments on prisoners in Ravensbrück. An inadvertently toxic preparation of sulfanilamide also had a central influence on the US Food and Drug Administration. A preparation called Elixir Sulfanilamide contained diethylene glycol as a solvent, which is toxic. This preparation killed over one hundred people, mostly children, and led to the passage of the 1938 Food, Drug, and Cosmetic Act.

The Sulfanilamide compound is more active in the protonated form, which in case of the acid works better in a basic environment. The solubility of the drug is very low and sometimes can crystalize in the kidneys, due to its first pK_a of around 10. This is a very painful experience so patients are told to take the medication with copious amounts of water. Newer compounds have a pK_a of around 5-6 so the problem is avoided.Many thousands of permutations of the sulfanilamide structure have been created since its discovery (by one account, over 5400 permutations by 1945), usually by the substitution of another functional group for a single amide hydrogen.

Preparation

Certain sulfonamides (sulfadiazine or sulfamethoxazole) are sometimes mixed with the drug trimethoprim, which acts against dihydrofolate reductase.

When R is usually a hetrocyclic group. They are made from aniline by the same process as used is the making of sulpahniamide taking RNH_2 in place of NH_3.

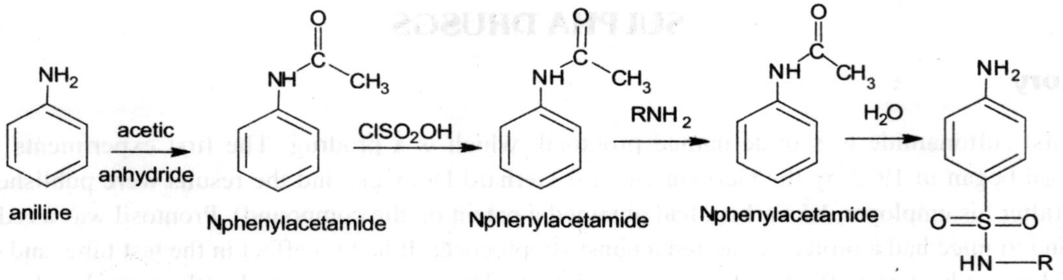

aniline —(acetic anhydride)→ Nphenylacetamide —($ClSO_2OH$)→ Nphenylacetamide —(RNH_2)→ Nphenylacetamide —(H_2O)→ NH_2–C_6H_4–SO_2–NH–R

A few of the more widely used sulpha drugs are

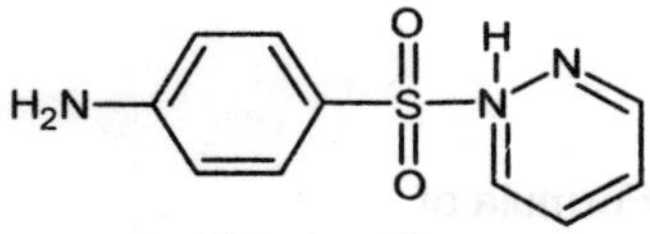

sulphapyradine

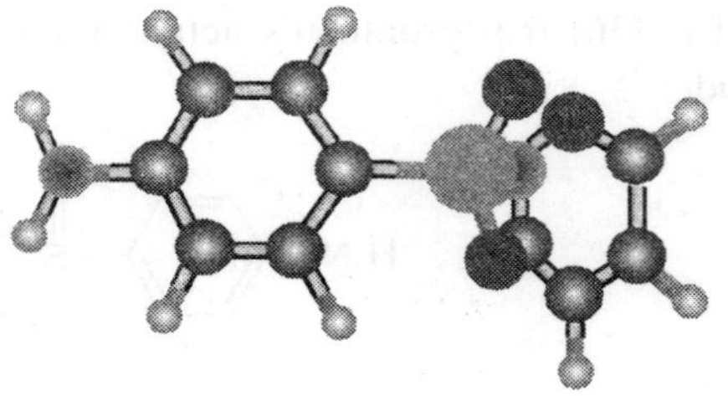

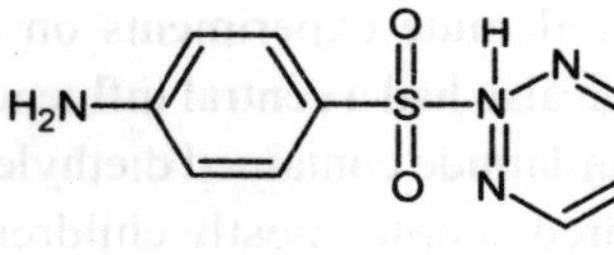

sulphadiazine

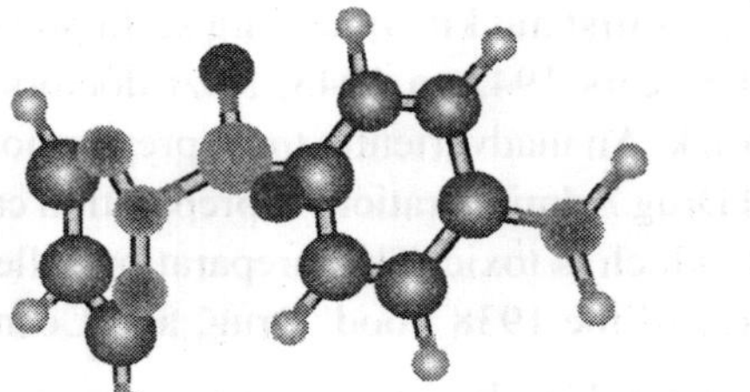

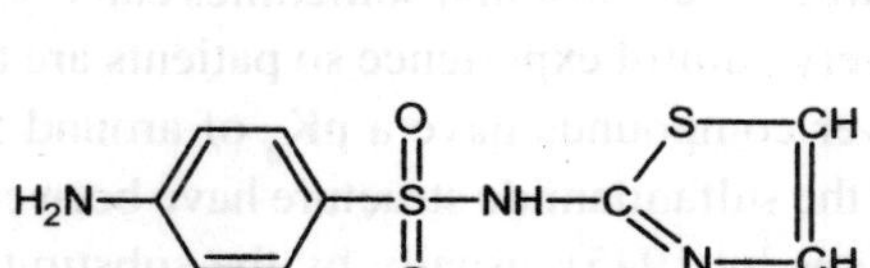

sulphathyazole

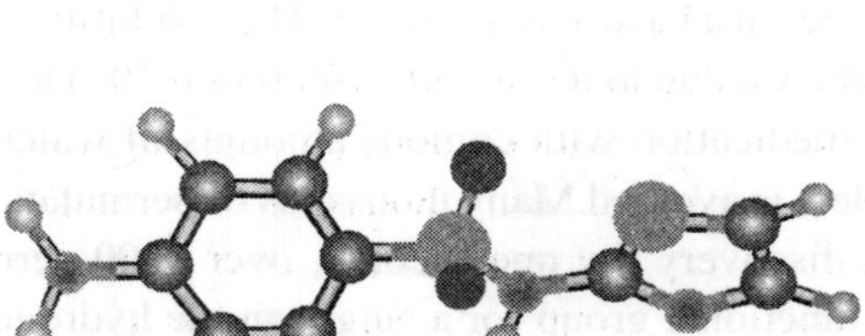

sulphaguanidine

> The detail description of these compounds are the subject of bio-chemistry which is out of syllabus of this book can be considered in future endeavour.

Sulphapyridine has a specific curative effect for pneumonia. Sulphaguanidine is useful for the treatment of intestinal infections. Sulphadazine and sulphathiazole are effective for the general microbial infections.

All through the sulpha drugs have been replaced by more powerful and less toxic antibiotics (penicillin auromycin etc) certain drugs of this class are still in the use. The following have been introduced and is found to be effective in small doses.

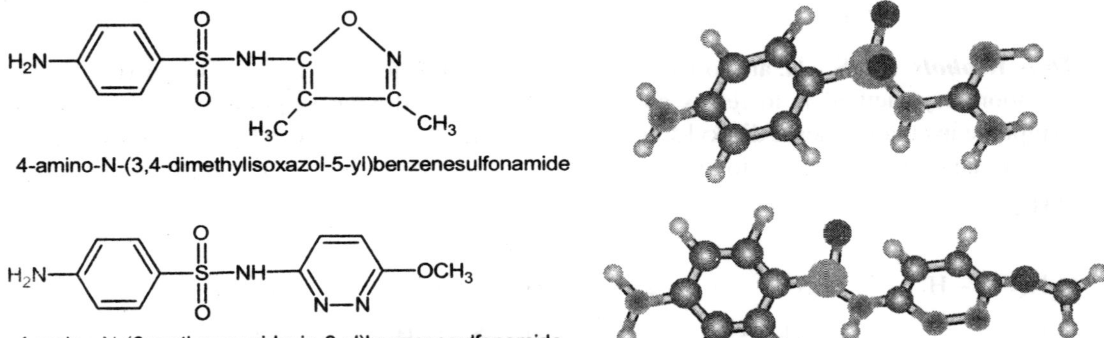

4-amino-N-(3,4-dimethylisoxazol-5-yl)benzenesulfonamide

4-amino-N-(6-methoxypyridazin-3-yl)benzenesulfonamide

The medicinal effect of the sulpha drugs has been cleavage of the molecule at the N – C bond

MORE ORGANO SULPHUR COMPOUNDS

Occurrence

Organo – sulphur compounds are widely distributed in nature; three a -amino acids of proteins viz methionine, cystine and cystine contain C – S bonds. Insulin has cystine units in its giant molecule. Thiamine (vitamine B1) contains sulphur in the ring. Co enzyme A, CoA – SH, is our system. Thyols, R – S – H, are responsible for the foul smell of petroleum garlic owes its odour to allyl hydrosulphide.

Methynethiol occurs in radish and propanethiol in onion. Allyl isothiocyanate gives black muster seed oil its characteristics smell.

Uses

Thiokol A oil resistant rubber [$-CH_2-CH_2-S_x-CH_2-CH_2-]_n$ is made from $Cl-CH_2Cl$ and sodium polysulphide Na_2S_x well known sleep inducing drugs many valuable direct vat dyes called thiozole dyes, are sulphur compounds. Saccharin, the sweetest material is the imide of orthosulpahmide – benzoic acid. Thioglycolic acid $HS-CH_2-COOH$ is used for cold waving for human hair.

Nomenclature

Organo – sulphur compounds in which bivalent sulphur replaces oxygen, are commonly named by adding the prefix Thio – (theior = sulphur Greek) Analogue e.g. thio ether $R-S-R$ thio urea $H_2N-CS-NH_2$ thio acetic acid $CH_3-CO-SH$, thiocyanates $R-S-CN$ etc. In IUPAC names of thio alcohols or thiols $R-S-H$, thiol is a suffic e.g

$H_3C-CH_2-HC(SH)-CH_3$ Butane thiol. Thiols were formerly called mercaptans e.g. ethyl mercaptans.

Thio Alcohols: $R-S-H$, are formally regardless alcohols $R-O-H$ are of water $H-O-H$. the classic laboratory method is to reflux an alkyl halide (or hydrogen sulphate) with excess sodium hydrosulphide in ethanol. Some dialkyl sulphide or thioether $R-S-R$ is also formed. These have much higher boiling point, their separation is easy.

$$C_2H_5Cl + Na^+HS^- \rightarrow \underset{\text{Ethanethiol}}{C_2H_5-S-H} + Na^+Cl^-$$

$$C_2H_5-S-H + HS^- \rightleftarrows C_2H_5-S-C_2H_5 + H_2S$$

$$C_2H_5-S + C_2H_5-Cl \rightarrow C_2H_5-S-C_2H_5 + Cl^-$$

Pure alkanethiols are prepared from thio urea and halide.

$$\underset{\text{thiourea}}{H_2N-C(=S)-NH_2} \rightleftarrows \underset{\text{carbamimidothioic acid}}{H_2N-C(SH)=NH} \xrightarrow{R-X} H_2N-C(S-R)=NH \xrightarrow[\text{warm}]{\text{aqNaOH}} R-S-H + NaX + H_2O + NH_2CN$$

Thiol in low yield is obtained by heating an alcohol with phosphorous pentasulphide

$$5R-OH + P_2S_5 \xrightarrow{\text{heat}} 5R-SH + P_2O_5$$

in industry an alkane with H_2S is passed over Lewis acid catalyst e.g. $AlCl_3$ or BF_3 at elevated temperature.

The addition is anti Markowinikoff's and the mechanism free radical. In ultra violet light similar liquid phase reaction at low temperature is also allowed to follow; some dialkyl sulphide is a co-product.

$$\underset{\text{alcohol}}{H_2C{=\!=}CH-R} + H_2S + \underset{\text{Allkanethiol}}{R-CH_2-CH_2-SH}$$

Alcohol and hydrogen sulphide are reacted in vapour phase over a dehydrated catalyst e.g. thoria or alumina, for making thiols.

$$\underset{\text{Allkanethiol}}{R-SH} + H_2S \xrightarrow[\text{heat}]{Al_2O_3} \underset{\text{thiol}}{H_3C-SH} + H_2O$$

Properties

Methanethiol is a gas, others are volatile colourless liquids, with foul smells. Those with 12 or more carbon atoms are practically odourless. Thiols are slightly soluble in water 1.5 gm of C_2H_5SH dissolves in 100 gms of water at room temperature (C_2H_5OH) is miscible with water) sulphur forms alcohols is also due to lack of inter molecular association by hydrogen bonds. But as polar compounds (C_2H_5SH has $\mu = 1.5D$) thiols associate by dipole – dipole attractions and boil much higher than alkanes of similar molecular weights e.g. $CH_3-CH_2-CH_3$ and $CH_3-(CH_2)_2-CH_3$ at -42° and -1°C respectively. Thiols (methane and ethane thiols mainly are removed from crude petroleum as they poison catalysts ion refinery process and also produce on combustion SO_2 which corrodes the automobile engine.

Reactions

(1) Salt Formation: An electron repelling alkyl group replacing a hydrogen of H_2S a weak bronsted acid ($K_a = 6 \times 10^{-8}$) makes thiols (K_a = about 10^{-11}) weaker still nevertheless they are stronger acids than parent alcohols C_2H_5OH has Ra = about 10^{-17}) and react vigorously with Na or K and dissolves readily in aqueous NaOH as salts called thiolets or mercaptides. Alkathiols were in the past removed sodium plumbite. Na_2PbO_2. Covalent heavy metal mercaptides of Cu, Hg, Ag Pb etc. are soluble in ether, benzene and isopropyl alcohol but insoluble in water sodium ethyl mercurithiosalycylate (thiomerosal) is an effective germicide

$$H_3C-SH + Na-OH \text{ € } H_3C-S-Na + H_2O$$

$$2C_2H_5-S-H + Pb(CH_3COO)_2 \longrightarrow \underset{\text{lead ethyl mercaptide}}{(C_2H_5-S)_2Pb} + 2C_3COOH$$

(2) Oxidation: Unlike alcohols thiols are easily oxidized. Due to lower electronegativity the S – H bond is weather (90 Kcal) than the OH bond in alcohols (118 Kcal) secondly sulphur can accommodate more than 8 electrons. Hence thiols are oxidized to disulphide even by weak oxidizing agents such as iodine in ethanols dimethyl sulphoxides). They also suffer autoxidation by air in presence of aqueous alkali.

$$R-S-H + I_2 \rightarrow R-S^+ + HI + I^-$$
$$R-S^+ + R-S-H \rightarrow R-S-S-R + H^+$$
Disulphide

$$2R-S-H + \frac{1}{2}O_2 \xrightarrow{OH} R-S-S-R + H_2O$$

Since HI is a strong oxidizing agent, the following reaction is reversible

$$2R-S-H + I_2 \rightleftharpoons R-S-S-R + 2HI$$

With excess chlorine or bromine water, oxidation proceeds further.

$$R-S-H + Cl_2 \xrightarrow{H_2O} HCl + R-SO_2-Cl$$
Alkyl Sulphonyl chloride

Powerful oxidizing agents e.g, HNO_3, $KMnO_4$, H_2O_2 etc. produce alkanesulphonoc acids via disulphides.

$$R-SH \xrightarrow{[O]} R-S-S-R \xrightarrow{[O]} R-SO_2-S-R \xrightarrow{[O]} R-SO_2-O-SO_2-R \xrightarrow{H_2O} R-SO_2-OH$$
alkane sulphonic acid

***(3) Reduction**:* The weaker C – S bonds, In thiols is more susceptible to catalytic cleavage by hydrogen than the C – O bond in alcohols, for example, a thiol in ethanoic solution forms an alkane on reduction with hydrogen and Raney nickel.

$$CH_3-(CH_2)_2-CH_2-SH + H_2 \xrightarrow{Ni} CH_3-(CH_2)_2-CH_3 + H_2S \text{ or } (NiS)$$
1 – butane thiol → n – butane

Crude petroleum is desulphurised in modern refinery plants with hydrogen and nickel catalyst. Mercaptans which have little commercial use are thus converted into useful alkanes R – H

$$R-SH + H_2 \xrightarrow{Ni} R-H + H_2S$$

***(4) Formation of Esters**:* In contrast with alcohols, direct esterfications of thiols with carboxylic acids in presence of a strong acid catalyst in unfavourable apparently because less electronegative sulphur has less attraction for the positive carbon of protonated carboxylic acid. Esters are prepared by reacting thiols with acyl chlorides or anhydrides.

$$R_1-CO-Cl + R-SH \longrightarrow H_3C-S-CO-R_1 + HCl$$
acyl chloride → thio ester

$$(R'-CO)_2O + R-S-H \rightarrow R-S-CO-R'-COOH$$

***(5) Addition of Carbonyl Group**:* Thiols form thioacetals and also thio – ketals on reacting with aldehydes and ketones in presence of a strong acid catalyst. The intermediate hemimecaptals are not stable in all cases.

Mercaptans, like acetyls, are stable in presence of bases. The reaction is faster than with alcohols due to the weaker S – H bond and lower electronegativity of sulphur.

$$R_1-\overset{O}{\overset{\|}{C}}-H + R-S-H \longrightarrow R_1-\overset{H}{\overset{|}{\underset{OH}{\underset{|}{C}}}}-S-R \xrightarrow[-H_2O]{RSH} R_1-\overset{H}{\overset{|}{\underset{S-R}{\underset{|}{C}}}}-S-R$$

aldehyde thiol hemimercaptal mercaptal

***Uses**:* Alkane thiols are mostly used in the synthesis of rubber industry for controlling chain length in the emulsion polymerization; e.g. 1 – dodecanethiol $C_{12}H_{25}SH$ for Gr – S rubber. Some are used as warming agents for leaks in domestic fuel gases; herbicides, pesticides, etc are also made form alkane thiols.

***Thio Ethers**:* Alkyl sulphides, R – S – R occur in traces in petroleum, methiomine, an essential amino acid, is thio-ester. Oil of garlic contains alkyl sulphide $(CH_2 = CH - CH_2)_sS$ Mustard gas by Germany. Thio esters are named either alkyl sulphides or alkyl thioalkanes e.g. $C_2H_5 - S - C_2H_5$ is ethylsulphide or ethylthioethane.

Preparation

(1) From Alkyl Halides: An alkyl halide and aqueous ethanoic sodium sulphide produce a simple alkyl halide and sodium thiolate may yield unsymmetrical ether on heating.

$$2R-X + Na_2S \xrightarrow{\text{heat}} R-S-R + 2NaX$$

simple ether (thio)

$$R-X + R'-SNa^+ \rightarrow R-S-R' + NaX$$

Mixed thio esters

(2) From alkanes: In ultra violet in presence of a organic peroxide, an alkanethiol adds on to an alkane against Markowinikoff's Rule, to give a thio ether. By free radical mechanism.

$$H_3C-CH_2-SH + H_2C{=}CH-CH_2-C{\equiv}N \longrightarrow H_3C-CH_2-S-CH_2-CH_2-CH_2-C{\equiv}N$$

ethanethiol but-3-enenitrile 4-(ethylthio)butanenitrile

(3) From Ether: Heated with phosphorous pentasulphide, ether forms thio-ether in low yield

$$5R-O-R + P_2S_5 \xrightarrow{\text{heat}} 5R-S-R + P_2O_5$$

(4) From Thiols: By passing thiols over sulphide alumina catalyst at about 300°C thio esters are obtained.

$$\underset{\text{thiol}}{2R{-}SH} \xrightarrow[300\,°C]{\text{catalyst}} \underset{\text{thioether}}{R{-}S{-}R} + H_2S$$

Properties

Thio – ethers are volatile, oily liquids of unpleasant odour, lighter than and also insoluble in water. They boil disproportionately higher than oxygen analogous, for e.g. methyl ethyl and n – propyl sulphides boil respectively at 38°C, 92 and 142°C. Where as the corresponding ether at 23°C, 34.5° and 90.5°. Lone electron pairs on sulphur are more liable than oxygen (it does not however attract a portion). Thio ethers thus form stable tri and tetra covalent compounds such as sulphoxides.

$R_2S == O$ and sulphones, R_2SO_2, which have no oxygen analogues.

Reactions

(1) *Oxidation*: strong oxidizing agents e.g H_2O_2 glacial acetic acid. HNO_3, $KMnO_4$ etc convert alkyl sulphides to sulphoxides, at ordinary temperatures a sulphoxides to sulphone at 100°C. the later reaction is slower due to greater steric hindrance and the presence of one lone pair (instead of two) on sulphur. With theoretical amount of 30% H_2O_2 mostly a sulphoxides results.

$$H_3C{-}\ddot{S}{-}CH_3 + H_2O_2 \xrightarrow[CH_3COOH]{25\,°C} \underset{\text{dimethyl sulfoxide}}{H_3C{-}\overset{\bullet\bullet}{\underset{\|O}{S}}{-}CH_3} + H_2O$$

DIMETHYL SULPHOXIDE, DMSO

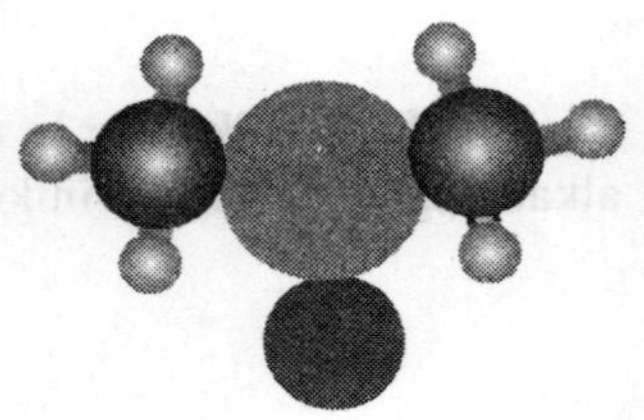

Dimethyl sulfoxide, also known as ***methyl sulfoxide***, dimethyl sulphoxide, dimethylsulfoxide, methylsulfinylmethane or sulfinylbismethane, is a sulfur-containing organic compound. It is a clear, colourless hygroscopic liquid. When it is pure it has little odor, but impure samples smell strongly of dimethyl sulfide. DMSO belongs to the class of "dipolar aprotic solvents" which includes also dimethylformamide, dimethylacetamide and *N*-methyl-2-pyrrolidone. It is readily soluble into a wide range of organic solvents such as alcohols, esters, ketones, chlorinated solvents and aromatic hydrocarbons. It is also miscible in all proportions with water.Dimethyl sulfoxide is a by-product of wood pulping and is frequently used as solvent in a number of chemical reactions. In particular DMSO proved to be an excellent reaction solvent for S_N2 alkylations: it is possible to alkylate indoles with very high yields using potassium hydroxide as the base and a similar reaction also occurs with phenols. DMSO can be reacted with methyl iodide to form a sulfoxonium ion which can be reacted with sodium hydride to form a sulfur ylide. The methyl hydrogens of DMSO are somewhat acidic in character (pK_a=35) due to the stabilization of the resultant anions by the sulfoxide group.It is important to use caution in handling dimethyl sulfoxide, due to its strong solvent power and the ease with which it penetrates the skin. It is especially important to avoid contact with the mucous membranes.In certain conditions, dimethly sulfoxide can produce an explosive reaction when exposed to acid chlorides.

Production

Dimethyl sulphoxide is manufacture by oxidising dimethyl sulphide (from waste sulphide liquir of paper industry) with air in presence of oxides of nitrogen. It is miscible with water, it has high dielectric constant of 45 and is an ionising solvent in amny reactions. Excess oxidising agents produces sulphones by further oxidation of dialkyl sulphoxides e.g.

$$H_3C-\underset{\underset{O}{\|}}{S}-CH_3 + H_2O_2 \xrightarrow{100\ °C} H_3C-\overset{\overset{O}{\|}}{\underset{\underset{O}{\|}}{S}}-CH_3 + H_2O$$

dimethyl sulfoxide dimethyl sulfone

The electron withdrawing oxygen in sulphoxides makes them waker nucleophiles than thioethers and also slows down their oxidation. Sulphones are crytalline solids resistant to further oxidation. The α-hydrogen atoms in sulphones (and sulphoxides) are liable and acidic both undergo claisen condensation.

Physical Properties

Molecular formula	C_2H_6OS
Molar mass	78.13 g/mol
Appearance	Clear, colourless liquid
Properties	
Density and phase	1.1004 g/cm^3, liquid
Solubility in water	Miscible
Solubility in ethanol, benzene, chloroform	Miscible
Melting point	18.5 °C (292 K)
Boiling point	189 °C (462 K)
Viscosity	1.996 cP at 20 °C
Acid dissociation constant	35 (in DMSO)
Structure	
Dipole moment	3.96 D
Dielectric constant	45

Uses

DMSO was discovered in 1867, but was not used commercially until after WWII. Other than its use as a solvent, both in organic synthesis and industrial applications (polymer chemistry, pharmaceuticals and agrochemicals), DMSO also makes a very good paint stripper: it is able to remove many paints from both wood and metal in a small amount of time. It is thought to be much safer than many of the other chemicals used as paint strippers, such as nitromethane and dichloromethane.In organic synthesis, DMSO can also be used in a number of oxidation reactions[1], the Pfitzner-Moffatt oxidation and the Swern oxidation. DMSO is also employed as a rinsing agent in the electronics industry and, in its deuterated form (DMSO-d6), is a useful solvent in NMR due to its ability to dissolve a wide range of chemical compounds and its minimal interference with the sample signals. In cryobiology DMSO has been used as a cryoprotectant and is still an important constituent of cryoprotectant vitrification mixtures used to preserve organs, tissues, and cell suspensions. It is particularly important in the freezing and long-term storage of embryonic stem cells and hematopoietic stem cell, which are often frozen in a mixture of 10% DMSO and 90% fetal calf serum. As part of an autologous bone marrow transplant the DMSO is re-infused along with the patient's own hematopoietic stem cell.

Use of dimethylsulfoxide in medicine dates from around 1963, when a University of Oregon Medical School team, headed by Stanley Jacob, discovered it could penetrate deeply through the skin and other membranes without damaging them and could carry other compounds deep into a biological system. Some people report an onion- or garlic-like taste after contact with the skin. This is likely due to catabolic processes which reduce DMSO to dimethyl sulfide. In the medical field DMSO is predominantly used as a topical analgesic, a vehicle for topical application of pharmaceuticals, as an anti-inflammatory and an antioxidant. It has been examined for the treatment of an extraordinary number of conditions and ailments. The Food and Drug Administration (FDA) has approved DMSO usage only for the palliative treatment of interstitial cystitis. Because it was first used several decades ago, it is currently not possible to patent it because of the 17 year length of pharmaceutical patents. This largely explains why little research is done into its possible therapeutic roles, because without the possibility of future patent protection, pharmaceutical companies rarely fund research on a drug. (Another substance having obvious therapeutic effect but not FDA approved for use as a drug is melatonin.) Also, DMSO is commonly used in the veterinary field as a liniment for horses.Dimethyl sulfoxide is the most powerful readily available organic solvent. It dissolves a great variety of organic substances to the highest loading level, including carbohydrates, polymers, peptides, as well as many inorganic salts and gases. Loading levels of 50-60 wt. % are often observed with DMSO (compared with 10-20 wt. % with typical solvents). Therefore it plays a vital role in sample management and High-throughput screening operations in drug design

(2) Formation of Sulphonium Salts: Alkyl sulphides form sulphonium salts on warming with an alkyl halidfe in ethanol. The reaction is reversible; on heating strongly the salts dissociates into the components.

$$C_2H_5 - S - C_2H_5 \quad + \quad C_2H_5 - I \rightleftharpoons \quad (C_2H_5)_2\ S + C_2H_5I$$

Di ethyl sulphide — tri ethyl sulphonium iodide

If the alkyl groups are different, e.g $(CH_3)(C_2H_5)S^+I^-$ the salt can be resolved into δ-and i-isomers it has a non- planner tetrahedrol confugaration and is symmetrical life quaternary ammonium salts. The optical isomers are quite stable. With moist silver oxide, the salts form hydroxides which are strong bases. On heating strongly, alkanes result as in the cases of quaternery ammonium hydroxide, e.g.

$$(C_2H_5)_3S^+I^- + AgOH \rightarrow (C_2H_5)_3S^+OH^- \rightarrow (C_2H_5)_2S + CH_2 = CH_2 + H_2O$$

triethyl sulphonium iodide; tri ethyl sulphonium hydroxide; Di ethyl sulphide

*(2) **Cleavage of C - S bonds:*** The C - S bond in thioether is more difficult to break than the C - O bond in the aqueous acetic acid however cleaves the bond

$$R - S - R' + BrCN \xrightarrow{65°C} R - S - CN + R' - Br$$

Alkyl thiocyanate

$$R - S - R' \rightarrow R - S - Cl + R' - Cl$$

Alkyl sulphenyl chlroide

Alkane sulphonic acids: R – SO_3H: These are derivatives of sulphuric acid, of which one (OH) group has been replaced by an alkyl group; the simplest is methane sulphonic acid. $CH_3 - CO_2 - OH$ Alkene sulphamic acids $R - SO_3H$ and alkanesulphenic acid R – SOH are unstable, they are readily oxidised to sulphonic acids. In alkanesulphonic acids sulphur is directly linked with carbon but in sulphuric acid esters oxygen joins the two, e.g. methyl hydrogen sulphate

$CH_3 - O - SO_2 - OH$ and dimethyl sulphate, $CH_3 - O - SO_2 - O - CH_3$. Aliphatic sulphonic acids are named by adding the alkane name as a prefic to sulphonic acid mentioning them as one word; e.g. ethanesulphonic acid $C_2H_5 - SO_3H$. Higher members are usually called alkylsulphonic acids for simplicity e.g. $C_{12}H_{25}SO_3H$ is lauryl sulphuric acid, not also present, it is a sulpho acid e.g sulpho-acetic acid. $HO_3S - CH_2 - COOH$.

Preparation

In the laboratory alkanesulhonic acid is prepared

(1) by refluxing a primary alkyl halide with excess aqueous sodium or ammonium sulphide e.g,

$$C_2H_5 - I + Na_2SO_3 \rightarrow C_2H_5 - SO_2 - ONa^+ + NaI$$

Ethyl Iodide; sodium ethane sulphonate

The free acid, a hydroscopic syrupy liquid, may be obtained in ethanolic sodium by passing dry HCl gas into group combines with sulphur to form a greater covalent bond C – S, rather than with lone electron pair of the silphite ion —O—S—O— (with S=O:). Sodium chloroacetate and sodium sulphite similarity yield disodium sulphoacetate.

$$Cl - CH_2 - COO - Na^+ + Na_2SO_3 \rightarrow Na^+ - O_2S - CH_2 - COO - Na^+ + NaCl$$

(2) From alkanes: Though n - hexane and higher memebrs produce alkenesulphonic acid with hot fuming sulphuric acid, the yield is low due to sulphoantion with sulphuryl chloride to form alkanesulphonic acids.

(3) From Alkenes: Aqueous sodium bisulphite adds to the terminal alkene in presence of oxygen or oxides of nitrogen at room temperature against Markowinikoffs rule to yield sodium alkenesulphonate by free readical mechanism.

$$H_3C—CH{=}CH_2 + NaHSO_3 \longrightarrow H_3C—CH_2—CH_2—S(=O)_2—O—Na$$

prop-1-ene (propane-1-sulfonato-kO)sodium

$$—O—S—OH + O{=}O \longrightarrow O—S(=O)—O + HOO$$

$$R—CH{=}CH_2 + O—S(=O)—O \longrightarrow R—CH_2—CH_2—S(=O)—O—$$

(4) From thiols: Alkanesulphonic acids are easily prepared by axidising thiols with a strong oxidising agent e.g HNO_3 via, disulphide.

$$H_3C—SH \xrightarrow[-H_2O]{O} R—S—S—R \xrightarrow{O} R—S(=O)_2—O—H$$

disulphide alkanesulphonic acid

Properties

Lower alkenesulphonic acids are hydroscopic liquids or low melting solids (CH_3 - SO_3H) melts at 20ºC, n - C_5H_{11} - SO_3H at 16ºC, n - $C_{16}H_{33}$ - SO_3H at 53ºC) freely soluble in water abnd polar solvents. They can be distilled undecomposed only under high vacuum. These culphonic acids are strong acids, much stronger than corresponding carboxylic acids, because the anion is stable due to response (distribution of negative charge is on three oxygen atoms); the positive charge on sulphur greatly stabilizes the negative charge on oxygen. They are almost as strong as HCl or H_2SO_4.

$$H_3C—S(=O)_2—OH \rightleftharpoons H_3C—\overset{+}{S}(O^{\sigma-})(O^{\sigma-})(O^{\sigma+}) + H$$

The density of alkenesulphonic acids fall with rise in molecular weight as the alkene character predominates (for CH_3 - SO_3H d = 1.484) and for n - C_6H_{13} - SO_3H d = 1.014) solubility in water also decomposes with increased carbon chain length.

Reactions: In contrat with the aromatic sulphonic acids e.g. benzene sulphonic acid C_6H_5 - SO_3H, the sulphonic acid groups - SO_3H isn alkane sulphonic acids cannot be replaced by OH CN NH_2 etc groups by reacting with NaOH NaCN $NaNH_2$

(1) *Salt Formation*: Alkane sulphonic acids from salts are usual with inorganic oxides hydroxides and carbonates. In many cases sodium salts are formed with aqueous NaCl

$$R - SO_2 - OH + NaCl \rightleftharpoons R - SO_2 - O - Na^+ + HCl$$

Most metallic salts including Ba, Ca, Pb, and Ag salts are soluble in water where as sulphates of these metals are insoluble or sparingly soluble in water.

(2) *Replacement of OH group by chlorine*: The OH group of sulphonic acids like the OH carboxylic acids can be repalced by chlorine on reacting with PCl_5 $SOCl_2$ or chloro sulphuric acid Cl - SO_2 - OH; alkane sulphonyl chlorides thus result. These are insoluble in water and are hydrolysed slowly

$$\underset{\text{Alkane Sulphonic acid}}{R - SO_2 - OH} + PCl_5 \rightarrow \underset{\text{alkane sulphonyl chloride}}{R - SO_2 - Cl} + POCl_3 + HCl$$

$$R - SO_2 - OH + Cl - SO_3H \rightarrow R - SO_2Cl + H_2SO_4$$

(3) *Formation of Esters*: Direct esterification with alcohol is unsatisfied from a sulphonic acid esters are prepared from a sulphonyl chloride and an alcohol in presence of aqueous NaOH (which makes the alcohol stronger nucleophilic by generating alkoxide ions R – O –)

These esters like alkyl sulphates,produce ethers on heating with alcohols (except esters of carboxylic acid)

$$\underset{\text{Sulphonate ester}}{R - SO_2 - O - R'} + R'' - OH \rightarrow \underset{\text{ether alkane}}{R' - O - R''} + \underset{\text{sulphonic acid}}{R - SO_2 - OH}$$

Alkane sulphonic acid as sodium salts are used as household synthetic detergents; these are not precipitated by hard water (their Ca and Mg salts are water soluble) many of these are sodium alkyl aryl sulphonates e.g. $R - CH_4 - SO_3 - Na^+$.

(R derived from petroleum) containing 12 or more carbon atoms for optimum detergency, the group including the aromatic portion should easily dissolve oil and grease as the hydrocarbon R – H or $C_6H_5 - R$.

CHAPTER

26

Nitrogen Derivatives of Aromatic Hydrocarbons

Nitration

Nitro compounds can be converted into nitro – derivatives, by the substitution of nitro groups, for hydrogen of the nucleus. The process is known as 'Nitration' is quite important in organic chemistry and many may be brought out by the following reagents.

(a) A mixture of Conc. nitric acid and conc. Sulphuric acid

(b) A mixture fuming nitric acid and conc. sulphuric acid

(c) A mixture of conc. nitric and acetic anhydride

(d) Nitrogen peroxide in the presence of anhydrous aluminium chloride (catalyst)

Conc. sulphuric acid or acetic anhydride in the above cases is used as a dehydrating agent

The number of nitro group which can be introduced in the nucleus depends upon.

(a) The concentration of nitrating agents

(b) The temperature maintained during nitration

(c) The mixture of substance undergoing nitration

In general nitration is facilitated if the nucleus is already substituted by hydroxyl ot alkyl group i.e. toluene and phenol are readily nitrated.

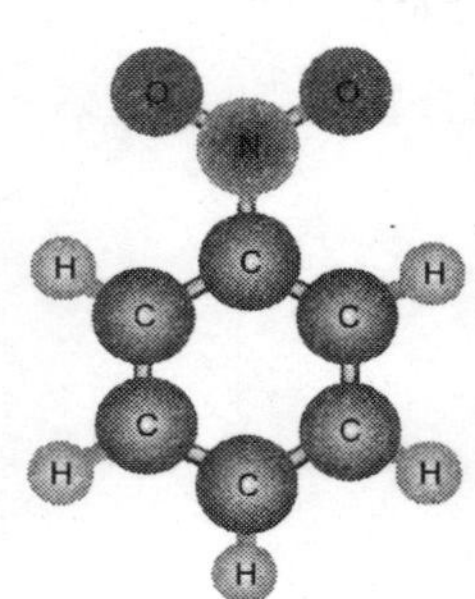

NITRO BENZENE

Nitro-benzene is also known as 'Oil of Mirbane'

Preparation

Nitrobenzene is prepared by the action of mixture of concentration of nitric acid and sulphuric acid on benzene.

$$C_6H_6 + HNO_3 \longrightarrow C_6H_5NO_2 + H_2O$$

benzene → nitrobenzene

Sulphuric acid removes water which would otherwise dilute nitric acid and decrease the speed of the reaction.

Laboratory Preparation

A well cooled mixture of concentrated nitric acid (12 parts) and concentrated sulphuric acid (16 parts) is slowly added to the pure benzene (10 parts) placed in a flask after addition flask is well shaken. The flask may be cooled under tap water, if the temperature rises above 50 – 50°C, at higher temperature some dinitrogen is formed when all the acid mixture is added, the flask is heated on water bath or heating mantle for about half an hour shaking well in intervals.(fig. 26.1 The contents of the flask are transferred into a separating funnel. The lower layer of the spent acid is first removed. The upper layer in the separating funnel is next washed by shaking it with water and then with dilute sodium carbonate solution and finally again with water. Since nitrobenzene is denser than these solutions, it collects at the

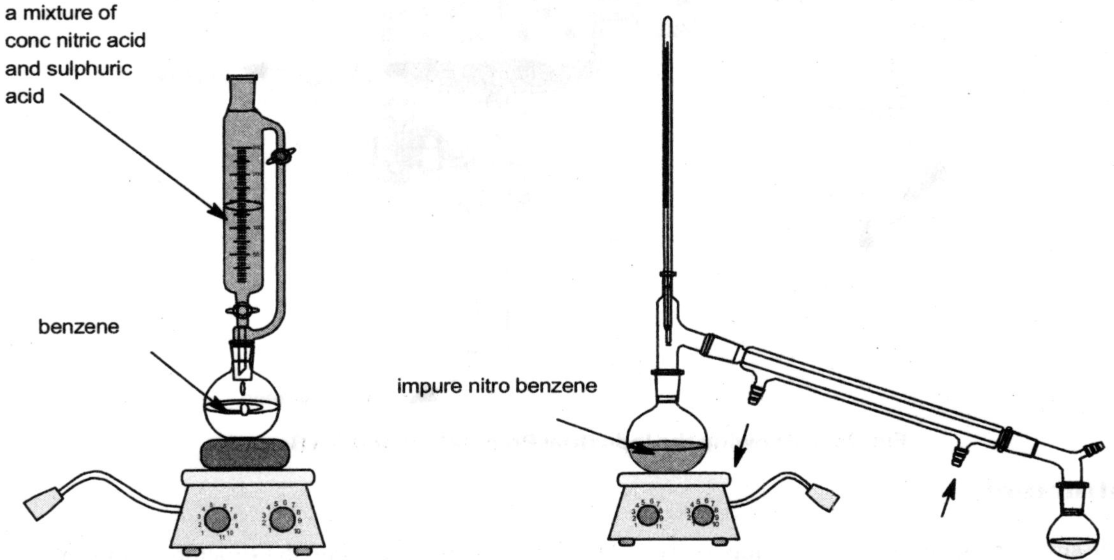

Fig. 26.1. Preparation of Nitro-benzene

Fig. 26.2. Distillation Nitro benzene

bottom. The nitrobenzene layer is run off into a small dry flask containing some anhydrous calcium chloride and shaken gently. Calcium chloride lumps absorb water and nitrogen – benzene is thus dehydrated (Fig. 26.2) it is taken in a distillation flask with condenser. Pure nitro benzene is collected between 209 - 210°C.

Industrial Preparation

On the large scale nitro-benzene is prepared by a similar process, provision being made for mechanical stirring and adequate cooling and heating of the mixture. The nitration of the mixture is carried in an iron vessel (see fig. 26.3).

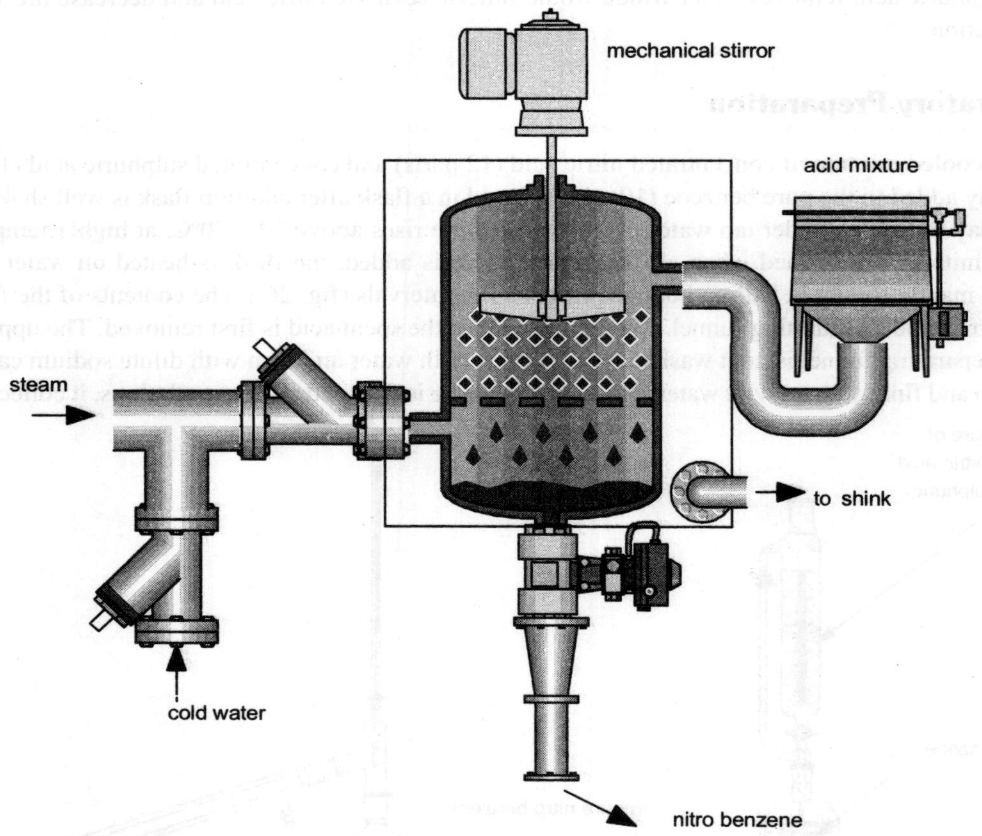

Fig. 26.3. Showing the Industrial Preparation of Nitro Benzene.

Structure

The structure of aromatic nitro compounds will be illustrated by discussing the structure of nitrobenzene. It is basically the same as that of benzene excepting that the unused *p* orbital of the ring carbons can interact with the unused *p* orbitals of Nand O atoms of the nitro group as indicated above. This is made possible because these *p*. orbitals are aligned and hence they can interact to give a larger delocalized p orbital encompassing the six ring carbons as also the Nand O of the nitro group.

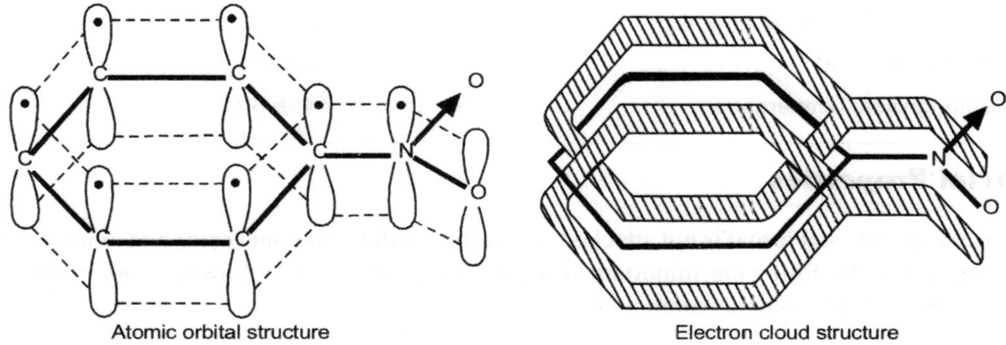

Fig. 26.4. Showing the Structure of the Nitrobenzene

The electronic structure of nitrobenzene can be represented as a- resonance hybrid of the following four canonical forms.

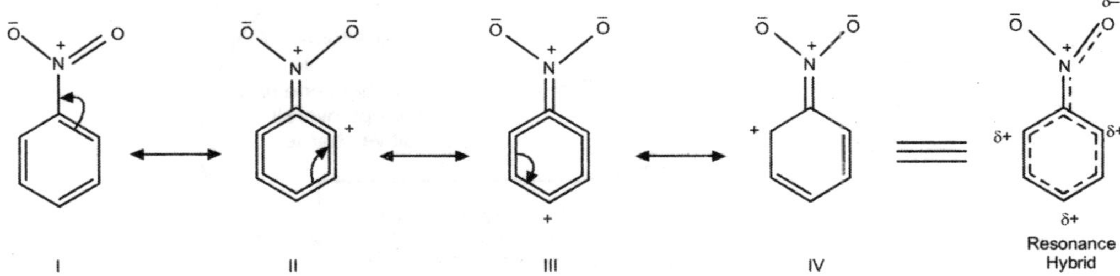

Fig. 26.5. The Ring formation and the resonance Hybrid of Nitrobenzene.

Physical Properties

Empirical Formula	$C_6H_5NO_2$
Molecular Weight	123.11
Freezing Point	5.7°C
Boiling Point @ 760 mmHg	210.8°C
Specific Gravity (20°C/4°C)	1.204
Vapor Pressure mmHg @ 44.4°C	1.0

Heat of Combustion @ 20°C, Kg-cal/mol	739.9
Heat of Vaporization @ 760 mmHg cal/gm	79.0
Refractive Index @ 20°C	1.5562
Solubility in Water, @ 15.6°C	0.19%
Flash Point, TOC	87.8°C
Auto Ignition Temperature	495.6°C

Chemcial Properties

Nitrobenzene is quite stable and is not attacked by caustic alkalies ammonia acids and oxidizing agents, it however, shows the following important reaction which shows the following important reactions which are typical nitro aromatic compounds.

A Reaction of Benzene Ring

We have already seen that aromatic compounds are resonance hybrids. Thus benzene can be represented as

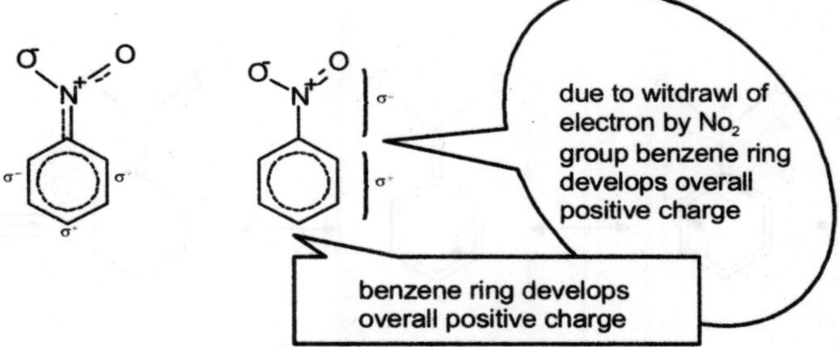

The electron withdrawal by NO_2 as a result of resonance and inductive effect from the benzene deactivates it to the attack to electrophiles. Further a partial positive charge develops on the ortho and

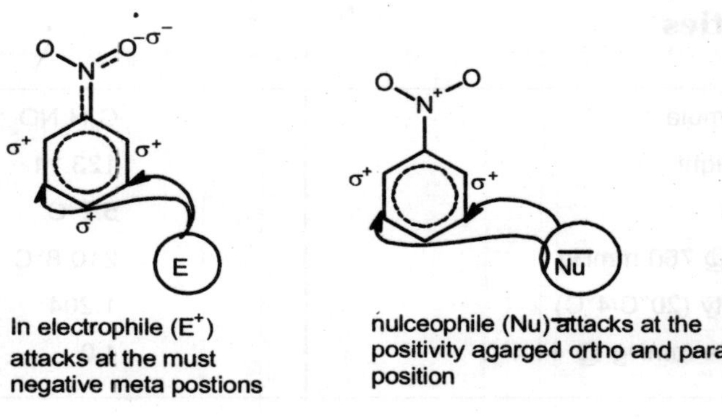

para position. Thus an electrophile seeking to attach at the most negative positions, will avoid the meta positions which beers no positive charge.

It may also be noted that the presence of the electron withdrawing $-NO_2$ group activates the ring for the attack of the nucleophiles at the ortho and para positions which bears the positive charge. This makes the nucleophilic substitution possible in ortho and para positions in the benzene ring of nitro benzene.

(I) Electrophilic Substitution: As already been observed, the $-NO_2$ group in nitrobenzene exerts deactivating and meta – directive effects on electrophilic substitution in the benzene ring. Thus the second $-NO_2$ group cane be introduced in it at higher temperature and in meta position.

$$C_6H_5NO_2 + NO_2 \xrightarrow{HNO_3/H_2SO_4} C_6H_4(NO_2)_2 + H^+$$

The third nitro group can be introduced in the ring under extreme difficult condition

(II) Nucleophilic Substitution: The ortho and para positions of benzene ring are activated and bear a positive charge, due to the presence of $-NO_2$ group in the ring. Thus when nitro-benzene is heated with aqueous potassium hydroxide in the presence of air o – nitro phenols are produced.

$$C_6H_5NO_2 + OH \xrightarrow{aq\ KOH} \text{o - nittrophenol} + H^-$$

$$2H + O_2 \longrightarrow 2OH^-$$

When a second $-NO_2$ is present in the benzene ring of nitrobenzene in the ortho or para position, it undergoes nucleophilic displacement. Thus

$$\text{1,2-dinitrobenzene} + aq\ KOH \longrightarrow \text{2-nitrophenol} + KNO_2$$

$$2H + O_2 \longrightarrow 2OH^-$$

$$p\text{-}O_2N{-}C_6H_4{-}NO_2 + 2NH_3 \xrightarrow{\text{alcoholic}} p\text{-}O_2N{-}C_6H_4{-}NH_2 + NH_4NO_2$$

B. Reactions of the Nitro Groups

The nitro group, being made of nitrogen and oxygen only show no other reaction except those of reduction.

Reduction of Aromatic Nitro Compounds

(1) Conversions to 1° Amines: Aromatic nitro groups on ($-NO_2$) reduction with powerful reducing agents are converted to primary amino groups ($-NH_2$). Thus aromatic primary amines are produced.

$$Ar{-}N^{+}(=O){-}O^{-} \xrightarrow[\text{reducing agent}]{\text{powerful}} \underset{1^\circ \text{ amine}}{Ar{-}NH_2}$$

(a) Metal – Acid Reduction: Iron or tin with hydrochloric acid e.g.

$$\underset{\text{nitrobenzene}}{C_6H_5NO_2} + 6[H] \xrightarrow{Fe/HCl} \underset{\text{aniline}}{C_6H_5NH_2} + H_2O$$

The reduction of nitrobenzene with iron and hydrochloric acid is used for the industrial production of aniline.

Pour a few drops of nitrobenzene into a test-tube, and add a few c.c. of glacial acetic acid and then a little zinc dust from the edge of a knife. When the first reaction is over, warm gently for . Add a little water, and decant the clear liquid. The solution aniline acetate together with zinc acetate

$$C_6H_5NO_2 + 3Zn + 7C_2H_4O_2 \rightarrow \underset{\text{Aniline acetate}}{C_6H_5NH_2C_2H_4O_2} + 3Zn(C_2H_3O_2)_2 + 2H_2O.$$

Add caustic soda to the solution until the zinc hydroxide redissolves, the liquid, which now contains free aniline, into a solution 1 hypochlorite. The violet colour, which is developed, is characteristic of aniline. Somewhat different result is obtained if the reduction occurs in a neutral solution, *e.g.* by the action of zinc dust, or the alumino-mercury couple, and water. Nitrobenzene is converted into *phenylhydroxylamine,*

$$C_6H_5NO_2 + 2H_2 \rightarrow C_6H_5NH(OH) + H_2O$$

This compound doubtless forms an intermediate stage in the production of aniline. Phenylhydroxylamine is a very reactive substance; on reduction it yields aniline, and on oxidation it is first converted into *nitrosobenzene,* C_6H_5NO, and then into nitrobenzene. Nitrosobenzene is a yellow, crystalline substance, which on heating changes to an emeraldgreen liquid. With mineral acids, phenylhydroxylamine undergoes isomeric change to p-amino-phenol (see the chapter phenol and derivatives). It reduces Fehling's solution, and also separates iodine from potassium iodide and dilute sulphuric acid.

A totally different effect is produced by the action of alkaline reducing agents, such as sodium methylate, zinc dust, and caustic soda, or stannous chloride and caustic soda. Azoxy-, azo-, and hydrazo-compounds are formed. Nitrobenzene is converted in successive steps into azoxybenzene, azobenzene, and hydrazobenzene.

nitro benzene   azoxy benzene   azo benzene   hydrazo benzene

The reduction of nitrobenzene has been effected at the cathodes of electrolytic cells. By varying the conditions different products are formed. If the acid is too concentrated, some of the phenol hydroxylamine undergoes rearrangement to the isomeric p-amino-phenol, $HO'C_6H_4'NH_2$. The action of acid on hydrazobenzene is to form the isomeric compound benzidine, $H_2N'C_6H_4'C_6H_4'NH_2$. Both changes are irreversible. In alkaline solution the main products are azo benzene and hydrazobenzene, the latter being easily oxidized again to azobenzene. The following scheme shows how the reduction goes :

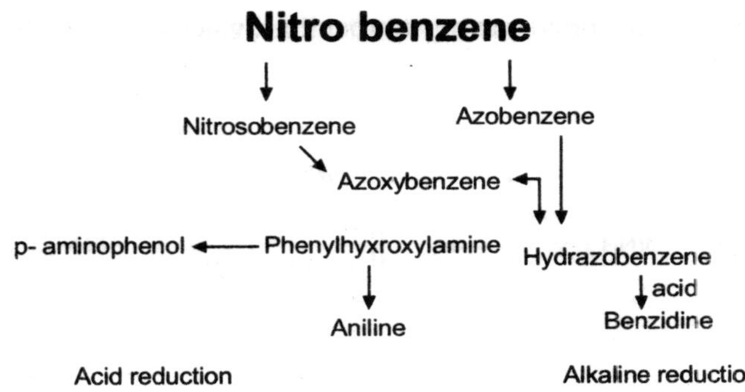

(b) Catalytic hydrogenation: **Hydrogen in the presence of a metal catalyst platinum or Raney Nickel**

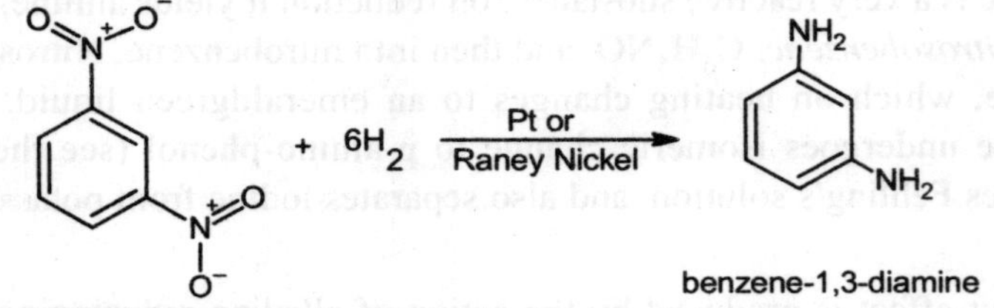

$$\text{1,3-dinitrobenzene} + 6H_2 \xrightarrow[\text{Raney Nickel}]{\text{Pt or}} \text{benzene-1,3-diamine}$$

(c) With Stannous Chloride ($SnCl_2$) or Titanous Chloride ($TiCl_2$):

This reaction is used for the quantitative determination of $-NO_2$ group

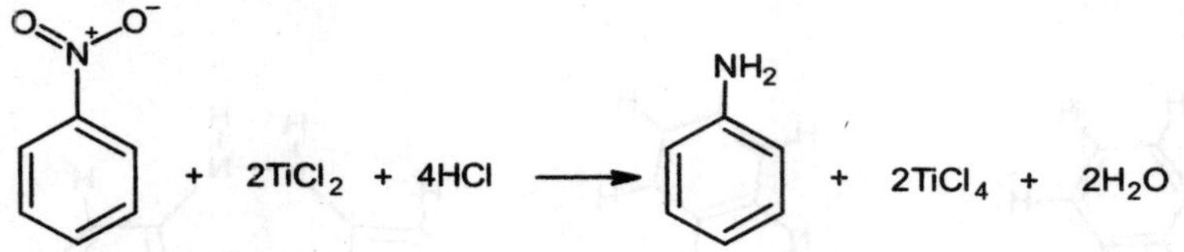

$$C_6H_5NO_2 + 2TiCl_2 + 4HCl \longrightarrow C_6H_5NH_2 + 2TiCl_4 + 2H_2O$$

(d) With Lithium Aluminium Tetrahydrate, $LiAlH_4$: **When p nitro benzene reacts with Lithium Aluminium Tetrahydrate, in the presence of ether. Aniline is produced.**

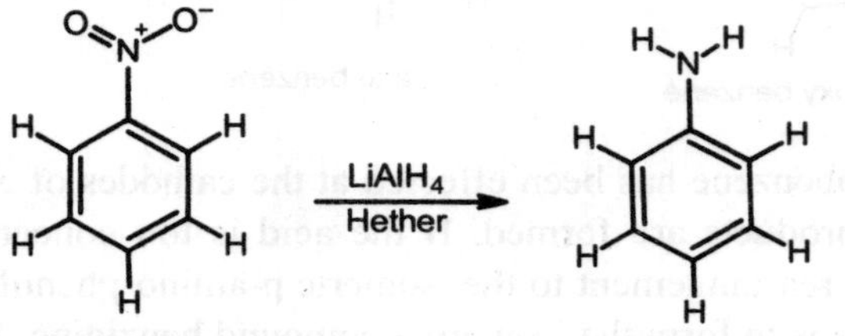

$$\text{nitrobenzene} \xrightarrow[\text{Hether}]{LiAlH_4} \text{aniline}$$

(2) Selective Reduction: **If two or more nitro groups are present in the aromatic ring it is possible to reduce one of them without the others. Such selective reductions may be carried by proper choice of a reducing agent. For e.g.**

(*a*) m – Dinitrobenzene is converted into m – nitroaniline, when reduced with sodium or ammonium sulphides.

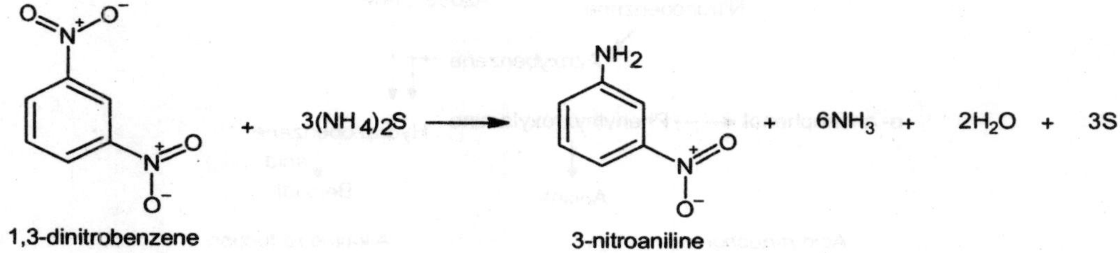

$$\text{1,3-dinitrobenzene} + 3(NH_4)_2S \longrightarrow \text{3-nitroaniline} + 6NH_3 + 2H_2O + 3S$$

(*b*) 2, 4 – Dinitro benzene may be reduced at either of two $-NO_2$ groups with suitable reducing agents.

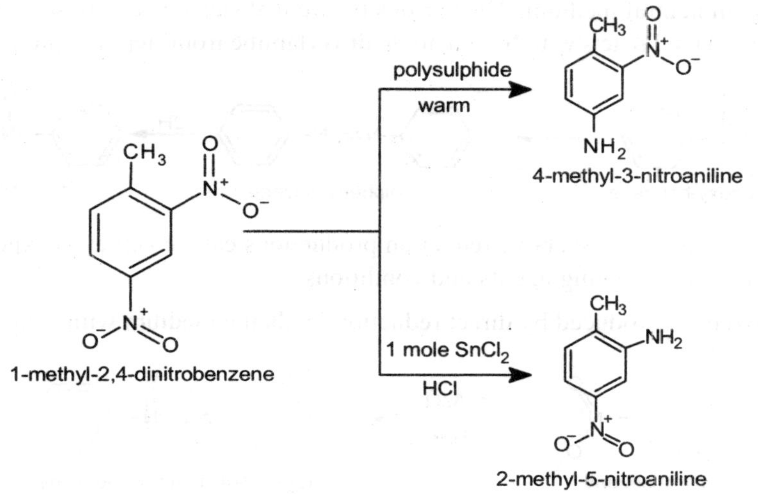

polysulphide
warm

4-methyl-3-nitroaniline

1-methyl-2,4-dinitrobenzene

1 mole $SnCl_2$
HCl

2-methyl-5-nitroaniline

(3) Reduction of Nitrobenzene under different conditions:

The reduction aromatic nitro compound is a complex process. In fact, it involves the formation many intermediate and side – products before the final stage – the production of an aromatic 1° amine – is reacted. For e.g., the electrochemical studies of the of reaction of nitrobenzene indicate that the reduction proceeds by the following steps.

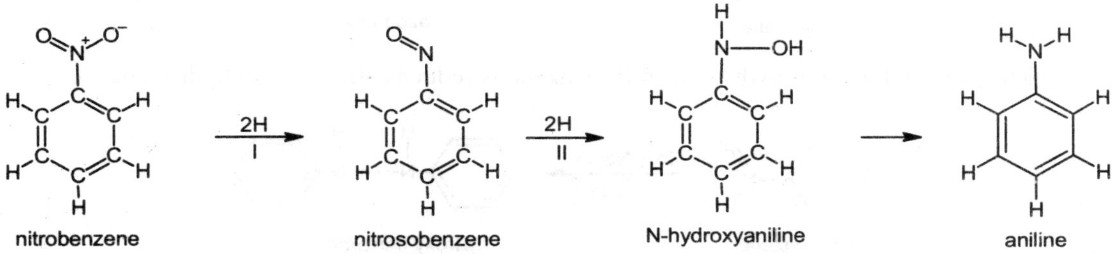

nitrobenzene —2H (I)→ nitrosobenzene —2H (II)→ N-hydroxyaniline → aniline

However, the nature of the product actually obtained depends on whether the reduction is carried out in acid, neutral or alkaline medium as also on the reagent used. It is noteworthy that all the products under different conditions can be aniline by reducing agents.

(*a*) Reduction in Acid Medium: The reduction of nitrobenzene with a metal and mineral acid gives aniline.

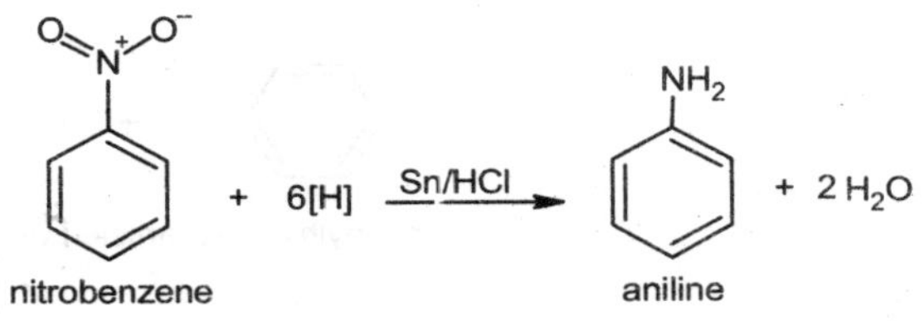

$$\text{nitrobenzene} + 6[H] \xrightarrow{Sn/HCl} \text{aniline} + 2H_2O$$

Here the step II and III of the general scheme described above are kipped over and the end product, the aniline is obtained straightly. This is so because the intermediate compounds (nitroso benzene an N – phenylhydroxylamine.

(*b*) Reduction in neutral medium: The various theoretical steps listed are same here also. But the nitroso benzene from step I reacts with N – phenylhydroxylamine from step II to produce azoxybenzene.

(phenyNNOazoxy)benzene → diphenyldiazene —2H→ hydroxy benzene

The formation of most of the above reduction productions can be achieved experimentally by the appropriate choice of the reducing agents and conditions.

(*i*) A zoxy benzene is produced by direct reduction in alkane medium with sodium arsenite.

nitrobenzene —Na_3AsO_3, $NaOH/H_2O$→ (phenyl-NNO-azoxy)benzene

Azoxybenzene on further reduction with iron and heat yields azobenzene.

(*ii*) Azoxybenzene is formed by reduction with zinc and methanoic sodium hydroxide.

nitrobenzene —Zn + NaOH, CH_3OH→ azo benzene

(*iii*) with zinc and sodium hydroxide, nitrobenzene is reduced straightly to hydroxybenzene.

nitrobenzene —Zn + NaOH, CH_3OH→ hydroxy benzene

(*d*) Electrolytic Reduction: When nitro benzene is reduced by electrolysis in dil. Sulphuric acid solution, the reaction is interrupted after stage II. The N – Phenylhydroxylamine as soon as it is produced, under goes an intermolecular rearrangement to yield p – amino phenol.

N - phenylhydroxyl amine from step II —rearrange→ 4-aminophenol

Test for Nitro Group

*(i) **Azo dye Test:*** A little given substance is suspended in water and reduced with tin and hydrochloric acid. The solution thus obtained is cooled and a little sodium nitrite solution added. A red precipitate of the azo dye shows the presence of $-NO_2$ group in the original substance.

(ii) Fehling's Solution Test: The given substance is boiled with zinc dust and ethanol and filtered. The filtrate is then heated with Fehling's solution. A red precipitate appears which indicates the presence of $-NO_2$ group. In this reaction the nitro compound is first reduced to arylhydroxylamine. Which reduces the Fehling's solution to red copper oxide.

DINITROBENZENE

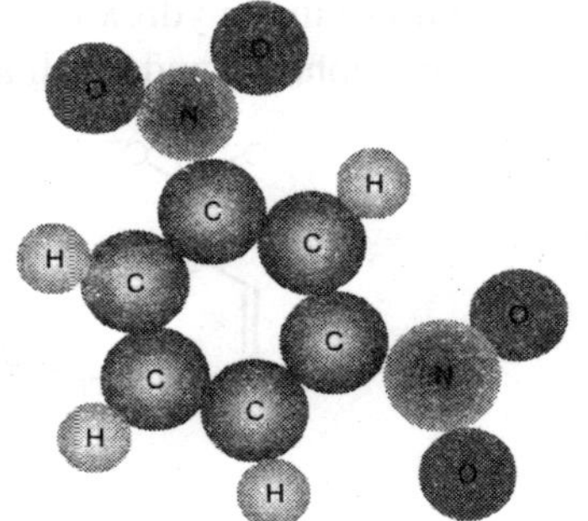

Introduction

Of the three isomeric dinitrobenzene (m,o,p) the meta derivative is by far the most important.

Preparation

It is prepared by a vigorous reaction between nitro benzene and sulphuric acid.

$$\text{nitrobenzene} + HNO_3 + H_2SO_4 \longrightarrow \text{1,3-dinitrobenzene}$$

About 7% of the o-isomer is also obtained which is removed during crystallization.

Physical Properties

Properties	m – isomer	o – isomer	p - isomer
Boiling point	302°C	319°C	299°C
Melting point	89°C	118°C	173°C
Density		5.79	
Vapour pressure	less than 1mm Hg	less than 1mm Hg	less than 1mm Hg
Flash point	150°C	150°C	

Chemical Properties

It is insoluble in water, but volatile in steam. It is very toxic, when reduced with ethanoic ammonium sulphide it is fist converted to nitroaniline and then to m – phenylenediamine.

1,3-dinitrobenzene $\xrightarrow[(NH_4S)]{[H]}$ 3-nitroaniline $\xrightarrow[(NH_4)_2S]{[H]}$ n - phenylene diamine

With sodium hydroxide and potassium ferricyanide, it under goes nucleophilic substitution yielding 2, 4 dinitrophenol and small amount of 2, 6 dinitrophenol.

2 1,3-dinitrobenzene $\xrightarrow[K_2Fe(CH)_6]{NaOH}$ 2,4-dinitrophenol + 2,6-dinitrophenol

Uses

Meta isomer of the dinitrobenzene is used in explosive Roburite and Securite are the two types for synthesizing m – phenylenediamine.

o – DINITROBENZENE

Preparation

(*i*) It may be obtained from o – nitroaniline by replacing the – NH_2 group by – NO_2, by diazotization and treatment with sodium nitrite in the presence of copper power.

2-nitroaniline $\xrightarrow[HBF_4]{NaNO_2}$ diazonium fluroborate $\xrightarrow{NaNO_2}$ 1,2-dinitrobenzene (*ii*) It may be obtained from o – nitroaniline with Caro's acid (H_2SO_5) and then to oxidize the mono – nitroso derivative with hot dilute nitric acid.

2-nitroaniline $\xrightarrow{H_2SO_5}$ 1,2-dinitrobenzene $\xrightarrow[\Delta]{\text{dil } HNO_3}$ 1,2-dinitrobenzene

Chemical Properties

o – Dinitrobenzene is a colourless solid; it resembles m – isomer in reduction. It differs from the later in that one of the – NO_2 groups can be replaced by –OH or – NH_2 by boiling with aqueous sodium hydroxide and ethanoic ammonia respectively.

2-nitroaniline $\xleftarrow[C_2H_5OH]{NH_3}$ 1,2-dinitrobenzene $\xrightarrow{NaOH}$ 1,2-dinitrobenzene

P – DINITROBENZENE

Preparation

Like the other isomer it may be prepared

(*i*) By diazotization of p – nitroaniline and treatment with sodium nitrite in the presence of copper powder.

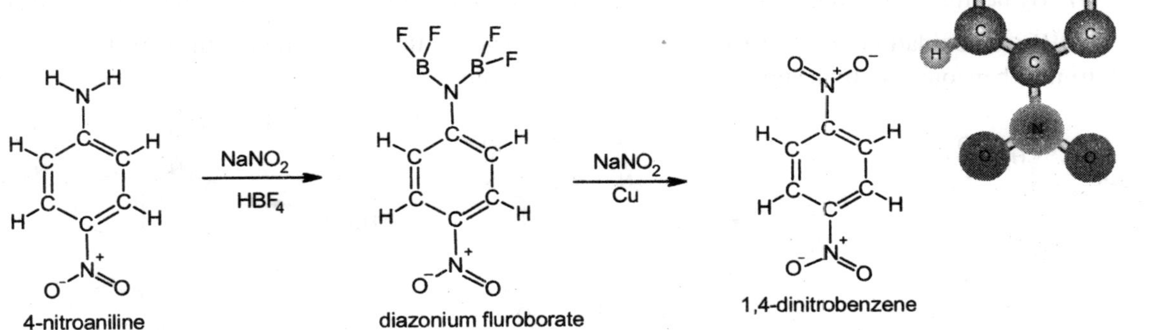

(*ii*) By oxidation of p – nitroaniline with Caro's acid and subsequent treatment of the p – nitrosobenzene with hot dilute nitric acid.

4-nitroaniline $\xrightarrow{H_2SO_5}$ 1-nitro-4-nitrosobenzene $\xrightarrow{HNO_3}$ 1,4-dinitrobenzene

Chemical Properties

It resembles in the chemical properties of the other nitro benzene isomers.

1, 3 TRONITROBENZENE(TNB)

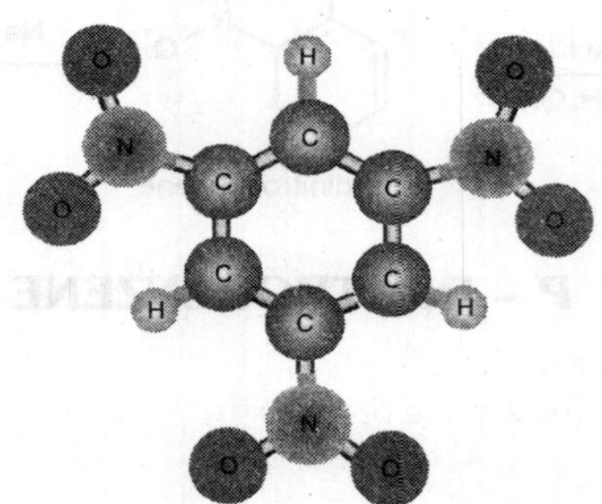

Preparation

(*i*) By heating in m – nitrobenzene with a mixture of fuming nitric acid and fuming sulphuric acid.

(*ii*) By the oxidation of T.N.T (tri nitro toluene). However, then knocking out the CO_2 from the 2, 4, 6 trinitro benzoic acid by heating it in acetic acid solution.

2-methyl-1,3,5-trinitrobenzene $\xrightarrow[H_2SO_4]{Na_2Cr_2O_7}$ 2,4,6-trinitrobenzoic acid $\xrightarrow[CH_3COOH]{\Delta}$ 1,3,5-trinitrobenzene

Laboratory Preparation

The crude trinitrobenzoic acid obtained by oxidation of 360 g. of trinitrotoluene is mixed with 2 L of water at 35°C in a 5-L. flask provided with a stirrer. Fifteen per cent sodium hydroxide solution is added, with continuous stirring, until a FAINT red colour is just produced. (See Notes.) The colour is then immediately discharged by means of one or two drops of acetic acid, and the liquid is filtered from unchanged trinibrotoluene. The filtrate is transferred to a 5-l. flask, and 70 cc. of glacial acetic acid are added. The mixture is then gently heated, with continuous stirring, when trinitrobenzene separates in crystalline condition, and floats on the surface of the liquid as a frothy layer. After about one and a half hours the evolution of gas ceases; at this point the crystals begin to stir into the solution. The heating and stirring is continued for three-quarters of an hour, when the mixture is allowed to cool, and the crystals filtered off. A sample of the filtrate should be tested for undecomposed trinitrobenzoic acid: if a precipitate is produced by the addition of sulfuric acid the process must be continued. After recrystallization from glacial acetic acid, the product melts at 121-122°C. The yield is 145-155 g. (43 to 46 per cent of the theoretical amount calculated from the trinitrotoluene).

2. Notes

During the solution of the trinitrobenzoic acid, the temperature should not be below 35°C, owing to the slight solubility of trinitrobenzoic acid in cold water. The heat of neutralization raises the temperature to 45-55°C, but the latter temperature should not be exceeded, since any trinitrobenzene formed at this point would later be removed with the unreacted trinitrotoluene.

Care must be taken that no more alkali is added than is just sufficient to produce the faint red colour. If an excess of alkali is added it produces a permanent colour, which is not removed by acid and colours the final product. When once the evolution of carbon dioxide sets in, the flame must be cut down so as to avoid the formation of a thick layer of froth which might foam over.

3. Other Methods of Preparation

1,3,5-Trinitrobenzene can be prepared by heating *m*-dinitrobenzene with nitric acid and sulfuric acid to 120°C; by heating 2,4,6-trinitrotoluene with fuming nitric acid in a sealed tube at 180°C for three hours; by heating 2,4,6-trinitrobenzoic acid or its sodium salt with water, alcohol, dilute sodium carbonate or other suitable solvent.

Physical Properties

MOLECULAR FORMULA	= $C_7H_5N_3O_6$
Formula Weight	= 227.1311
Composition	= C(37.02%) H(2.22%) N(18.50%) O(42.27%)
Molar Refractivity	= 50.71 ± 0.3 cm^3
Molar Volume	= 141.2 ± 3.0 cm^3

Parachor	= 411.3 ± 4.0 cm^3
Index of Refraction	= 1.637 ± 0.02
Surface Tension	= 71.9 ± 3.0 dyne/cm
Density	= 1.608 ± 0.06 g/cm^3
Polarizability	= 20.10 ± 0.5 $10^{-24}cm^3$
Monoisotopic Mass	= 227.017835 Da
Nominal Mass	= 227 Da
Average Mass	= 227.133136 Da

Uses

TNB is an explosive used in military shells, bombs, and grenades, in industrial uses, and in underwater blasting.

NITROTOLUENES

Introduction

Described in the chapter of benzene derivatives nitration of toluene is now going to be discussed in more details.

Preparation

Toluene is nitrated more easily than benzene because of the activating effect of the – CH_3 group. When reacted with a mixture of nitric acid and sulphuric acid, it yields a mixture of o and p nitro toluene.

$$\text{toluene} \xrightarrow[H_2SO_4]{HNO_3} \text{o nitrotoluene} + \text{p nitrotoluene}$$

toluene o nitrotoluene p nitrotoluene

These isomers may be separated readily by fractional distillation under reduced pressure m – nitro toluene can be separated in small amounts (about 4%) by direct nitration of toluene. When needed in large quantities it is obtained indirectly from p – nitrotoluene by (i) reduction (ii) acetylation of p- toluidine (iii) nitration of acetyl derivative and (iv) replacing the amine group by hydrogen.

1-methyl-4-nitrobenzene —Sn/HCl→ p-toluidine —$(CH_3CO)_2O$→ N-(4-methylphenyl)acetamide → N-(4-methyl-2-nitrophenyl)acetamide —H_2O, H^+→ 4-methyl-2-nitroaniline —$NaNO_2$, H_2SO_4→ [(hydroxythio){(E)-[(4-methylphenyl)imino]oxonio}nitroryl]oxidanide —C_2H_5OH, boil→ nitro toluene

The $-NH_2$ group in toluidine has to be acetylated to protect it from oxidation and to help ortho oreantation.

Physical Properties

Empirical Formula	$C_7H_7NO_2$
Molecular Weight	137.1
Physical Form @25°C	Liquid
Colour	Pale Yellow
Freezing Point	–2.9°C (–9.3°C)
Boiling Point	222°C
Specific Gravity g/cm^3	1.166 ± 0.06
@ 4°C/4°C	1.1742
@ 30°C/4°C	1.1531
@ 100°C/4°C	1.084
Pounds per gallon @25°C	9.7
Flash Point °F	223

Specific Heat cal/mole	48.4
Heat of CombustionK cal/mole	897
Heat of Vaporization@25°C cal/gm	73
Refractive Index @20°C	1.5466
Viscosity 15°C CPS	2.62

Chemical Properties

When toluene is nitrated nitrotoluenes are produced. The o – nitrotoluene and p – nitrotoluene and when oxidized and then nitrated, nitro benzoic acid is produced.

toluene → ($KMnO_4$) benzoic acid → ($H_2SO_4 + HNO_3$) 2-nitrobenzoic acid

toluene → ($H_2SO_4 + HNO_3$) o - nitro toluene + p - nitro toluene

o - nitro toluene → ($K_2Cr_2O_7$) o - nitro benzoic acid

p - nitro toluene → ($K_2Cr_2O_7$) p - nitro benzoic acid

Phloroglucinol a reagent for estimating furfural and pentose, is synthesized from toluene, when it is nitrated.

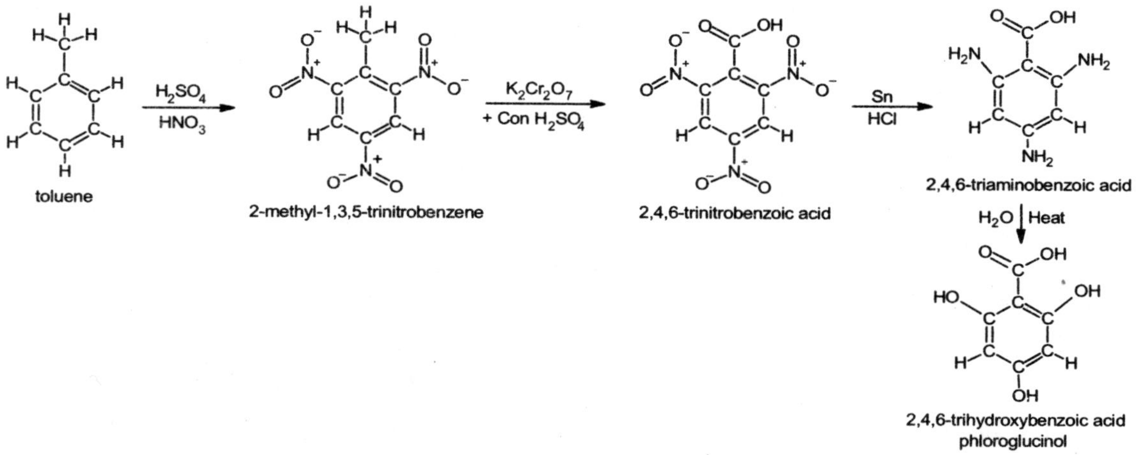

H_2SO_4 / HNO_3 → ; $K_2Cr_2O_7$ + Con H_2SO_4 → ; Sn / HCl → ; H_2O | Heat ↓

toluene

2-methyl-1,3,5-trinitrobenzene

2,4,6-trinitrobenzoic acid

2,4,6-triaminobenzoic acid

2,4,6-trihydroxybenzoic acid
phloroglucinol

(The detail study of phloroglucinol is out of syllabus of this book, it is in biochemistry)

Uses

Phloroglucinol is used as plant hormone and as a reagent for the detection of pentoses.